A Personal Journey into the Quantum World

God's Silent World

Quantum Physics, Cosmology, Evolution of Life, Consciousness, Afterlife

Jean Paul Corriveau

iUniverse books may be ordered through booksellers or by contacting:

iUniverse
1663 Liberty Drive
Bloomington, IN 47403
www.iuniverse.com
1-800-Authors (1-800-288-4677)

ISBN: 978-1-4401-4825-5 (sc)
ISBN: 978-1-4401-4829-3 (hc)
ISBN: 978-1-4401-4826-2 (ebook)

Library of Congress Control Number: 2009910656

Printed in the United States of America

iUniverse rev. date: 12/09/2009

The universe is driven by thoughts and quantum energy.
Nature is the product of interpretations of them.

Life's origin is nonphysical.
Therefore life after physical death is plausible.

Contents

Preface

The initial intent for this book was solely to ponder the origin of the universe and the origin of life. Did the universe have a beginning? If so, did God create it? Was there divine intervention in the creation life? Is there life after death? So Chapter 7 (Did the Universe Have a Beginning?), Chapter 9 (What Created Life?), and Chapter 10 (Is There an Afterlife?) were the first chapters I tackled.

I soon realized that an understanding of the makeup of the universe was necessary in order for me to comprehend its origin and life. So I started to ponder the nature of the elements of physics: matter, space, and time. This led me to write Chapter 1 (What Is Matter?), Chapter 2 (What Is Space?), and Chapter 3 (What Is Time?). These profound questions about matter, space, and time quickly led me into the world of quantum physics.

As a result, my emphasis shifted toward the quantum world away from God Himself. My endeavor became a journey into the quantum world. In the end, I became convinced that quantum physics is at the root of everything in physics, and that it's also the entry door into the world of thoughts, consciousness and God.

Journey into the Quantum World

My journey into the quantum world started one July summer day in 2007 when I drew a straight line on a blank sheet of paper and thought about what that action of drawing a line meant physically. *What makes the motion possible?* I wondered. I quickly realized the answer isn't at all obvious, and so the first eight chapters at times bring back to the surface that question of motion in space whose answer proves to be so elusive. Unexpectedly, this question led me into the world of quantum physics, special and general relativity, and cosmology.

Moreover, the notion of time as a dimension preoccupied me a great deal through many chapters because it remained unclear to me what time is. The topic of gravitation also perplexed me a lot. Finally, once I realized that there was more to the universe than its physical aspects, I began considering the nature of thoughts and consciousness as well as where God fits in all this. As a result, what I first thought would be a short personal essay of a few pages ended up being this book!

The image that most people and even some physicists have of quantum physics is that it makes nature seem very weird and apparently illogical. There's some truth to that. Quantum physics is weird simply because it deals with things so incredibly small that its domain is beyond our imagination. However, quantum physics isn't illogical, as everything in nature is based on logic.

What drove me while writing this book is the belief that quantum physics isn't entirely beyond people's imagination. To demonstrate this, for the most part this book makes use of knowledge of physics and mathematics no more advanced than what was known during the times of the celebrated scientist Isaac Newton in the seventeenth century! I'll demonstrate that some elementary aspects of quantum physics can be deduced by a high school-educated person using simple logic. Logical thinking alone can take us a long way.

God's Silence

Although most of this book deals with physics and nature, God was never far from my mind. God manifests Himself is very subtle ways, and this book deals with many subtle aspects of nature such as quantum physics. This domain of physics deals with the infinitely small scales in nature, billions of time smaller than the smallest virus! Using logical arguments, I conclude in this book that God manifests His presence at the quantum scale. But before coming to that conclusion, I found it necessary to spend many chapters helping you as the reader discover the world of quantum physics and other domains of physics and science. I needed to spend so much time on these domains because the discoveries are so very subtle.

The subtitle of the book suggests that God's world is silent. There are a number of reasons for this:

1. God doesn't reveal His secrets so easily. God is completely silent about the blueprint of His creation. Only through application of both scientific and mathematical methods, along

with a fair amount of intuition and inspiration, have scientists been discovering the workings of the universe—and there's still a lot more to be learned.

2. As mentioned previously, God's manifestation occurs at the quantum level where there's no sound at all. Arguments suggesting His manifestation at that level will be provided in various places in this book. It's not a coincidence that people are asked to remain silent when prayers are given during a religious service. It's also no coincidence that most of my inspirations while writing the book occurred while in the complete silence of the night.

3. God didn't have a "hands-on" approach to the creation of the universe, meaning that God didn't create its makeup directly. This book isn't for those who believe that the universe, as we know it today, was created in seven days or that the earth is 6000 years old, as there's plenty of evidence that the earth is many millions of years old. God's involvement was and still is silent and very subtle. This is why it took so long for the earth to form and for life to take root. God isn't concerned about how long it takes to create something. As this book will show, God isn't confined by time. To God, there's no difference between a single second and one billion years! A couple of sections of Chapters 4 and 7 will show that it's nature that keeps a tab on time, not God.

4. Silence is a sign of strength—loudness, a sign of weakness. The most powerful commands are the ones that are executed without anyone having to utter a single word. Such an orderly and silent execution is only possible if all parties involved are in perfect harmony, which is an evolutionary process based on nature's ability to very slowly and silently sort things out. This sorting out will be discussed at length in Chapter 9.

5. As a corollary to the point above, silence is necessary for good communication! This seems a total contradiction, but it isn't. We have to know when to be silent in order to understand what someone else is saying. Also we learn only when we are silent, even if the silence may last only a fraction of a second. It's like when we swallow, as we can't breathe and swallow at the same time. Just the same, we can't learn and speak at the same time. In this book, I theorize that nature has an ability to learn. Consequently, nature learns in silence. There's some truth to that, as when we leave the city and its noise and go into the forest or anywhere else in nature, most times it's fairly quiet. In Chapter 9 I'll suggest an association between evolution and nature's ability to learn.

As a fourth point, I make a correlation between orderly execution and silence. Orderly execution is the reverse of randomness, so silence leads to the reverse of randomness. Conversely, randomness leads to disorderly execution. This is quite easy to grasp. For instance, clearly Beethoven's Ninth Symphony was an orderly execution of sounds. Replace the musicians with monkeys and you get random noise. So if there's a correlation between orderly execution and silence, then there's a correlation between silence and intelligence. As I was writing this book, it became apparent to me that nature manifests some intelligence. But nature is humble and silent about that.

A Brief Overview

Clearly a study of what God created—the Universe—is far too vast an undertaking to fit in a single book. Therefore, I limited my effort to covering some of the most fundamental questions about physics, cosmology, life, consciousness, and the afterlife. The book is split into ten chapters. The first eight chapters deal with physics ranging from what matter, space, time, and gravitation are to how the universe was created and its expansion.

This book starts off discussing what's most present in our daily life: matter. When we touch a billiard ball, what are we really touching? In Chapter 1: What Is Matter?, I use a simple argument to demonstrate that matter is made of quantum particles. Moreover, these particles are composed of waves. In the process, I prove by using a totally unique approach that Einstein's famous formula $E = mc^2$ has its roots in quantum physics. People's understanding of Einstein's equation is that it expresses the immense amount of energy released when an atomic bomb explodes. But that equation also expresses other things that people don't suspect. (Read all about that in this first chapter as well as in Chapter 5)

In Chapter 2: What Is Space?, using again a simple argument, I show that particles travel in space in quantum steps. I explain that a fourth dimension is necessary to account for motion. I also wonder about the constitution of space itself, including: is it discrete or continuous? Physicists don't yet know, but most will say that it's discrete. I beg to differ.

In Chapter 3: What Is Time?, I show that time moves in quantum steps. Is time the dimension that accounts for motion? Is time discrete or continuous? Physicists don't yet know, but most say that it's discrete. Based on a simple argument, I'll show on the contrary that that time is continuous, not discrete.

In Chapter 4: Why Is Time Relative?, I explain why time ticks at a different rate depending on a person's frame of reference. You may have heard of the thought experiment of the two twins used to portray time dilation as predicted by Einstein's Special Theory of Relativity. One twin remains on the earth while the other travels at a very high speed in outer space. When that twin comes back many years later, he finds that he has aged more slowly than his twin who remained on the earth.

Einstein's Special Theory of Relativity proves that the faster a body moves the slower is its time clock, so the slower it ages. Conversely, the slower a body moves, the faster is its time, so the faster it ages. Why is this so? To my knowledge, this effect has never been explained satisfactorily to the average person. In this fourth chapter, I show that the answer lies in quantum physics, and I provide a very simple, unique explanation based on my own discovery that at the quantum level, there's a one-to-one relationship between the energy of a particle and its time clock.

In Chapter 5: What Makes Motion Possible?, based on the finding of the previous chapter, I propose a mechanism that drives the motion of particles. I explain that space is structured in a way that enables motion, somewhat like the valves in our veins that enable blood to flow. In the process, I'll argue that space is possibly continuous, not discrete. I'll also introduce you to other aspects of quantum physics such as the Heisenberg uncertainty principle, De Broglie duality principle of quantum particles, and Planck's quantum energy equation $E = h\upsilon$. I'll show you that there's nothing weird about what these physicists have discovered. However, the answer to the question of motion won't be complete until we reach Chapter 8: How Did the Universe Start?

In Chapter 6: What Is Gravitation?, I propose my own theory of gravitation based on my own understanding of time relativity from Chapter 4 and of motion from Chapter 5. My theory isn't a crackpot one because I show its equivalence to Einstein's space–time theory (general relativity). However my theory is based on energy, not time. This chapter proposes an explanation for the cause for gravitation. Contrary to popular belief, Einstein's General Theory of Relativity describes the effect of gravitation, not the cause.

In Chapter 7: Did the Universe Have a Beginning?, I prove using a very simple argument that the universe did have a beginning, and that it started from a single point. I also wonder what created that single point of origin, and I propose an answer. Be ready for a surprise: I show that the universe has a *creator*! Contrary to what some scientists theorize, the universe did *not* create itself.

In Chapter 8: How Did the Universe Start?, I provide a brief overview of the events that followed the Big Bang that came out of that single point and dispel a few myths about the Big Bang itself. An attempt is also made to determine how God may have been implicated. Then I propose that not only has the universe been expanding for many billions of years, but that this expansion has been accelerating—a fact verified by physicists. Unexpectedly, a study of that expansion led me to determine how space is constituted and to discover Newton's formula for gravitation using a totally unexpected approach. The study of the expansion of the universe led me to my own simple proof of existence of what physicists call dark energy, which, as you'll see, is a mysterious kind of energy.

Dark energy led me to discover that the constitution of space is what allows for motion of quantum particles. Space isn't void of structure. I provide an explanation for the motion of matter such as a baseball in which again the answer lies in quantum physics. Quite unexpectedly, the equation $E=mc^2$ is involved in the mechanism that enables motion. How can that be? This chapter explains.

What's the fate of our universe? Will it eventually collapse into itself or expand until it fades away? Did you know that the universe was initially a black hole? I'll show that the universe preserved to this day, and probably will forever, some properties of a black hole. Could this mean that the universe is contained within a black hole?

The last two chapters deal with life, its origin, its evolution, consciousness, and the afterlife. I have some intriguing surprises for you in those last two chapters. Especially if you're not interested in physics, you might find these chapters to be the most captivating because they aren't about physics! Nonetheless the subjects of life and the afterlife implicate quantum physics. These last two chapters explain how so.

In Chapter 9: What Created Life?, I calculate the odds of our human DNA (deoxyribonucleic acid) having been created solely by chance, and find that they are *literally* out of this universe! Nonetheless, I propose using a simple model that it's quite plausible that life did evolve via natural causes by nature cooperating in a way that some people might call intelligent design. I also argue at length about the possibility of different levels of consciousness in nature at the biological level, at the molecular level, and possibly even at the quantum level. I show that living cells most likely possess some level of consciousness without which complex life-forms could never have formed by chance alone.

Most scientists believe that consciousness plays no role whatsoever in the evolution of life. However this isn't so certain because it's scientifically impossible to prove whether consciousness exists or not as it's beyond the realm of science. Most scientists claim that evolution is entirely explainable by random DNA mutations. This is counterintuitive. Yet, using a simple argument, I explain why evolution is indeed largely driven by random mutations. Nevertheless, I propose that evolution isn't entirely without intelligence and consciousness.

I also wonder what separates the living from an inanimate object. In the process, using simple logical arguments, I conclude that life on earth appeared in a moment of spontaneous *creation*—and the laws of evolution took over from there. The theories of Evolution and Creation should be viewed as complementary theories rather than conflicting ones. Finally, using a simple argument I show that thoughts manifest themselves at the quantum level.

In Chapter 10: Is There an Afterlife?, I propose based on the findings in Chapter 9 that two universes exist, the physical one in which we live and another one that is a universe of consciousness, that is, a system of thoughts from the past and present. After death, only our thoughts remain into that universe that some might call Heaven. I propose a possible mechanism that might enable entities from the beyond to communicate with us. I demonstrate that there's a similarity between the creation of thoughts in the brain and the event just prior to the Big Bang! By coming up with our complex brain, evolution has come full circle back to the Creator of the universe. The universe appears to have been created for the purpose of being discovered! I also ponder some more about the implication of the Creator in our world.

Who Is This Book For?

This book is for anyone with an analytical and inquisitive mind-set, a spirit for wonder, and an eagerness to wonder about things and ask some deep questions. This book is loaded with carefully constructed logical arguments that will require you to have a focused mind. The arguments are driven by intuition, logical thinking, and deductions. I took to heart Albert Einstein's famous quotation, "The only real valuable thing is intuition." As a result, as I wrote this book, instead of asking myself "How do the universe and nature work?", I asked, "How do the universe and nature *think*?"

Many intriguing questions are raised—questions that are just as important as answers. As Einstein once said, "Imagination is more important than knowledge. Knowledge is limited. Imagination encircles the world." Some of the ideas in this book are my own ideas that you won't likely come across anywhere else!

A fair amount of intuition and inspiration guided me along, but obviously there are a number of facts I had no way of dreaming up on my own. So I used the Internet to obtain specific facts or data I needed about the laws of thermodynamics, biology, Darwin's theory of evolution, and DNA structure, the complex molecular structure that contains the characteristics of all life-forms on earth.

This book will slowly bring you toward discoveries. Most quantum physics books meant for the general public inform the reader of "precooked" findings made by physicists, but don't necessarily show how those findings were made or, most importantly, how readers can discover them on their own. In contrast, this book is a journey of discoveries. Because my approach is unique, you're not likely to find many of my ideas and analogies in other books.

Physics is a very complex subject, yet I tried to make this book as easy to read as possible. Most times, I use analogies to explain physical phenomena rather than delving into detailed precise explanations that would be above most people's heads anyway.

Nevertheless, I must inform you that many times, I couldn't avoid using mathematical algebraic manipulations in pursuing my discoveries. A few times, I had to resort to first-year university calculus. But mathematical tools and methods beyond that level were never required.

For the sake of keeping the book reasonably easy to read, I've placed most of the difficult mathematical derivations in appendices. Reading these appendices isn't necessary. However, I couldn't remove entirely the algebraic manipulations from the body of the book as some mathematical derivations are so much part of the arguments that they couldn't be displaced. So it would be helpful if you have an elementary knowledge of algebra—but it's not strictly necessary as you can still get the gist of the arguments.

Because this book is a personal quest for answers, some of my findings are new ideas, or new approaches not found anywhere as far as I can tell. For instance, I conclude that the universe is one of thoughts and energy and nothing else. The rest are details interpreted by nature and our brains.

I use the word *nature* a lot in this book. Keep in mind that this word has different meanings depending on the context. For instance, there's nature at the quantum level, nature at the molecular level, nature at the biological level, and nature at the astronomical level.

To most scientists, some of my ideas that attempt a connection between thoughts and nature may seem crazy. But who knows? It's impossible for us as humans to have a totally unbiased objective view of the workings of the universe while we are *deeply embedded* inside it. The only way we may have a complete and unbiased view of the universe is to be elevated out of it. This is clearly impossible for any living beings, as only God is elevated out of the universe, as will be argued in Chapter 7. Sometimes far-out ideas should be considered rather than discarded straightaway. Once again, Einstein said it best: "If an idea does not seem absurd at first, then there's no hope for it." I hope you as my reader will keep that in mind.

Jean Corriveau

Acknowledgments

This book has been written entirely by me in my spare time. Besides the editor and the production team, there were five people who read my manuscript and offered their feedback. I chose three reviewers without knowledge of quantum physics: my brothers François and Luc Corriveau and my sister-in-law Kerry Clermont. My approach intrigued them, but a couple of them advised me to tone down the level of mathematical difficulty. I've followed their advice by placing the scary mathematical work in appendices.

My work was reviewed by two persons with established knowledge in physics who made sure that the work didn't contradict current accepted knowledge. They are my brother Gilles Corriveau, a retired meteorologist and physicist, and Allan Dill, an instructor of physics and mathematics at the Saskatchewan Institute of Applied Science and Technology in Canada. My brother Gilles carefully read the manuscript and advised me on numerous occasions to make corrections on the parts dealing with quantum physics and Einstein's General Theory of Relativity. I followed through on some of his recommendations, but certainly not all of them because I wished for a work that remained of my own creation, especially where I offer my own views on gravitation and the expansion of the universe. I also wish to thank Allan for the many hours he spent reading the manuscript and for his detailed annotations where corrections on my part were called for.

Finally, I wish to thank Jean Vouillon, instructor of Multimedia at the École Technique et Professionnelle of the Collège Universitaire de Saint-Boniface, where we both teach, in Winnipeg, Canada. He designed the book cover, the three-dimensional pictures of the spinning photons in Chapters 5 and Chapter 8 and the 3-D image of the planet in space density in Chapter 6.

I also wish to thank the following people for giving me permission to quote them or use information posted on their Web sites:
- Louis Savain for his Web site rebelscience.blogspot.com.
- John Baez, mathematician, for his Web site math.ucr.edu/home/baez.

- Frank L. Lambert, Professor Emeritus (Chemistry), Occidental College, Los Angeles, for his Web sites www.entropysimple.com and www.entropysite.com.
- The webmaster of the Web site www.ornl.gov that hosts the Human Genome Project.
- Dr. Matin Durrani, editor of the Web site physicsworld.com.
- Kurt Johmann, computer scientist, for his Web site www.johmann.net.
- Dawn O'Driscoll, Syndication Account Manager of Telegraphmediagroup, for the Web site www.telegraph.co.uk.
- The webmaster of the Web site www.care2.com.
- Ross Heaven of the The Four Gates Foundation for his Web site http://www.thefourgates.com.
- Sorin Brabete, news editor for the web site news.softpedia.com.
- Gordon Ramel, for his web site www.earthlife.net.
- Edward R. Winstead, Managing Editor of the Web site www.genomenewsnetwork.org that hosts the Genome News Network.

This book would never have come to be without a spiritual e-mail that I received in early July 2007 from my sister-in-law Kerry Clermont. I forget the content of that message, but I recall that it sparked inside my mind a sudden strong desire for reflection and a wonder of God's creation of the universe and nature. The desire was strong and relentless, and it just wouldn't fade away. Dear reader, you have in your hand the result of that desire for questions and answers.

On the one hand, I've used metric units; I trust that my U.S. readers will not have too much trouble with that. On the other hand, I've used American spellings, which I hope will not alienate my Canadian readers.

Chapter 1: What Is Matter?

On September 10, 2008, numerous physicists conducted an experiment in Geneva using a newly built particle accelerator. This machine is designed to force quantum particles to collide and split into yet smaller particles. The main purpose of the experiment was to try to duplicate the conditions at the time of the Big Bang. A secondary purpose was to discover the most elementary quantum particle in nature. Quantum physics defines *quantum* as the smallest discrete quantity of some physical property that a system can possess. So the investigation of what matter is made of appears to be the best place to start our journey into the quantum world.

Over the past hundred years, physicists have discovered numerous quantum particles. It's not my intent in this chapter to delve into the details surrounding these particles, just to show you that it's very easy to deduce that matter is made of quantum particles. Moreover, I'll demonstrate that Einstein's famous formula $E = mc^2$ can easily be derived from a very simple and elementary understanding of quantum particles.

Let's get started.

What is matter?

Suppose we take a rock and cut it in half, then in half again, then in half yet again, and so on represented as follows:

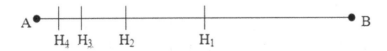

The H_i represent the cuts. How many times do you think we can cut the rock until we have nothing left to cut? Would there be an infinite number of cuts? No! If there were infinitely many cuts, then that would imply that there's always matter left to cut, *forever*. We might think that at one point, there would be nothing left to cut. Indeed, it is so. Just think in reverse to be persuaded. If the rock had infinitely many pieces, nature would have required an infinite amount of matter (and events) to put it together. Consequently, it would have taken nature forever to create it. That's impossible!

So the rock has only a finite number of pieces, perhaps something like this:

What's in between them? Vacuum—that is, empty space? Possibly, depending on what we mean by *empty*! If the rock is made of discrete pieces of matter, then there has to be some force that keeps them together, otherwise the rock would fall to pieces (actually it would "evaporate" spontaneously). If a force holds the pieces together, the rock can't be penetrated by other pieces coming from other objects such as the molecules inside my hand holding that rock. When the rock is cut, what happens is that the *forces* between the pieces give way, not the pieces themselves. We essentially force the pieces to separate from one another.

Nowadays we know that those "pieces" of matter are atoms and that they are made of smaller particles inside such as protons, neutrons, and electrons. So when we touch a rock, we don't really touch it. The electrons and protons that form part of our bodies are attracted or repelled by other electrons and protons in the rock we think we are touching, but there's *never* an actual contact. The sense of hardness is due to electromagnetic forces between atoms because at the quantum level, there's no such concept as hardness or softness of a surface.

Around the beginning of the last century, over a span of ten years two physicists conducted experiments that not only proved the existence of atoms, but proved in 1897 the existence of the electron, a particle with a negative electric charge, and proved in 1909 the existence of a positively charged nucleus at the center of the atom. These physicists were J.J. Thompson and Ernest Rutherford respectively.

However, the idea of atoms making up all matter wasn't new. Around 1803, John Dalton developed the first useful atomic theory of matter. He imagined the atom as a sphere full of an electrically positive substance mixed with negative electron. What a stroke of genius this man had!

But it really does not take a genius to be persuaded that matter is made of atoms. In the fifth century BC, the Greek philosophers Democritus and Leucippus proposed that matter was made up of tiny, indivisible particles they called *atoms*, or in Greek *a-tomos*. The reason they assumed this is because nothing can come from nothing. My argument that I developed above goes along the same line of thought. Moreover, in the next few paragraphs my argument will go beyond that of the Greeks.

These atoms are matter, so they have a size. Then it's reasonable to expect that each atom can also be cut into smaller pieces, then those smaller pieces can also be cut, and so on. At one point, the last piece left *cannot* be cut. Why is that so? Is it because its composition is so hard that it can't be cut, not even by nature? No. If nature could put any two pieces together, it can split them apart, although this might require great force—a force much greater than I can apply. So why is it that at one point, *nature* cannot cut the leftover piece any further? Is it that the piece left has no dimension? Yes, precisely: it has no size because if it did have a size, it could be halved. But this leads to a paradox. As the cuts are halves, then the size (noted s) of the piece left over after n cuts is given by the simple formula

$$s = \frac{S}{2^n}$$

where S is the size of the uncut piece that we started with. So if after n cuts, there's nothing left to cut, then $s = 0$. So

$$0 = \frac{S}{2^n}$$

This formula implies that $S = 0$ (i.e., that the uncut piece had no size to start with). But this is surely nonsense! So what's up? It's that the formula works for all values of $n > 0$ up to a *last* cut (call it the cut number w). Say that the particle had size (*s*) *before* that last cut w:

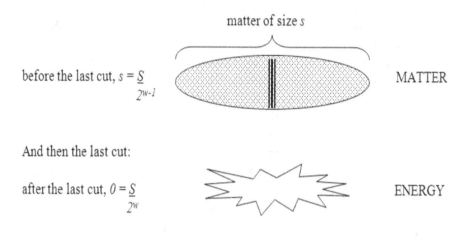

matter of size s

before the last cut, $s = \dfrac{S}{2^{w-1}}$ MATTER

And then the last cut:

after the last cut, $0 = \dfrac{S}{2^{w}}$ ENERGY

After that last cut, there's no size, but there's still something: energy! We enter at this point into a different world inside the matter: the world of *waves*. Waves are the carrier of energy, but have no size. Waves have only three properties: frequency, amplitude, and ability to carry energy. Physicists will say that waves have a fourth property, a spin, but I'll postpone discussing that one until Chapter 5. We can disturb the shape of a wave and make it change direction, but any attempt to cut it will only alter its shape.

Also energy (i.e., massless energy) travels at the speed of light as will be seen in Chapter 4. I'll come back to that last cut in Chapter 5 in the section titled "The Quantum World" and speculate about what type of particle that last piece of matter might be. You'll be surprised at what it is!

To help you understand how it's possible to pass from having a tangible piece of matter to simply intangible energy, let me use the following analogy. Suppose a balloon is filled with air. It has a size. If we attempt to cut it in half, we won't end up with two smaller balloons as the balloon will just burst and we'll end up with nothing. So it's possible to end up with nothing out of something! But actually there's something left: the energy that made the balloon explode. I have a hunch that a similar effect occurs with matter. When the very last cut is made, the matter explodes into energy created by the waves generated by the explosion. The famous equation $E = mc^2$

can be derived from this balloon analogy with the mathematical manipulation below.

Derivation of the Famous Equation E = mc²

Einstein's famous equation seems so bizarre. How can matter become energy and vice versa? When I was a child, it sure boggled my mind, but actually there's nothing bizarre about it as the basic idea is intuitive. The equation can easily be derived starting with the following analogy. Suppose we have a balloon with hydrogen gas inside it. Hydrogen atoms hit the wall of the balloon and bounce back. If we were to burst the balloon, it's the overall energy generated by the force of each atom against the wall that would constitute the energy of the explosion (and the bouncing back may be ignored as it won't happen). My idea is that simple.

Now with this analogy in mind, let's go back to the balloon in the previous section that represented the last piece of matter just before the last cut burst it. I wish to derive a formula for that energy generated after the cut of that *last* balloon of matter. But first I need to figure out the force applied on the surface of that balloon. Let's use the analogy of the balloon of hydrogen. Suppose an atom is at distance *d* from the wall of the balloon. It picks up some speed *v*, and hits the wall of the balloon:

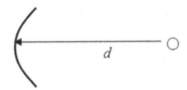

In case you don't know, Isaac Newton lived in England between 1642 and 1727, and in my opinion was the most brilliant scientist the world has ever known. The force on the wall is given by his Second Law of Motion, which is expressed by the formula $F = ma$, where m is the mass of the atom, and a is the acceleration applied to it.

Before continuing onto Einstein's famous equation, I wish to explain what that equation $F = ma$ is all about. It's a simple yet profound formula because it manifests itself in different ways:

1. It says that if you wish to move an object, you have to apply a force to it.

2. It says that if a force is applied to an object, the object will accelerate. For instance, if you push a table, the table should accelerate. But that is wrong: we all know that the table won't accelerate because the legs of the table apply friction on the floor. This friction is itself a force against your own force to move the table. Newton's equation assumes *no friction*.

3. If two objects fall from the same height (for instance, off a bridge), they'll reach the bottom at the same time regardless of their mass. So theoretically a feather and a hammer will fall at the same time (i.e., at the same speed). However, in reality we again know that this won't happen due to friction from the air sliding against the objects.

By definition acceleration is given by the change in speed over a time span:

acceleration = change in speed/time span

So if a car changes speed from 30 kilometers/hour to 50 kilometers per hour in 4 seconds, then the acceleration is $(50 - 30)/4 = 20/4 = 5$ kilometers/second/second $= 5$ kms/seconds^2. Notice the units of acceleration: kms/s^2.

Now back to Einstein's equation. The atom at a distance d from the wall of the balloon will pick up some speed v and hit the wall of the balloon. So the atom's acceleration is given by $a = \Delta v/\Delta t$, which represents a differential of speed over time (the symbol Δ is used to indicate a differential; for instance, suppose my weight was 63 kilograms last year, and it's now 73 kilograms, the differential is

Δweight = 73 − 63 = 10 kilos). Using the symbol *Δ*, the formula *F* = *ma* may be replaced by

$$F = m \ (\Delta v / \Delta t)$$

As we assume that the atom had an initial speed of 0, then *Δv* = *v* − *0* = *v*, giving us

$$F = m \ (v / \Delta t) \ (1)$$

where *v* is the speed of the atom hitting the wall, and *Δt* is the time it takes it to get there. But also *v*•*Δt* = *d*, with *d* being the distance that the atom is away from the wall and the dot being a symbol for multiplication. So *Δt* = *d/v*. Plugging that expression for *Δt* into formula (1), we get

$$F = m \ \left({}^{v} / _{(d/v)} \right) = m(v^2 / d)$$

The resulting equation, *F* = *m(v²/d)*, is equivalent to this one by placing *d* on the other side:

$$F \cdot d = mv^2 \ (2)$$

The term *F*•*d* (force multiplied by distance) turns out to be the definition of *energy* produced by a force applied over a distance. Indeed, a force applied over a distance requires some energy! So *E* = *F*•*d*, where *E* stands for the energy of the atom hitting the wall of the balloon. This is very interesting because then the formula *F*•*d* = *mv²* may be written as

$$E = mv^2 \ (3)$$

In the present analogy, the variable *v* is the speed of the atom, but in the reality of our "balloon" of matter, that "atom" is rather the last quantum particle of matter before the last cut (after which that matter bursts into energy). When the matter is cut a last time, the energy bursts at the speed of light because massless energy always goes at that speed at the quantum level. How do I know that? It is because energy at the quantum level is electromagnetic in nature. It was Michael Faraday, a scientist in the early nineteenth century, who had the incredible intuition that electromagnetic energy traveled at the speed of light. All physicists of the time thought he was crazy because at that time, it was believed that electromagnetism and light were two entirely different things. But then one generation later, physicist genius Clark Maxwell developed mathematical equations that proved Faraday right!

Back to my analogy: so the quantum particle hits the wall at the *very* moment that it becomes energy. This means that the speed *v* equals *c*, the symbol used for the speed of light. Therefore the formula (2) becomes Einstein's famous formula

$$E = mc^2$$

And voilà!

Now this has got to be the shortest derivation (and the most intuitive one) of that famous formula. As this formula has been proven to be correct in many different ways over the last hundred years, my derivation of it confirms that at the quantum level, that last piece of matter, the "balloon", is made of waves simply because that is what comes out of the balloon after the last cut. The conclusion is that at the very most basic level—the quantum level—*matter is composed of waves*! Real hard tangible matter doesn't actually exist. Isn't that shocking?

Waves (such as waves on water) easily change shape all the time. The same happens at the quantum level. As a result, matter is actually made up of wobbly quantum particles that don't have a well-defined shape as there's a little bit of fuzziness about the shape of quantum particles. As an analogy, quantum particles are somewhat like very soft Jell-O that jiggles all the time. In the next chapter we'll see another factor that contributes to the fuzziness of quantum particles.

This view of matter seems in total contradiction to our everyday experiences. When I touch a rock, sensors in my hand tell me that I am touching a *solid* rock with a well-defined contour. But this is a deception. Those sensors are made of cells, which are made of molecules, which are made of atoms, which are made of smaller particles, which are finally made of waves. These waves are electromagnetic ones that generate electromagnetic forces. These forces are present inside the rock and inside the atoms of the cells making up the sensors of my hand and they give me the feeling of hardness at our macroscopic level. But as we delve into smaller scales, the hardness of matter slowly diminishes all the way to the quantum level where everything is fuzzy and wobbly.

The famous equation $E = mc^2$ has implication behond the creation of matter. As we'll see in the section "Motion Finally Explained!" of Chapter 8, the equation $E = mc^2$ is also implicated in the mechanism that enables the motion of quantum particles—essentially the mechanism that makes motion possible altogether.

From the thought experiment above that led us to discover the famous equation $E = mc^2$, another equation can be discovered for the kinetic energy of an object. In our analogy of an atom, the equation (2) is for the energy of the atom as it travels the distance d. Now imagine that this atom is inside a beach ball. At any one time all atoms within the ball will travel an average distance of d. The sum of the energy of all atoms is given by

$$E = F \cdot D = MV^2$$

where D is the sum of all the distances d, the variable V is the velocity of the baseball, and M is the mass of the ball.

However not all atoms move in the direction of the ball. Consider this picture of the ball moving to the right:

Direction of the ball

Atoms move in all kinds of direction. For simplicity, the picture shows only six atoms going to the right, with only three to the left. However, because atoms move in all kinds of directions, on average half the atoms go one way, and the other half go the other way. Now do the atoms going to the left keep going in that direction? They certainly don't! If they did, the ball would split apart. The molecular forces keep particles going to the left to change direction. Of course, the opposite also happens: some particles going to the right may turn around and go to the left. The important result is that the ball moves to the right. So, the speed V of the ball is the result of the *average* of the sum of the speed of all atoms inside the ball.

Because atoms buzz around the baseball is all directions, the actual amount of energy that accounts for the motion of ball is on average half of what is expressed in the equation above. At any moment, on average half the particles move in the opposite direction from the motion. So, the actual amount of energy contributing to the motion of the ball is half of the entire energy, so

$$E = \frac{F \cdot D}{2} = \frac{MV^2}{2} \quad (4)$$

If you took an elementary high school course in physics, you'll recognize this to be the formula for the *kinetic energy* of a body. This is a measure of energy of the body due to its motion. My derivation of that kinetic energy formula has got to be the simpliest one possible, and also the most intuitive one. Note that kinetic energy doesn't include the energy "tangled up" inside each quantum particle. In other words, it does *not* include the energy $E = mc^2$.

While we are on the subject, let me solve one puzzling effect that motion has over energy consumption. Have you noticed that when you drive a car at 100 kilometers per hour, it uses much more fuel than when you drive at 80 kilometers per hour? The fuel consumption is disproportionate compared to the increase in speed. Why is that? It simply has to do with the fact that energy consumption is proportional to the *square* of the speed of the car. So, if you increase the speed from 80 to 100 kilometers/hour, the increase in speed may be only 20 percent but the fuel consumption jumps to $m(v+v \times 0.20)^2/2 = m(v \times 1.20)^2/2 = (mv^2 \times 1.44)/2 = (mv^2/2) \times 1.44$.

So, the energy consumption increased by 44 percent—more than twice the increase in speed. If the car was to double its speed, it would consume four times the amount of gas. Go easy on the gas accelerator!

Nature's Sensitivity to Energy

Understand that these waves are *extremely tiny* simply because they "show up" only after no matter at all is left, so they are smaller than subatomic particles. They are the smallest things in the universe (as we'll see in the section "The Quantum World" in Chapter 5).

The realization that waves are the very makeup of matter is a crucial deduction that led me later to conclude that, deep down, nature's sensitivity occurs only at the level of those waves and of emitted waves—essentially light. Nature doesn't really feel matter. Rather it feels its energy (such as electromagnetic energy) because energy is carried by waves making up the matter.

This is the reverse of our senses. For instance, the tip of my finger easily feels a rough surface, but can't feel the tiny bumps of a very smooth surface. Nature is sensitive to the tiniest things, with nature's view being that big things take care of themselves. By *big things*, I mean anything beyond subatomic level, or certainly anything above the atomic level. Nature doesn't care much about what happens beyond the quantum level. The implications of this are:

1. Beyond the quantum level, things are *abstractions* as far as nature is concerned. We'll discuss that in the section "Reality Versus Abstraction" of Chapters 2 and 9 in particular.

2. Every single *physical* phenomenon has its roots at the quantum level as we'll see for instance in Chapter 6. In fact, in Chapter 9 I'll theorize that even life has its roots at the quantum level.

This gives us a clue already that nature's point of view is definitely below our human macroscopic level. This doesn't mean that humans can't feel energy—of course we do! It's the *mechanism* of energy transfer that happens at the quantum level, and it happens at the speed of light. We humans just feel the resulting effect. For instance, when we exercise, our muscles get hot. That is the result of energy transfers that happen deep inside our muscle cells. These cells are made of atoms, and it's at that level or even at a lower one that energy transfer takes place.

So anything elevated out of the quantum level is in a different reality of nature, or as the term I use in various sections of Chapter 9, at a *different level of consciousness*. This realization of nature's sensitivity to energy is important, and we'll come back to it many times throughout this book. To be more precise, nature is sensitive to energy *transfer*, not energy itself. The reason for this will be further explained in a later chapter. For instance when a particle moves around, there's a transfer of energy, and nature feels that rather than the particle itself.

Creation of Matter

We're only in the first chapter, and already we are discovering quantum physics! There's a lot more to come. Just a couple of sections ago, I hinted at the relationship between matter and its energy. Matter contains energy because it's made of particles that move around, emit energy, and absorb energy, with the particles being made of those tiny waves mentioned earlier. Atomic bombs demonstrated to the whole world that matter can be transformed into a massive amount of energy. This is reflected by the famous formula $E = mc^2$. But what about a manifestation of that equation the other way around: energy that transforms itself into matter? That's the equation $m = E/c^2$.

Note that energy is equivalent to light because energy travels at the speed of light. So energy is a kind of light. This is what physicists Faraday and Maxwell discovered.

But hold it! When I heat up a pot of pea soup, the energy doesn't spread into the soup at the speed of light. Energy in the soup can't spread at the speed of light because there are many billions of atoms to heat up. One atom receives energy and spreads it to nearby atoms, and so on. Energy spreads at the speed of light at the *quantum* level only, not at the atomic or molecular level, and certainly not at the cellular level.

Because the speed of light, *c*, is so incredibly large, the equation $m = E/c^2$ says that it would take an immense amount of energy to transform itself into matter. The immense energy has to be concentrated in a very small area of space—so concentrated that it changes form and becomes matter. It's as if that small area of space has to be become *saturated* with energy.

Here's an analogy based on when we put sugar in a coffee cup and stir. The sugar dissolves and becomes energy that our bodies will use. However, if we put in too much sugar, it reaches the point where it no longer dissolves. But what condition leads to a state of saturation in the case of light (energy)? Waves of emitted light from the *highest* possible energy coming from opposite direction and also at right angles cause the waves to "shatter" and form curled-up waves. These waves are then looped together, thereby keeping the energy within a tiny region of space. These curled-up loops form what we call *matter*.Why do I suggest that matter is the result of curled-up, tangled-up waves? It's because of the thought experiment in the first section of this chapter where I deduced that matter at the quantum is like a balloon. Here is the picture:

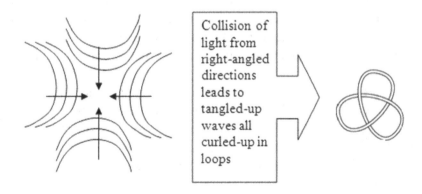

Collision of light from right-angled directions leads to tangled-up waves all curled-up in loops

There are at least two conditions this scenario of resulting curled-up loops has to meet to be considered plausible:

1. Matter does not lose its mass. For instance, an atom of hydrogen always weighs the same. The tangled-up waves are closed in in such a way that those waves preserve their energy like a well-working spring, which ensures that no energy is dissipated and lost. This *non-emitted* energy is what constitutes matter. Any other type of energy is emitted rather than being matter. Matter is essentially energy that is "condensed" and "trapped" in loops. That "balloon" making up matter that I was using at the beginning of this chapter is made of those curled up waves. When the balloon is cut, the curled up waves burst out and untangle themselves, thus producing energy as per the formula $E = mc^2$. It's very much like a spring that's let loose: its energy is released.

2. Matter has to feel relatively solid. Now, *solidity* is very much a relative term. To a fly, air is solid, but not to us humans. So maybe it's better to describe matter as something *tangible*. As was discovered by physicist Clark Maxwell in the 1850s, the light that meets and shatters at right angles is made of *electromagnetic* waves. So when the tangled-up waves are formed into a balloon, they are also electromagnetic waves. Because they are bent waves, the electromagnetic forces shield the tangled waves giving it a sensation of solidity. Take a copper wire, for instance. What makes it solid? It's the electromagnetic forces that bind the atoms together. If it were not for electromagnetic forces, matter couldn't exist.

This energy saturation, mentioned a few paragraphs ago, is only possible with light of the highest frequency possible, because at the highest frequency present, the highest energy is emitted, and the formula $m = E/c^2$ indicates that an immense amount of energy is indeed needed to create even the smallest amount of matter. But think about it. Is it really the highest frequency that creates matter? No because the highest frequency in nature has so much energy that if it could for a moment create matter, the curled-up waves of that matter would be so full of energy that they would explode into energy immediately. The highest frequency waves in nature are thought to be 10^{40} Hz, and they don't produce matter. So it isn't the highest, but rather simply "high enough" energy waves that can produce matter. How high does the frequency have to be? Physicists are not sure and this is one reason why they conducted that experiment in Geneva using the newly built particle accelerator that I mentioned at the very beginning of this chapter.

A quick search on the Internet revealed to me that this scenario of light (waves) colliding as high frequency does actually happen, and it does produce matter! Here is an example: two photons p colliding may produce an electron e^- and a positron e^+ (an electron of positive charge)

$$p + p \rightarrow e^- + e^+$$

The resulting matter created (electrons) depends on the frequency of the photons in collision. Note that a photon is a massless quantum particle that carries the energy of light.

So some time after the Big Bang, matter started to form this way. Details about the Big Bang and the creation of matter will be given in Chapter 8. This scenario of high frequency energy crashing at right angles has some truth as will be debated in the section in Chapter 5 titled "The Quantum World."

Note that because matter is made of tangled-up waves, it's inevitable that matter's properties are dictated by those waves. Therefore the mass and the electric charge of a particle are determined by the values of the frequency, amplitude, and spin of the tangled-up waves that the particle is made of. In other words, the way that the particle "shakes" determines its mass and electric charge. Of course this applies at the quantum level only. If you make a baseball vibrate, its mass will not change.

A last note worth noting: *mass* and *matter* are not the same concept. *Matter* is what I described above: the curled-up waves. *Mass* is the entire energy content of a body. As matter is made of energy, mass therefore includes matter. One type of energy that's part of the mass is the kinetic energy of particles inside the body. Consider a baseball:

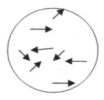

The bouncing back and forth of particles inside the ball contributes to kinetic energy. Understanding the difference between the two concepts—mass and matter—is important. For instance:

1. Gravitational force acts upon mass, not just matter.

2. As an object increases its speed, Einstein's Special Theory of Relativity dictates that its mass increases. How mysterious! But this is simply because it gains more energy as it speeds up, hence it gains more mass. But its matter does not actually increase, as it's clear there's no new matter created and no new atoms.

To put it simply, matter is any *tangible* substance, and mass is the gravitational force acting on it. So then gravitational forces act upon *energy* as well as matter! We'll investigate these two effects in later chapters, in particular Chapter 6.

Chapter 2: What Is Space?

After I reflected on possible answers to this question, it came apparent to me that it's difficult to have a complete discussion of space without introducing time. On the surface, this seems obvious, for when a particle travels through space, it has a speed which is in units of distance over time. What needs to be understood in both this chapter and the next one is the concept of time as it's not as obvious as it seems at first. Also it became apparent to me that the way we view matter influences the way we view space.

As illustrated in the previous chapter, I'll assume that matter is made of quantum particles. These particles could be anything that has a mass and possibly emits energy. In the previous chapter, I showed that deep down; matter is actually made of waves—light. But I choose in this chapter to view matter as made of quantum particles rather than waves. Viewing matter as waves would lead to different arguments about what space might be, as will be seen in the section "Motion Finally Explained!" in Chapter 8.

After developing my arguments about what matter and space are, it became clear that matter, energy, and space are very closely related in that it's impossible to talk about one without involving the others. In addition, this chapter can't provide a complete answer to the nature of space because space is tied in with other things besides time such as gravitation. I'll provide a more complete answer in Chapters 6 and 8.

Discrete or Continuous Space?

Suppose we cut a piece of ham in half repeatedly. From the previous chapter, we know that at one point, we'll be left with nothing to cut. But there's something left: space. So we can imagine "cutting" space now. The same idea of splitting in halves used in the previous chapter is applied here. Is it possible to "cut" space forever? That seems ridiculous. Most physicists believe that space is discrete, not continuous—meaning that space cannot be cut forever.

Philosophers have difficulty with this idea of a discrete space. Discrete means that there are only a finite number of spots of space. It's like a classroom with only a finite number of seats. Therefore, a particle would have to move in space from one spot to the next one, making what's called a quantum step. The consequence of a discrete space is that there's an interval of *nothingness* between the steps. What would that nothingness be? Is it a vacuum? No, because vacuum means the absence of particles or energy, not the absence of *space*. Here, by the very definition of discrete, nothingness has to mean the absence of space—a "void" so to speak.

Unfortunately, this presents a dilemma. How can the particle accomplish a quantum step when there's nothingness to step over? Also, as the particle is made of energy, its jump over the void would inevitably fill that void with energy. So any void in space would disappear, which implies that space would no longer be discrete. Is space continuous after all then?

But not so fast: we need to ponder about the fabric of space quite a bit more before jumping to any conclusion. We'll come back to the idea of a void in space in Chapter 6. (Note that a complete discussion of the fabric of space requires technical details that will be covered in other chapters, in particular Chapter 8.)

Philosophers favor the concept of space being continuous. But continuity presents a dilemma as well. Suppose that a particle proceeds to move from point A to point B. How is that going to happen in a continuous space? Continuity is the opposite of discrete. A continuous space between these two points A and B means that there are *infinitely* many points between them (the exact mathematical definition of continuity is a bit more complex than that).

A way to view the impasse is this: Before reaching B, the object has to reach halfway H_1. Before reaching that point, it has to reach halfway H_2 between A and H_1. Before reaching that point, it has to reach halfway H_3 between A and H_2, and so on. Here is the situation:

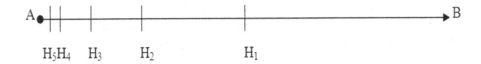

It's clear that as we split space in halves, we get closer and closer to where point A is located. But, as we assume a continuous space, we can always—and *forever*—divide the remaining segment AH_i in half because there are infinitely many points in any interval of space between point A and H_i. This is a consequence of the meaning of continuity. So we'll *never* even reach the first point right *next* to A. Consequently, it's impossible to jump from A to the spot right next to it in space because there's *no* next spot.

With a discrete space, we can pinpoint the next spot, but with a continuous space, that's impossible. We could have flipped the intervals on the other side toward point B and concluded due to symmetry that it's possible to start from point A, but impossible to reach B. So depending on how we look at it, on the one hand we can start from point A, but on the other hand, we can't! As that is a contradiction, something is very wrong with this scenario.

The fallacy is the assumption that the halfway points can be pointed out precisely. As we assume that space is continuous between A and B, the continuity implies that it's therefore impossible to *precisely* know where those points are. This is a mathematical implication of what continuity means. Well, if it's *mathematically* impossible to locate a point precisely, you can be sure that it's *physically* impossible to do so either. Consequently nature can't pinpoint precisely any location in space! If there's a little fuzziness in space, can space really be continuous? Note that physicists still don't agree on the answer to this question.

Skipping through Space

Both a discrete space and a continuous space present dilemmas. I find a continuous space more plausible for two reasons: (a) I'll conclude in the next chapter that time is continuous, and (b) I'll explain in the section "Motion Finally Explained!" in Chapter 8 that matter *moves* in space as fuzzy, shapeless waves, not as particles. Nonetheless, it remains that matter is *made* of particles. So we'll continue to view matter as particles for the next few chapters.

How can a particle manage to reach point B from point A assuming space is continuous? Clearly motion is possible from point A to point B: I can draw a line with a pen from A to B without any problem. So particles within the pen *do* move from point A to point B. How do they do it in a continuous space? There isn't much of a choice: the only way to reach point B is for the particles within the pen to skip *infinitely* many points so that there will be only a *finite* number of steps to execute in order to reach B.

Although this seems ridiculous, there's simply no other way! Does this suggest that space is discrete after all? No—it simply implies that the *motion* of particles is discrete. But what's the mechanism that makes motion discrete? This is such a deep question that it can't be answered in a simple paragraph. In fact, I'll have to come back to this question many times throughout this book in order to come up with a satisfactory answer in the section of Chapter 8 described above.

The idea that the pen skips space while going from point A to B is so bizarre that I bet that just a second ago, you took a pen, and moved it on a line, but you didn't feel the pen skipping. The reasons are simple:

1. The discrete steps that the pen takes across the line are incredibly smaller than what the nerves of your fingers can feel. As nerves are made of cells, the skipping is therefore much smaller than the size of a cell. Now if we look into a microscope that magnifies everything a few thousand times, we can see the inside of the core of the cell, and maybe even chromosomes wiggling around. Even at that many thousand times of magnification, we still wouldn't see any discrete, jerky motion. Nothing skips around.

    So it's clear that the skipping in space that we're considering here is *billions of times* smaller than what can be seen, even with the most powerful microscopes. The skipping is even smaller than what we can imagine!

2. Based on the first point, we are clearly here down to the level of subatomic quantum particles, perhaps even smaller. Something has to skip space, and it has to be those quantum particles. But this incredibly tiny quantum skipping isn't the only reason that we can't feel the pen skipping. The pen is made of billions of quantum particles. These particles don't skip space all at the same time, or even in the same direction. They aren't *synchronized*, and so they move all over the pen. So the effect of skipping of one particle is immediately nullified by the skipping of another particle, and so on. Consequently the pen itself as a whole doesn't skip while it's moved along by my hand.

This reality of the particles of the pen skipping in space presents a difficulty: when a particle skips from one step to the next one, it has to skip over an interval of *infinitely* many points in space (remember that we decided to assume that space is continuous). This is obviously a problem.

If there are infinitely many points to skip, the particle will never reach the second step—actually it won't even be able to *start* crossing the line at all between two steps! I already illustrated in the previous section the reason for that, which has to do with the continuity of space. So, there can't be infinitely many points in an interval (i.e., within a quantum step). Maybe there are only a finite number of points in an interval, but then we are still stuck with the question of what lies between those new points.

The other idea mentioned earlier was that *nothingness* separates the steps. The intervals are made of nothing, of void. But that presents a problem too: what does it mean for a particle to skip nothingness? Either way of looking at space presents a dilemma. It seems that motion is physically impossible, yet ... it does happen. What then makes motion possible?

Just out of curiosity: what would happen if *all* quantum particles of the pen were synchronized—that is, what if they all skipped at the same time, and in the exact same direction? To begin with, there wouldn't be any overlapping of skipped intervals. Electromagnetic forces that hold the pen together go normally in all kinds of directions, but with all particles synchronized, these forces would be nullified. The consequence might be that the pen would simply disappear into space (although its particles would still be around). Some physicists suggest that the pen would disappear into another universe—a parallel one, but I very much doubt that possibility as the pen would simply disappear into the fabric of space. We'll investigate the fabric of space in later chapters.

A Fourth Dimension Is Necessary

Whether there's space or nothingness between the two quantum steps (intervals) that a particle takes, it remains impossible to reach point B. So what should we think of this? How do we get around this impasse? I believe that this is the turning point where another dimension *has to exist*. The obvious candidate is that a time dimension has to exist (remember I told you earlier that a discussion of space leads to a discussion of time). Essentially, space needs time! But this is a matter of *human* interpretation of what time is.

Later in this book, I'll discuss whether the time dimension really exists or is simply imaginary. Here I'll just note that Einstein's space–time theory proved the existence of the time dimension. However, Einstein never explained what time is. I'll propose answers later, mainly in Chapters 4, 5, 8 and 10.

Whatever the answer, clearly a fourth dimension is needed to account for motion. If there's no time dimension, then some other dimension has to exist to account for motion. In this chapter and the next three, I'll assume the existence of the time dimension to explain motion.

In the section "The Energy Dimension" of Chapter 8, I'll attempt to explain motion using some sort of energy dimension. It's then that we'll discover the real mechanism for executing motion. However, let's stick with a time dimension for this and the next three chapters because it's easier to understand motion within that context, and it will also allow me to discuss Einstein's space–time theory.

Here is my first attempt to explain motion: nature cannot keep track of where the particle is *exactly* in space given an *exact* precise time. By the time it finds (more precisely, it feels) the particle, it's too late—the particle has moved, and it has moved over the troublesome quantum interval (quantum step) in the *time dimension*. I explain this concept further in the next section. But the particle gives the time dimension a chance to locate it when the particle emits energy. More precisely, at the *same* time that this energy is emitted, the time dimension "lifts" the particle in time, thereby avoiding the troublesome space interval that we debated in the previous sections. Here's the picture:

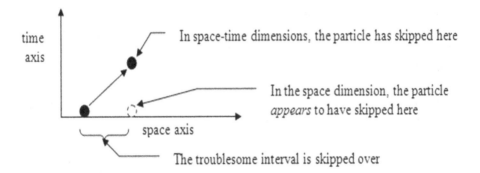

The convenience of a fourth dimension is that we don't have to deal with the question of whether space is discrete or continuous because the fourth dimension makes the particle skip over the troublesome interval. To better understand this concept of lifting the particle in time, imagine a simpler world of a single dimension in space. A particle at position *x* needs to go to position *y*, but a line of particles lies in between:

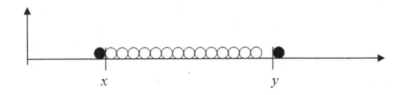

The particle at *x* can't reach particle *y*. However, if a second dimension in space is invented, then it can *jump* into that second dimension and over the line of particles:

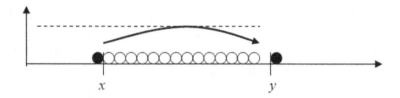

This analogy should clarify why the space dimension needs another dimension so that the particle can avoid the troublesome interval by skipping it. There's one important difference though between the space and time dimensions. In the space dimension, the particle has control over where it lands. In this simple example, it lands where *y* is on the same axis (or horizontal dimension). With the time dimension, the control of movement is reversed: it's the time dimension that decides where the particle will move onto the time dimension axis. The particle doesn't really jump over an interval—instead it more flies over it in the direction of time.

By the way, this notion of motion over an interval of space–time is purely conceptual in that the particle does not really *physically* move in the time dimension. There will be more on that point of view in the section "Motion in Space–Time" in Chapter 5 where I'll attempt to give a more physically realistic explanation behind the quantum motion of particles.

However, for the sake of keeping the explanations simple, this chapter and the next two won't delve into those details. So as you read on, keep in mind that the motion of particles doesn't happen in reality the way that it will be explained from here on until the reality is revealed in Chapter 5. Accordingly, for simplicity's sake, we'll remain at the conceptual level for now.

When comes to motion in the space–time dimensions, nature is somewhat like a producer filming a movie that has twenty-four *frames* per second. Again, it doesn't quite happen this way, but this difference isn't significant because in Chapter 4, I'll provide a mathematical proof that this analogy actually works out to nearly the same effect as if it were the reality. The film misses whatever happened between each frame. Except that in nature, it's *time* that makes the particle miss space (actually this isn't quite true as we'll see later). This reasoning suggests that time is a necessary dimension without which matter wouldn't be able to move at all over the troublesome intervals!

As you may have begun to realize, space–time motion is very unclear and confusing in a physical sense. What does it mean to move in time? To my knowledge, physicists have never clearly answered that question. Instead they've devised countless equations for motion in space–time that work perfectly well. It remains to be seen whether there will be a plausible physical explanation of motion in time; for now, let's just stick with the official explanation.

Note that it's the time dimension that makes the particle move through time and not the object that moves itself through time. This will be important to remember later, in particular in the section "Motion in Space–Time: The Plot Thickens!" in Chapter 5 where I'll argue that objects don't really move alongside the time dimension. But although this detail is important, for the moment we can ignore it as otherwise the present discussion will become far too messy.

Here's a more complete picture to illustrate a particle skipping through space with the help of the time dimension:

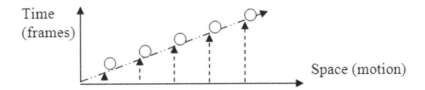

Of course space has three dimensions, but to simplify the argument, only one is shown here. The point of origin is simply where we start taking *frames* of the moving particle. Each dot represents the particle at a certain point in time and space.

Note that I assume that the particle moves through an empty space (not to be confused with nothingness), *empty* meaning that there are no other particles around. If the space between points A and B contains matter, then I suspect the particle won't go to point B on a perfectly straight line as it would be affected by the presence of the particles across the line AB. But as it would travel in the empty space between those particles (not through them, of course), the particle will still skip around as above—it would just require more skips and more time.

Space–Time Fuzziness

During periodic interval of times of say Δt fraction of a second, nature loses track of where the particle is. Then it finds it again. In other words, there's a little *fuzziness* in space. This fuzziness is in part due to the fact that

1. It takes nature a tiny bit of time to detect the particle.

2. Nature doesn't view the particle as a particle, but as internal, compacted, tangled-up wobbly waves (or simply as energy) as discussed in the previous chapter. Nature nonetheless presents those waves to us as particles. But they are fuzzy particles in the sense that quantum particles move along like a wave, but give off their energy like a particle (the section "Motion Finally Explained!" of Chapter 8 will prove that this is so). This is very much the way that a photon manifests itself. Recall that photons are just like everyday light.

We'll delve into these two factors in the sections "Emitted Energy Versus Internal Energy" and "Quantum Steps Taken by Particles" in Chapter 5 where I'll explain the real physical reason for the skipping of particles.

It's this fuzziness that allows the particle to skip around in space. This fuzziness happens in the interval of the continuous space between particules (or interval of nothingness, whichever it may be) talked about earlier. Let's say that this happens at every Δs fraction of a meter interval. In other words, the particle (the pen as referred before) not only skips space, but time makes the particle skip *time* as well at every Δt fraction of a second right when the particle is about to go through the troublesome space interval! By skipping that time that it takes to skip over that troublesome interval between two steps, the object does not have to go over that interval.

This idea seems out of this world, but there's no other way that the object can bypass the troublesome interval. As I mentioned before: anyway you look at it, a fourth dimension is necessary to account for motion.

Now, how is it that nature loses the interval Δt of time? There are two reasons:

1. It has to do with the confusion as to how to locate the particle at a precise given time. Nature has to know where the particle is *now*. Unfortunately nature can't pinpoint where the particle is "now." The reason is that if nature can pinpoint it, it would take a very tiny moment to do so, so that by then the particle would no longer be located at the point "now." Consequently, the "now" is always *undetermined* (you'll see in chapter 4 another argument for the trouble with the "now"— an argument based on time relativity). This is equivalent to saying that it's impossible to determine the exact position of a moving particle in time, and it has important implications. One of the following takes place:

    a. Nature can locate a particle in space *precisely* only when it's at *rest* in time. By rest, I mean that time took a snapshot of the particle. If the particle is at rest in the snapshot, then time has no way of knowing what *speed* the particle is going at that very spot. So it certainly can't know where it's going. An analogy would be the police taking an instant picture of a speeding car. They could know exactly where the car was at the time of the picture, but from the picture alone it's impossible to know the speed.

    b. If nature knows the *exact* speed at which the particle is going, then it can't know the exact location in space of that object simply because speed takes place over time. So it has to view the particle within an interval of space and an *interval* of time— not at a precise time. An analogy would be the speed detection devices that police use. When the signal hits the speeding car, the signal takes a short time to come back to the device. By then the car has moved even though the device knows exactly the speed of the car at the time of detection.

Putting the factors in points (a) and (b) together, we now have a mathematical way to express what I meant earlier about fuzziness: either a particle's place in space is fuzzy or its speed is fuzzy. So a measure of the level of fuzziness of the whereabouts of the particle might be expressed by simply multiplying these two factors:

Fuzziness = $\Delta s \cdot \Delta v = \Delta s \Delta v$ = fuzziness in space multiplied by fuzziness in speed

The idea is very much like multiplying two factors of a number such as $3 \times 5 = 15$. Note that the dot in $\Delta s \cdot \Delta v$ signifies multiplication and is usually left out by default. As most people will know, the symbol $\times$ also signifies multiplication, but mathematicians prefer the dot.

That particular multiplication seems to make sense because both length interval and speed contribute to the uncertainty (fuzziness) of where the particle is. Nature acts like an *observer* seeing particles going by. The better that nature can see a moving particle, the less fuzzy its movement will appear.

Note that the unit for Δs is meter, and Δv is meters per second. This leads to the units of $m \cdot m/s = m^2/s$ for the fuzziness formula $\Delta s \cdot \Delta v$. So the fuzziness is over a plane (m^2), not a cube (m^3). We would expect it to be over a cube because the particle travels in a three-dimensional space. Indeed, but the fuzziness occurs within one skip—one quantum step. That is the *smallest* trajectory physically possible so it *cannot* have a curve in it. Consequently, it occurs in two-dimensional space.

This begs the question: does a quantum particle even know that it moves in a three-dimensional space? Of course it knows, doesn't it? After all, it's obvious that we live in a three-dimensional space. Yes, it's obvious to us humans, but a quantum particle isn't a human. As the tiniest kind of particle in the universe, inevitably it has a different view of space than we do.

Remember that just a thousand years ago, humanity thought that the earth was a flat 2-dimensional surface. It was only by deductions that the Greeks (then much later, the Europeans) discovered the truth. We'll come back to this topic of two-dimensional motion of quantum particles in the section titled "Motion Finally Explained!" in Chapter 8.

2. As was argued in Chapter 1, nature's sensitivity is at the level of the extremely tiny tangled-up waves that make up the particle. Nature doesn't really feel matter: it feels energy. So if it has a hard time feeling matter, it has a hard time feeling the particle, and therefore certainly has difficulty locating it or guessing how fast it's going.

    It's like someone who is lost in a forest. The person can easily feel the trees, but has great difficulty have a feeling for the forest as a whole. Also there's confusion about the *semantics*: a forest is made of trees, but a tree isn't a forest. Likewise, a particle is made of waves, but a wave isn't a particle. Because of this realization regarding energy, we'll see in the section "Another Formula for Fuzziness" in Chapter 5 another way to express that fuzziness formula $\Delta s \Delta v$—one that involves energy.

The above findings suggest that the fuzziness is both over the *behavior* of a particle and its *meaning*. Note that although the notion of fuzziness applies to all matter, its effect is felt mostly at the subatomic level, and less at the atomic level or above (this will be demonstrated in the section "Quantum Steps Taken by Particles" in Chapter 5). There's clearly next-to-nothing fuzzy about a basketball thrown through a hoop. At that level, fuzziness is negligible due to the fact that at our macroscopic level, the value of fuzziness of all particles making up the ball cancel each other out, which is a good thing as otherwise it would be difficult to feel motion at our level — in fact the basketball might even disappear for a while, then reappear. What a bizarre world that would be!

The term *particle* here and everywhere in this book is used to indicate a very small quantum particle at the subatomic level, or perhaps something even smaller that emits energy. It's important to realize that the principles described here do *not* apply to large-scale objects such as a pen, but only to the small particles inside the object. Of course because a large object such as a pen is made of such small particles, its motion will be dictated by all its particles overall.

So far we've got around a continuous space by saying that the object jumps in the time dimension at every quantum step, thereby avoiding both the troublesome continuity and discreteness dilemmas. In other words, the time dimension allows us to account for motion whether space is continuous or discrete.

A Model for Space

In regarding space, I've imagined a model that might just satisfy both quantum physicists and philosophers. My discussion so far should have persuaded you that subatomic particles change their state in quantum, discrete steps. They emit energy in the same manner (more about that idea in Chapter 4). So it's natural to suspect that they move in quantum steps. This is what most physicists believe, and as I've explained so far, it has to be happening that way.

A particle moves through space skipping from one spot to an adjacent one. I'll call those spots *floats*. It seems that space is granular this way in which each circle represents a float (i.e., a spot where a particle jumps on). This could be analogized to people stranded in the ocean where there are floats fixed from the bottom of the sea

around them. If they want to move, they go from one float to another. From that, I imagine the following representations of space.

First, space is in a form of a grid in which each circle represents
a float:

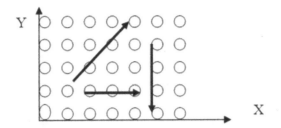

Only two of the three dimensions are shown to make it easier to
imagine. If a particle moves parallel to the X axis, this move is executed the way explained before. By symmetry, it does likewise when
the particle moves parallel to the Y axis. But if the particle moves in
any other direction, we see that it has to execute wider skips from
one float to another, as shown by the diagonal arrow. There's absolutely no good reason that the skips should be different in length
regardless of the path taken by the particle, simply because in the
absence of anything, space has to be nice and even—there's no reason to believe otherwise.

Therefore, space can't be structured as a rigid grid. So then how
can the steps be all the same length regardless of the path of the particle? The only way is that the particle applies *"tension"* on space so
as to attract the floats at equal distance all around the particle as it
moves along. This is more or less illustrated below with a circle
around the particle, supposing that the particle is at position (x, y):

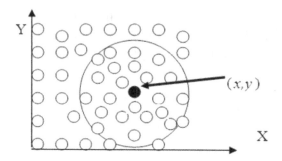

The floats are stirred up a little, but eventually the grid would be transformed into a structure in which all spaces between the floats are equal. This is only possible with the particle in the middle as follows:

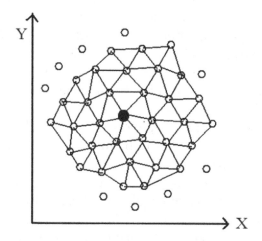

The grid would be structured similar to equilateral triangles adjacent to each other. This ensures that the distance between each couple of floats is the smallest possible (i.e., the term Δs used earlier). This is why equilateral triangles make sense. This strongly suggests that space is made of this structure. In other words, space is *not* void of structure. A proof that space has a structure will be provided toward the end of Chapter 8.

This structure is particularly a consequence of space being filled with light for which the floats are adjacent to one another (as will be demonstrated in Chapter 4). Because there's light everywhere in space, the structure of space is something like this:

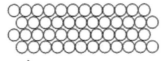

On the one hand, when light travels in space, it doesn't really skip, but rather follows adjacent floats:

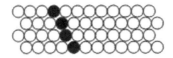

On the other hand, as will be shown in Chapter 4, particles (unlike light) follow floats that are not adjacent such as follows:

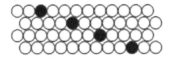

I'll demonstrate in Chapter 4 that light and particles do indeed skip as shown here.

Flexible Space

We realize now that space has to be a little bit flexible, meaning it's not structured as a rigid grid. This implies that the floats are never guaranteed to map onto the X and Y axes evenly because they may be disturbed by a passing particle. Here's an example of an uneven mapping on just the X axis for simplicity (but the same applies to the Y and Z axes):

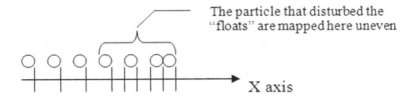

The particle that disturbed the "floats" are mapped here uneven

X axis

We see that the intervals on the X axis aren't all the same. And because the particle can come from any direction—and there can also be other particles around—the floats could be all over the place (i.e., no longer in a nice and regular grid or equilateral triangles).

This suggests that the X axis has to be able to map floats in *any x* location. This implies that an interval on the X axis has to be continuous. It follows that the entire X axis has to be continuous (regardless of whether space is discrete or continuous). The X and Y axes have to be able to map any possible intervals no matter how small. In other words, for any (x_1, y_1) location, there always has to be values of x_2 and y_2 so that the intervals $|x_1 - x_2| < \varepsilon$ and $|y_1 - y_2| < \varepsilon$, for any $\varepsilon > 0$, however small it may be. This is the mathematical definition for continuity, and it implies that the X and Y dimensions have to be continuous.

In conclusion, space is continuous, but *not* the paths the particles take through that space as particles skip through space. A way to visualize this is a hockey player skating on the ice. He makes skate lines as he moves along, but doesn't have to touch the ice in every single part in order to move. However, if there were a great many players on the ice, the ice would better be covered *everywhere*. This is the idea of continuous space with quantum paths of the particles.

Another example would be to take a bicycle gear and roll it onto a surface. The teeth of the gear will leave discrete quantum dots on the surface, yet the surface itself is continuous:

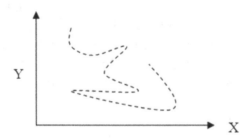

Nice, but hold it—the argument above suggesting that space might be continuous seems erroneous! The XYZ axes of space are *imaginary* axes as will be demonstrated in the section "Imaginary Spaces Axes and Real-Time Axis" of Chapter 5. The mapping made on the

X axis as described is used purely for mathematical purposes, perhaps to figure out the equation of the path of the particle. We can *imagine* XYZ axes anywhere we need them, and in any orientation: on a chemistry table, on the blackboard, in the air, etc. If we just take the earth as an example, there are XYZ axes anywhere we need them (here showing the X and Y only for simplicity):

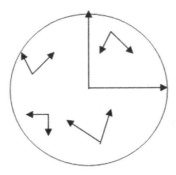

We can even imagine XYZ axes for the entire earth if that's what we need.

The *domain* of values of the axes may be continuous, but *space* itself might not be. At the moment, most physicists believe that space is discrete, not continuous. But the jury is still out on that question.

Reality Versus Abstraction

While I am on the subject of imaginary axes, I might mention I've read things by a few people on the Internet who reject the Special Theory of Relativity on the basis that it implies that a particle has as many relative velocities as there are possible *frames of reference* (another term to mean imaginary XYZ axes)—in fact, an infinite number of frames. That is true. But these frames of reference are an abstract conception. When it comes to abstractions, infinity is easily conceivable. Take for instance a family on cruise ship. The child riding his tricycle on the deck of the ship is inside a frame of reference which is the tricycle. The child is also inside another frame of reference—the ship. The lighthouse near the shore line is another frame of reference. The Earth is another frame of reference that the child is in, and so on. In each of these frames, the child moves with different velocities.

These are relative velocities because they are measured relative to a frame of reference. Consequently, relative velocities don't give the absolute velocity of a body. We may even imagine negative velocities, if we wish. Another argument against relativity is this one I obtained permission to reprint from the Web site rebel-science.blogspot.com:

> How does a particle "know" about its motion relative to another frame of reference so as to know how to move relative to them? [...] No particle in the universe can make use of the relative because it has no access to it. It follows that the universe does not use the relative.

I thank the author Louis Savain for this quote because it illustrates well a common misunderstanding about Einstein's Special Theory of Relativity. I wish to analyze all three statements of that quote.

This first statement is correct: a particle does not "know" its motion relative to another frame of reference simply because the particle doesn't need to care where it is in time or space relative to another particle. But this doesn't mean that relativity doesn't exist.

To easily understand this, imagine automobile A going at speed V_A on a road when automobile B approaches it at speed V_B. The driver of automobile A will see the automobile B approaching at the relative speed V_A+V_B. Nobody can deny that this is a relative speed, and the driver knows that this is a relative speed, but the automobile surely does not! In fact, the particles making up the automobile don't "know" that the other automobile is even there.

The second statement is incorrect, as a particle does have access to the relative. For instance, both automobiles "feel" each other coming at relative speed V_A+V_B. It's just that the automobiles aren't aware that this is a relative speed. Consequently, the third statement is also incorrect. The truth is that the universe uses relativity all the time as it's simply the result of the interaction of two entities, big or small. Using Einstein's Special Theory of Relativity proper terminology, these two entities are frames of reference.

Take the solar system as another example. If a scientist measures the speed of the earth within the frame of reference of the solar system, and then measures the speed within the frame of reference of the whole galaxy, the two speeds will differ. Thus, the earth has many more relative velocities. That is fine because these velocities are the result of frames *imagined* by humans. Nature has *no* sense of such frames; nature only cares about interactions. The earth "interacts" relative to the solar system, which "interacts" relative to the galaxy. So what's the real speed of the earth? There's no way of knowing because there's no absolute frame of reference in space (actually, there might be a sort of absolute frame of reference as will be suggested in the section "An Absolute Frame of Reference!" in Chapter 5).

As there's an obvious relationship between speed and energy, the previous paragraph seems to imply that a body has no absolute energy. This seems illogical and ridiculous. We can imagine a relative energy, but surely there has to exist an absolute energy to a body or a particle. There is, and we'll see in Chapter 5 how nature might sense absolute speed and energy.

But we don't need to wait until we reach that chapter to be persuaded that energy is absolute. As mentioned in the previous chapter, energy manifests itself at the quantum level. At that level, there's no abstraction (except for time as we'll see in a later chapter). Velocity, matter, and masses are abstractions because they are things constructed from energy. As will be explained later in this section, velocity is the result of a calculation, so it's clearly an abstraction.

Essentially anything above the quantum level is an abstraction. For instance, energy coming from a steam of vapor is an abstraction because it's made of matter moving around. Matter is an abstraction, and so is motion of particles. As we'll see later in Chapters 5 and 8, waves account for motion, not particles.

Other realities of life are actually abstractions as far as nature is concerned. Have you ever been to a three-dimensional movie where people wear special glasses? But is the picture really in three dimensions? No. So why do we see in three-dimensional something that's projected on a two-dimensional surface? Here's how: Two copies of the exact same pictures are sent to the screen, but in different colors and at slightly different angles—the same angles that separate our eyes. From birth, our brain gradually learned to see in three-dimensional. But in the first few days of weeks of life, we didn't see much other than fuzzy figures and colors.

Being able to see the objects around us is something that our brains had to *learn*. It's just like figuring out how to throw a basketball through the hoop. Note that our brains learn to view in three-dimensional using *two* senses: touch and sight. Touch allows us to learn how to perceive depth so as to realize that an object has three dimensions. Thus, the sense of touch is needed too. Accordingly, a person born blind can still "see" in three-dimensional by touching. At a three-dimensional movie, our brain perceives that something in two-dimensional is three-dimensional because it receives the same signals and in the same angles as in a real three-dimensional world.

Having said this, as our brains had to learn to see; maybe they also had to learn the *significance* of everything they see. The brain had to learn what velocity, acceleration, length, mass, etc, mean. The significance of something is the result of *interpreting* what has been learned. Once a brain learned to see in three-dimensional, it realized the significance of *length, height,* and *depth.* When viewing something in motion, it learned what *velocity* is. By holding objects, the brain learned what a *mass* is, and what the *force* needed to hold it is.

Keep in mind that these concepts are interpretation of what we learned of the physical environment. What's nature's interpretation of all this? You can be sure that nature has a different view of these concepts than we do. By nature, I mean also the "observer." Yes, there's an observer in the universe! Is it God? (We'll discuss this more in later chapters).

Our brain limits our view of nature. This is particularly true at the quantum level, so what we perceive as real may not actually be the true reality. We essentially live in a virtual universe—an idea that will be further developed in the section "Our Virtual Universe" in Chapter 10. This reminds me of another of Einstein's thoughts: "Reality is merely an illusion, albeit a very persistent one." It's indeed not at all clear what reality is. The limitations of our brain are somewhat like our social limitations. A society's point of view is shaped by many factors such as its cultural background. For instance, in some societies, a boy is considered a man only after he has passed through some ritual. The definition of manhood differs from one society to another. Because manhood is a relative concept, it's also somewhat abstract.

I mentioned earlier that to nature, velocity is abstract. This may seem ridiculous, but think about this: how do we measure velocity? Answer: by taking a distance divided by time. So velocity is the result of a *calculation*. It's also the reading on an odometer of a motorized vehicle. Nature doesn't perform any calculations or read anything, so nature has no concept of velocity. What nature detects though is motion, which is essentially a transfer of energy. Velocity is a measure of that transfer of energy.

The notion of size is also abstract. Nature has no sensitivity as to the size of an object. For an object to have a size, it has to have a surface. But like I said in Chapter 1, deep down matter is made of wobbly, fuzzy, tangled-up waves. As two objects don't really touch, their surfaces are not well-defined. You might attempt to refute my argument by saying that if size doesn't exist, then I should be able to insert as many coins in my piggybank as I want. Why is it that the piggybank becomes full if nature has no sense of size? My reply is that the piggybank becomes full because the waves making up each coin and the piggybank eventually interact with one another in such a way that no more waves can be inserted. It's all about waves, not size as far as nature is concerned. As a corollary to my argument, the notion of length is also abstract.

In my opinion, time is also abstract! Indeed, we measure time by means of motion of some object in *constant* velocity such as the hands of a clock. This implies that we measure time using *distances*. I just argued in previous paragraphs that both velocity and distance are abstract as far as nature is concerned. This should imply that time is also abstract as far nature is concerned. That then begs the question: does time really exist? Of course time exists, right? Probably, as its purpose is to ensure motion, as I discussed in the section "A Fourth Dimension is Necessary" in this chapter. But we'll see in later chapters that the time dimension is profoundly problematic when it comes to explaining motion. We'll continue this discussion of abstraction, mainly in Chapters 5 and 9.

Space "Tension"?

Let's go back to the graph depicted earlier in this chapter with the circled particle:

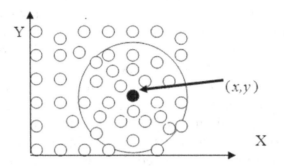

As shown, the presence of the particle causes the floats to try to remain at an equal minimal distance between one another. How far from the particle will this effect take place? Answer: it's a classic snowball effect. One float moving toward the particle causes another float to displace itself. It's like when a child removes a piece of bubble gum from a gumball machine—it causes all the remaining balls to move around a bit, especially those close to the ball that was removed. Keep in mind that these floats aren't particles: they are nothing more than places in space that particles can skip onto.

What causes this movement of the floats? I'm not referring to the movement of particles at this time, just the floats. A displacement of a float causes a decrease in the *tension* in the space around it that causes other floats to react:

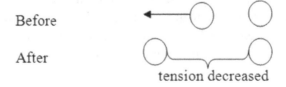

This is another classic case of cause and effect. What sort of tension is that? It certainly can't be tension in the way we understand it with matter, as with the tension in a rope that holds a box together. Tension is a reaction force to another force so to keep a balance between the two forces. When the tension is too high, the rope snaps. Quite obviously space is *not* matter, and those floats are not matter, so it's difficult to grasp what tension could mean in this situation. I'll propose a realistic physical interpretation in the section "Emitted Energy Versus Internal Energy" of Chapter 4 and in "Motion Finally Explained!" in Chapter 8.

Relationship between Energy and Space

The constitution of space remains puzzling. As noted previously, physicists continue to debate the continuity of space. Perhaps the notion of continuity isn't the same in physics as in mathematics. Perhaps we look at the behavior of particles in the wrong way. If they are indeed particles, then we have no choice but to question what's between them. But as far as nature sees it, there are no particles! As I said a few times in Chapter 1, matter is made of "balloons" of tangled-up waves that when packed together in a certain way, *act* as particles. This will be shown with mathematical formulas in the section "An Absolute Frame of Reference" in Chapter 5.

How do waves travel through space? First, waves transport energy. It seems that the nature of space is closely tied to what energy is because matter is energy—and waves are too. I have a satisfactory answer as to what space is that I'll suggest in Chapter 6 and in the sections "The Energy Dimension" and "Motion Finally Explained!" of Chapter 8.

One thing is certain: where there's matter or energy, there's obviously space. So there's a cause-and-effect relationship. Does that mean also that energy creates space? If so, then where there's space, there's also energy. The relationship goes both ways because what would be the point of having space without energy in it? It's like if we blow air into a balloon. Air fills the whole balloon so there's no empty space inside. I discuss in the section "Motion in Space–Time:

The Plot Thickens!" in Chapter 5 a little more about the relationship between energy and the motion of particles in space.

This isn't the end of the story. It turns out that to better understand space, we have to also understand time, or at least what we think time is. (Let's discuss that some more in the next chapter.)

The nature of space and time will occupy a great deal of discussion in this book. Coming next is the chapter about time, but we'll come back to the discussion of space later in various places, notably in the sections "Motion in Space–Time: The Plot Thickens!" of Chapter 5 and "The Energy Dimension" and "Motion Finally Explained!" of Chapter 8.

Chapter 3: What Is Time?

Throughout this book, you'll find that at times I argue that the time dimension exists, but later that it might not exist as a real dimension but rather as an imaginary one. Regardless, time is a dimension as Einstein's theory of space–time proved. Nonetheless, the "nature" of time remains unclear, but I'll attempt to clarify this in the next chapter.

We measure time with a clock whose hands change position at every tick. So a change of state of an object is evidence of time passing by. Why does a hand move? It moves when energy is transferred from an internal part of the clock, probably a gear, to the hand. Some parts interact with other parts better or faster depending on how the clock was manufactured. This is why no two clocks tick at exactly the same rate. Even clocks that use quartz oscillations aren't absolutely accurate.

We used the time interval symbolized by $\varDelta t$ in the previous chapter. This interval is the smallest unit of time needed to transfer energy—one that happens at the smallest subatomic level. The reference of that tiny interval $\varDelta t$ suggests that time is discrete, but doesn't imply it. I'll suggest later in this chapter what some physicists also believe: that the time dimension may be continuous. The transfer of energy between two subatomic particles is evidence of time passing, though it doesn't explain what time is. We'll explore that question soon.

As in the previous chapter, the discussion of time in this one will remain conceptual. The notion of lifting the particle in time over an interval of space is purely conceptual in that the particle does not really physically move in the time dimension. In the section "Motion in Space–Time: The Plot Thickens!" in Chapter 5, this notion of motion in time will be further investigated.

Discrete or Continuous Time Dimension?

Most physicists believe that time is discrete, as it moves along in quantum steps. However, many philosophers have difficulty with this idea, and wonder what to make of the interval between two quantums of time. We encounter here the same dilemma we did with our discussion of space. As with space, I concluded that a fourth dimension is necessary regardless of whether space is discrete or continuous. Recall from the previous chapter the case of the moving pen from inside that we considered the motion of a single particle:

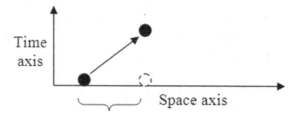

In the previous chapter, it was said that the XYZ axes are imaginary, and can therefore be placed anywhere and in any orientation. This same applies to the XYZT axes. So the axis above could easily have been oriented as follows:

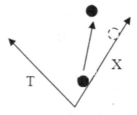

or we even could have placed the starting particle at the origin:

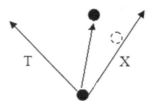

Another thing is that the illustrations I used earlier to illustrate the path of a particle through space–time are slightly incorrect. The correct picture as per space–time theory taught to university physics students is this one:

Before the particle moves:

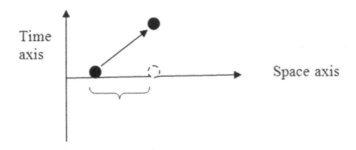

After the particle moves:

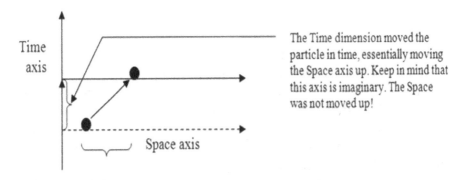

The Time dimension moved the particle in time, essentially moving the Space axis up. Keep in mind that this axis is imaginary. The Space was not moved up!

Only one of the three space dimensions is used here, but the same idea applies to the other two as well. The particle moves up in the direction of the time axis—something that applies to everything in the universe, including us. So the graph below that was used in the previous chapter isn't quite correct after the particle has skipped five times:

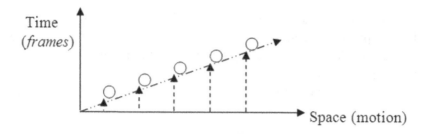

This one is correct:

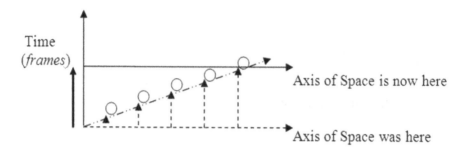

The axis of space has moved up with the time.

I've imagined a model almost identical to the one for space to consolidate the views of both physicists and philosophers. The universe has four dimensions, but to illustrate my idea, let's imagine only two dimensions, X (space), and T (time). According to Einstein's Special Theory of Relativity, the speed of a particle shortens the speed of time (we'll examine that in Chapter 4). In other words, the time intervals get shorter. Imagine the path of the particle as follows as it accelerates in space:

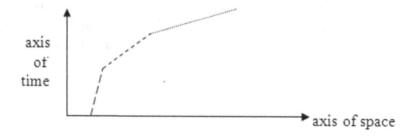

As can be seen, the time intervals shorten as the speed increases. Both the X and T axes have to be able to map all kinds of space and time intervals depending on the speed of the particle. The particle here speeds up in the second segment, and then accelerates much more in the third segment, thereby decreasing the time intervals. But this is only one particle: the universe has billions and billions of such particles, and the time axis has to map them all. So, can that axis really be discrete? I doubt it.

There's another way to see that the time dimension might very well be continuous. Recall that transmission of energy is evidence of time gone by. Let's continue with the example in the previous paragraph that considered only one particle. In the universe, there are billions and billions of particles that transmit energy at the speed of Δt fraction of a second. The time dimension has to be able to map all those intervals, but obviously those intervals don't happen in synchronicity. For instance, for four particles, it might happen this way:

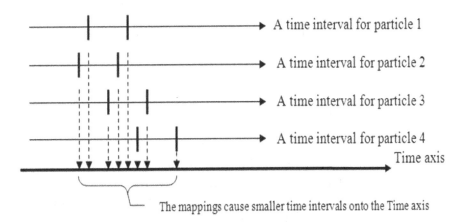

A time interval for particle 1

A time interval for particle 2

A time interval for particle 3

A time interval for particle 4

Time axis

The mappings cause smaller time intervals onto the Time axis

We can easily see that although the smallest interval of time for energy to take place for each *individual* particle is Δt, the time dimension has to map intervals much smaller.

A continuous dimension means that there's a point at every single possible location. The definition for continuity for time is the same as the one expressed for space in the previous chapter. For a given point t_1, there has to exist a point t_2 such that the interval $|t_1 - t_2| < \varepsilon$, however small $\varepsilon > 0$ may be.

If there were an infinite number of particles in the universe, this condition would be true, and the time dimension would be certainly continuous. But there are only a *finite* number of particles in the universe. Oops! There's no way of knowing when those intervals from particles take place. As when an interval of energy transfer takes place, it implies a time interval, then all possible time intervals on the time dimension (no matter how small) have to be "open" in case an energy transfer takes place in that interval somewhere in the universe. This is a strong argument for a continuous time dimension. That dimension is continuous while "motion" in it is discrete.

But hold it! You might point out two possible flaws in my argument for a continuous time dimension, as it assumes two things:

1. That all particles in the universe use only one and the *same* time axis. This seems to contradict what was said earlier about the XYZT axes being imaginary, and so for every particle of interest, we can imagine its own XYZT axes.

2. That the time intervals associated with all particles in the universe occur asynchronously (i.e., they can happen at any time).

However, in my opinion, there are no such flaws, as explained below.

To refute these two flaws, the rest of this section requires some knowledge from Chapter 4. For greater clarity, I advise you to jump ahead and read the first sections of Chapter 4 up to and including the section "The Correct Tick Formula." However, I'll provide now a quick summary of that section, as it essentially says that as a particle speeds up in space, its "motion" along the time axis slows down. Conversely, as a particle slows down, its "motion" along the time axis speeds up.

To refute the first flaw, consider now this picture of the universe with the same four particles above with their path shown as the dotted arrow:

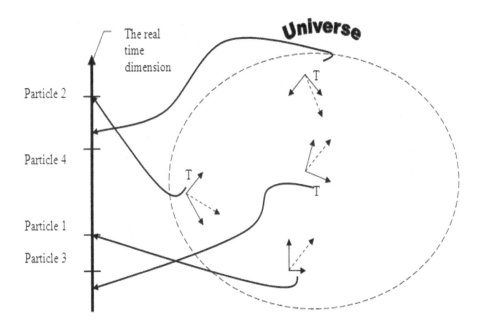

We can imagine each particle having its own time axis just like we've done so far. Also the dotted arrows are at different angles of motion in the space–time dimensions, indicating that the particles go in all kinds of directions in those dimensions. It was discussed earlier that the space XYZ axes are imaginary for each particle.

Is the time axis *imaginary* as are the space dimensions? Yes, but the big difference between the space and time axis is that the time axis of each particle has to map to a *same* time dimension axis shared by *all* particles of the entire universe! In other words, there's an *absolute* time axis, whereas the same can't be said of the space axes. In the picture, I showed the time axis as outside of the universe just because it's probably easier to visualize the idea explained here that way. Also by keeping it outside, I can remind you that the time axis is obviously not visible in the universe. In fact, the time axis isn't really an axis because time manifests itself everywhere in the universe and it doesn't really have an orientation.

Placing in the figure above the time axis outside the universe isn't just a means to convey the idea of an absolute time axis. As you'll see in section "The Time Dimension is Nonphysical" of the next chapter, time is just that—nonphysical. This implies that time resides outside of the physical world. Moreover, as said before, the time axis is an absolute axis in that everything maps to it. Indeed everything has a time associated with it. In section "The Clock of Universe" of the next chapter, it will be shown that there exists indeed an absolute time clock.

At first grance, my idea here seems to contradict Einstein's Space-time relativity which says that time is relative. No, there's no contradiction. An absolute time axis still allows for time relativity as is being discussed here. Read on, and you'll see what I mean. Also, later, we'll further investigate the relation between this absolute time axis and time relativity in section "Past, Present and Future" of the next chapter where I'll suggest a stunning possibility.

Now, why is there a need for an absolute time axis for the universe? This is rather obvious. Take for instance particle 1. You see in the picture above that in the time dimension, it's ahead of particle 3. Suppose that particle 3 all in a sudden begins to slow down, thereby increasing its time "tick," and meanwhile particle 1 starts accelerating at a very high speed, thereby decreasing its time tick. Here is the picture in space–time:

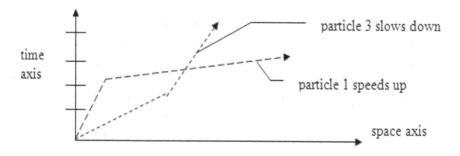

The graph shows that eventually on the time axis, particle 3 will catch up to particle 1 at the intersection of the two paths. If we focus only on the absolute time axis, the picture is:

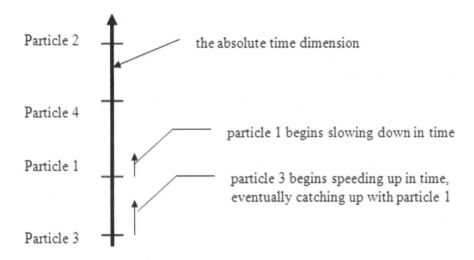

Particle 3 moves up faster in time than particle 1, so eventually particle 3 will go past particle 1. How could it do so if there were not a *common* absolute time axis? I don't see how that could take place otherwise. This refutes the first flaw.

The second flaw states that time intervals of all particles of the universe don't occur asynchronously. In other words, they are synchronous with respect to the time dimension. If so, then the absolute time axis is stepped evenly as follows, suggesting that the time dimension is discrete:

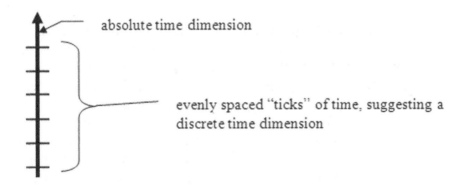

Remember that when a particle emits energy or moves, it is evidence that time went by. For the time axis to act synchronously as just depicted above, all particles would have to act in a synchronous way.

It's like in a military parade: all soldiers walk in step. If it works that way with particles too, this suggests that it's the time dimension that tells particles when they can emit energy or move in space. It doesn't make sense that all particles of the universe be synchronized. So the second flaw is refuted as well.

So there we are: both alleged flaws have been refuted. Accordingly, the time dimension could very well be continuous after all.

Another way to see that an absolute time dimension truly exists is that time (a) is nonphysical and (b) the concept of time is the same everywhere in the universe. Any sort of dimension with these two properties can be viewed as absolute. For instance, take the concept of weight. Suppose that every planet in the solar system was inhabited. Then we wish to make a graph of the weight of every person on the earth as well as on all the other planets. Whould I draw a separate graph for each planet? Yes, I could. This way, I'll know that a person with a bigger weight has a bigger mass—weight becomes synonymous to mass. However the *concept* of weight—the force applied by the gravitation of the planet onto the mass of a body— is the *same* for all planets. Consequently a person that weights 80 kilograms here on earth and another person weighing also 80 kilograms on another faraway planet weigh the same regardless of their mass or gravitational pull. Therefore they cannot tell the difference in their respective mass and gravitational pull.

So, there exists an absolute weight axis for the cosmos for the same reason that there exists an absolute time axis. So, I may draw a single and absolute graph with the vertical axis being for the weight for each person, and the horizontal axis mapping each person, and this would include all inhabitants of all planets in the cosmos. A change in the gravitational pull of the planet or a change in the mass of the person would make that person change position onto the absolute weight axis. Also if that person was to travel to a distant planet, his weight there would change, and so his position onto the weight axis would change too. Just the same with the absolute time axis: a person taking a trip in a speeding rocket would see its time clock slow down, and so his position onto the time axis would change.

To sum up, there's a lot of room in the XYZ space dimensions, but very little in the time dimension. That dimension is a lot busier as every single particle in the universe has to map to it. This is the essential difference with the space dimensions and suggests a continuous time dimension. There's something definitely different about the time dimension as we'll see again later.

Most physicists believe that time and space are both discrete, but I beg to differ. Based on my arguments presented in this chapter, I conclude that time is continuous. Because of the simple equation for motion, *distance = speed × time*, if time is continuous then space is continuous as well. Motion in those dimensions is discrete, however. In the section "Motion Finally Explained!" in Chapter 8, I provide another argument for continuous space.

What Makes Time "Tick"?

If a transfer of energy between two entities is evidence of time passing by, is the reverse also true? If a "tick" of time has taken place in the time dimension, does that imply that a transfer of energy had to have taken place somewhere in the universe? Can time go by without any energy having been transferred? Another way to put it: does the transfer of energy at the quantum level dictate the tick of time? We'll see in Chapter 4 that the answer is *yes*. This means that time requires energy in order to exist. An implication of this is that time cannot possibly have existed before the beginning of the universe. We'll discuss this issue in detail in Chapter 7 and 8. In Chapter 4, we'll also see that the smallest tick of time can be smaller than the Δt symbol used in this chapter.

Moreover, should time be no more than an imaginary mathematical dimension then it can easily be imagined to be continuous, meaning that the interval Δt may be as small as possible. If the time dimension is imaginary, then who is it imagined by? Is it by humans? Is it by nature? Is it by God? We'll find out later.

Remember that in the previous chapter, I said that if there's energy, there's space; if there's space, there's energy. With time, it goes both ways as well: if there's energy, there's time; if there's time, there's energy. Consequently, if there's space, there's time; if there's time, there's space. Space, time, and energy go together.

Consequently they were created simultaneously at the moment of the Big Bang.

Dimensions Versus Axes

The existence of time as an *attribute* of everything in the universe is acceptable, but the existence of the time *dimension* isn't so obvious. I'll detail my reasons later in the section "Motion in Space–Time: The Plot Thickens!" in Chapter 5 where I discuss a little further the "motion" of particles in time.

By the way, I'm in good company with my doubts about the existence of the time dimension. Kurt Gödel, one of the most brilliant mathematicians the world has ever seen, didn't believe that time existed as a dimension of the universe either. In the 1940s, he often debated this with Einstein at Princeton University where they both worked. We'll talk about Gödel's most brilliant accomplishment in Chapter 7.

While we are on the subject of dimensions, there's an important difference between a dimension and an axis. The space XYZ axis and the time axis are depicted often in this chapter and the previous one. An axis is anything to which units can be imagined and quantities measured. So an axis can be devised to measure an attribute, a property, a factor, or a parameter about something. An axis is somewhat like a ruler—a way to measure something like weights, temperatures, etc.

A dimension is more than an axis. In physics, a suggestion to imagine an additional new dimension needs to be considered only as a last resort when physicists are at an impasse. For instance, few physicists thought of time as a dimension until an experiment conducted in the late 1890s showed that the speed of light remains constant regardless of the direction from which we look at it. This flew in the face of common sense and eventually forced physicists to consider time as a dimension. For a physicist to speculate on the discovery of a new dimension is a very serious claim as physicists have to "get real."

On the contrary, mathematicians can invent dimensions at will, but with one condition: dimensions have to be *independent* from one another. That means that there's no inherent relationship between the values in one dimension and another dimension. This condition applies to both physical and mathematical dimensions. For instance I can move my hand in any direction in the XYZ axis in space. Of course when a car travels down a road, the (x,y,z) coordinates taken by the car are in a relationship dictated by the direction of the road, but the XYZ axes themselves are independent. Here are a few examples to explain the difference between a dimension and an axis:

1. Suppose an oven has just been turned on with the heat dial turned to a target temperature of H. A good mathematician could come up with a formula that gives the temperature h of any point (x, y, z) in space dimensions in the oven at any given time t in the time dimension and for any target temperature H. So the temperature of any point in the oven is said to be described by the 6-tuple (x,y,z,t,H,h), where h represents the temperature at the point (x,y,z,t). A formula would give the value of h depending on the parameters (x,y,z,t,H). There are five parameters, so five axes, and they're all dimensions except for heat. Heat is *not* considered by physicists to be a physical dimension, but instead a mathematical dimension. Personally, I consider time to be a mathematical dimension as well (more on that in later chapters).

2. Take the path of a ball across a field. At any point (x,y,z,t), the ball has a speed V. Can V be seen as a physical dimension or as a mathematical dimension? Neither one because its values can be derived from the other dimensions using Newton's formula of motion in a gravitational field. We could, if we wanted, draw a speed axis for the ball, but speed remains just a parameter or an attribute of the ball. How about the size and mass of the ball? Yes, they are dimensions because they obviously can't be derived from the other dimensions. But they are *mathematical* dimensions, not *physical* dimensions. Size is obviously related to the space dimensions, and mass (energy) is related to the space dimension (recall Chap-

ter 2) because there's no space where there's no energy, and vice versa. Because of that relationship, size and mass aren't physical dimensions.

3. Suppose a physician came up with a formula that calculates the health index of a person based on a (age), h (height), w (weight), and c (calories consumed per day). So the 4-tuple (a,h,w,c) would represent a point in a four-dimensional world for human health. None of these are real physical dimensions, of course. Also, it could be argued that only age and calorie consumption are mathematical dimensions. As the others (height and weight) are not independent of the age or calorie consumption, they are just parameters from which an axis can nonetheless be drawn.

Chapter 4: Why Is Time Relative?

The previous chapter falls very short of a complete explanation for what time is or how it works. Before Albert Einstein came along, people and scientists always believed that time was absolute—it ticked along at the same rate anywhere in the universe. In the early 20th century, Einstein stunned the scientific community when he proved that time ticks at *different* rates depending on the frame of reference that the subject is in! In other words, time is relative. What implication on the "nature" of time does this relativity have? Upon a long reflection about this, I realized that time may not be at all what people think it is. You'll find out later in this chapter what I mean.

At this moment, you may be reasoning that if time is relative, then there isn't an absolute time axis. Therefore I was wrong in the previous chapter to suggest that such an absolute axis exist. No, I am still correct, and I will clarify this in section "Past, Present and Future" toward the end of this chapter.

Discussion about time will resurface many times for the remaining of this book because it's implicated in just about everything in nature. But let's first tackle time relativity. As an illustration of Einstein's Special Theory of Relativity, you may have heard of the famous thought experiment involving the two twins. One twin remains on the earth while the other travels at a very high speed through space. When that twin comes back many decades later, he finds that he has aged more slowly than his twin who remained on earth. This effect is called *time dilation*.

How Einstein Derived Time Dilation

Einstein imagined a train going past an observer at speed V. In the train, there's a light at the ceiling throwing a beam straight down toward the floor. Supposing that the ceiling is a height of H above the floor, then $H = ct_T$, where t_T is the time it takes light to reach the floor in the train and c is the speed of light. Here is the picture:

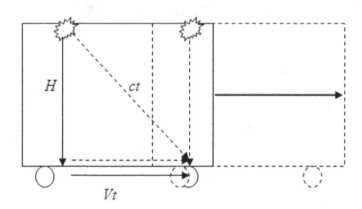

As the train moves a distance of Vt, t being the time as perceived by the observer outside the train, that observer will view the light travel downward diagonally. This is the line labelled ct. Regardless of the speed of the train; the light will always be seen at the speed of c by the observer. This isn't an illusion. Light always travels at speed c no matter the angle at which the light comes toward or away from the obverser. This is what physicist Clark Maxwell proved mathematically in the 1800s, and the effect was confirmed in the late 19th century by two physicists in the famous Michelson-Morley experiment that bears their names.

Einstein's thought experiment is brilliantly simple. Clearly the line H is shorter than the line ct, so $H < ct$. Given that $H = ct_T$ then $ct_T < ct$. The speed c cancels from both sides of this inequation giving $t_T < t$. That is, the time inside the train passes by more slowly than the time outside the train! This result defies our senses, but it's correct.

As you can see in the figure, a right-angled triangle is formed with sides of lengths $H = ct_T$, Vt and ct. Now, do you remember from high school the Pythagorean equation for a right-angled triangle $h^2 = a^2 + b^2$, where h is the hypotenuse and a and b, are the sides?

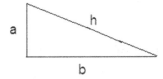

Let's use that equation here. The length ct is the hypotenuse, the lengths H and Vt are for the two right-angled sides. So the Pythagorean equation becomes

$$(ct)^2 = H^2 + (Vt)^2$$

or

$$(ct)^2 = (ct_T)^2 + (Vt)^2$$

or

$$(ct_T)^2 = t^2(c^2 - V^2)$$

Taking the square root on both sides gives

$$ct_T = t\sqrt{c^2 - V^2}$$

Dividing by c on both sides gives

$$t_T = t\sqrt{1 - (V/c)^2}$$

This formula says that the faster the train goes, the slower time inside it goes by. If the train could reach the speed of light, time would stop ticking!

An historical note worth making here: Dutch physicist Hendrik Lorentz discovered this exact same formula above that came to be called the *Lorentz transformation*. Many historians are in agreement that Lorentz was indeed the first to derive this equation, not Einstein. It's however Einstein who first interpreted this equation as time relativity. Nevertheless I believe that Lorentz should be recognized as the Father of Relativity Theory.

Derivation of Time Dilation from the Quantum Level

Einstein's Special Theory of Relativity proves that time dilates depending on the speed of the object or body: the faster a body moves, the slower is its time, so the slower is its aging. The converse is also true: the slower a body moves, the faster is its time, so the faster is its aging. Why is this so? In my opinion, Einstein's thought experiment of the previous section doesn't provide a satisfactorily answer.

In this chapter (and the next two), I show that the answer lies in quantum physics! Indeed taking the path of quantum physics allowed me to find out what time really is. Be ready for a surprise! But you'll have to be patient because a complete answer is possible only once I deeply wondered about the nature of motion and gravitation, the subjects of Chapter 5 and 6 respectively. So you'll have to read up until then.

In the meantime, in this chapter, I provide a very simple and unique explanation for time dilation based on my own discovery that at the quantum level, there's a one-to-one relationship between the energy of a particle and its time clock. We'll discover that at the quantum level, that the same time dilation factor that Einstein discovered from our macro level applies at the quatum level too! Given that Einstein discovered time dilation before the age of quantum physics, you can be sure that he came about it differently than the way I did. Here is how I derived time dilation.

The mechanism for the effect of time dilation (time relativity) puzzled me for a long time, but the idea is actually simple. There are two ways to look at it:

1. If we use the analogy proposed in Chapter 2 to explain time with reference to movies as the *frames* (or snapshots) taken by nature of an object moving by, then simply put, the *faster* an object goes, the *harder* it is for the time dimension to keep track of where the object is. So it loses "time" thereby decreasing the "tick" of time. This is just an analogy, not the completely correct explanation, but let's study it for now in the section below and see where it takes us.

2. The second view is the correct physical explanation of time, not just an analogy. Recall that in the section of Chapter 1 titled "Nature's Sensitivity to Energy" I said that nature only senses energy. Imagine a particle speeding by at speed V. Nature will detect that particle only when it senses its presence, and that will happen only when nature senses the particle's energy. But nature can't sense the presence of the particle *instantly* because it can detect the particle only *immediately* after it feels its energy.

As will be determined below, the energy is emitted at the speed of light, not instantly. Consequently, there's an inevitable *delay* between the motion of that particle and the moment that nature detects that motion. It's that delay that nature uses as the "tick" of time of that particle. It's that simple. So, the *faster* the particle moves, the *shorter* the delay of time, hence the *slower* its tick of time. We'll delve into this second idea in this chapter in the section titled "A Better 'Tick' Formula."

Keep these two different interpretations of time in mind because we'll be studying them carefully in the first six sections of this chapter. Let's begin with the first interpretation.

By the way, apparently Einstein's Special Theory of Relativity has been proven to be correct in laboratories. Therefore, time is truly a dimension. Great! However, his theory is silent regarding the *meaning* of that dimension. We'll investigate that as well in this chapter, Chapter 8 and in Chapter 10.

From the Planck Units to a Time "Tick" Formula

Remember from Chapter 2 that nature has a hard time detecting where an object is *now*. In the section of that chapter titled "A Fourth Dimension is Necessary," I introduced the idea of an object's necessary ability to skip some space interval Δs and time interval Δt in order to account for motion.

Guess what? It turns out that those intervals do exist! The physicist Max Planck discovered them at the very beginning of the twentieth century. He found that Δt is actually of value 5.4 x 10^{-43} second which he called the *Planck time*. The interval Δs is actually 1.6 x 10^{-35} meter and is called the *Planck space*. These numbers were adjusted to fit into quantum theory, which means that in reality these values vary depending on the speed of the quantum particle.

Now that we have a name for our two intervals, let the Planck time be noted by P_T, and the Planck space be noted by P_S. These are respectively the shortest unit of time (5.4 x 10^{-43} seconds) and the shortest unit of distance (1.6 x 10^{-35} meters).

The way to visualize the implication of these two values is this: At every P_T time, nature gives the time dimension a snapshot of the object. Suppose that an object travels at a speed of V meters per second. There are $1/P_T$ frames of time in a second. In a meter there are $1/P_S$ steps (this is also the number of floats) that the object travels through. So the object moves at a rate of $V \bullet (1/P_S)$ steps per second, and there are $1/P_T$ frames of time in one second. So how many frames of time of the object are taken in a single step as *perceived* by the observer (that's nature)? This many frames:

$$Number\ of\ frames\ per\ step = \frac{\frac{frames}{second}}{\left(\frac{meters}{second}\right) \times \left(\frac{steps}{meter}\right)}$$

or

$$Number\ of\ frames\ per\ step = \frac{(\frac{frames}{second})}{(\frac{steps}{second})}$$

or

$$Number\ of\ frames\ per\ step = \frac{1/P_T}{V \cdot (1/P_S)} = \frac{P_S}{V \cdot P_T} = \frac{P_S/P_T}{V}$$

It's interesting to note that $P_S = 1.6$ x 10^{-35} meters divided by P_T = 5.4 x 10^{-43} seconds gives us 3 x 10^8 m/s—the speed of light! In other words, $P_S/P_T = c$, where c stands for the speed of light. So the formula becomes

$$Number\ of\ frames\ per\ step = c/V$$

Actually the first frame is a "skip" move toward the first frame, so it doesn't count. Thus, the correct formula is

$$Number\ of\ frames\ per\ step\ =\ c/V - 1$$

This is a time dilation formula that says that the more frames that the object goes through, the more time it will go through. Likewise, the fewer frames that the object goes though, the less time it will go through. Given the formula above, it can be seen that

1. The *slower* the object moves, the *more* frames will be taken of it, and so the *more* time that it will go through. In other words, its time is *faster*.

2. The *faster* the object moves, the *fewer* frames will be taken of it, and so the *less* time that it will go through. In other words, its time is *slower*.

Note that if the object goes at the speed of light, no frame is taken: time has stopped for that object! This is precisely what Einstein's Special Theory of Relativity says.

Note that the formula

$$Number\ of\ frames\ per\ step\ =\ c/V - 1$$

doesn't seem totally correct. If the speed $V = 0$, then we have $c/0 - 1$ = infinity many frames per step. This result comes from the fact that if the particle doesn't move, it takes no step forward. However, at the quantum level, particles *always* move, as there's no such a thing as a still particle. So the formula doesn't apply in this case of $V = 0$.

When we think about it, the formula $c/V - 1$ is incorrect for another reason. The formula does work when the object goes at the speed of light, but for the wrong reason. The formula doesn't answer the question "Why does time slows down as an object speeds up?" but rather this one, "At what speed does an object have to go for time to stop?" This error will be fixed in the sections "A Better 'Tick' Formula." and "The Correct 'Tick' Formula."

Energy at the Speed of Light

Now that we know that the ratio of the Planck time and Planck space is the speed of light, this implies that massless energy transmits from one particle to the next at the speed of light. However, this doesn't imply that a *particle* jumps from one Planck step to the next at the speed of light. (This will be better demonstrated in the section "Emitted Energy Versus Internal Energy" later in this chapter). Remember that we're at the quantum level. None of this implies that a baseball moves through a field at the speed of light! All it says is that quantum particles within the ball *emit energy* at the speed of light. This agrees with what we discovered in Chapter 1, and what will be discovered in section "Motion Finally Explained" of Chapter 8.

Note that because P_T is a time value, it has to be subjected to the time dilation formula of the previous section. Accordingly, its time interval will shorten or increase depending on the speed of the quantum particle. It turns out that the value of P_S changes too, and at the *same* rate as P_T. This is because it's always true that $P_S/P_T = c$. It will be shown later in this chapter in the section "Emitted Energy Versus Internal Energy" that

1. The *slower* a quantum particle moves, the *longer* is its Planck time, and the *longer* its Planck space is too.

2. The *faster* a quantum particle moves, the *shorter* is its Planck time, and the *shorter* its Planck space is too.

Later in this chapter, we'll devise a formula for these Planck intervals.

A Better "Tick" Formula

Like I said before, the formula developed a couple of sections ago

$$Number\ of\ frames\ per\ step = c/V - 1$$

works only when time has stopped. The reason that it fails everywhere else in the time dimension is simply that it doesn't take into account that the particle travels in space–time and not just space. In my description of the events, I implied that the particle did travel in time, but I didn't introduce that into the formula. The correct picture is that nature observes the particle going by at the speed v, and all the while this observer also *moves along* the time dimension. It's that motion that now needs to be taken into account.

OK—this is the explanation as accepted by nearly everyone who learned special relativity theory, including all physicists. But I take a different view: I believe that the theory is conceptually correct, but physically wrong! In reality, the observer doesn't move along the time axis, and neither does the particle. We'll deal with my view in the sections "Motion in Space–Time: The Plot Thickens!" in Chapter 5 and in "Motion Finally Explained!" in Chapter 8. But for now, let's continue with the official view.

Time moves the quantum particle along the time axis while the particle moves itself along the space axis, say the X axis. So at the beginning, the picture is as follows, just as nature is about to begin observing the particle speeding toward it:

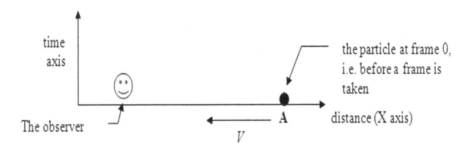

The particle starts at point A, moving at speed V. The step lands at point B. So the step is AB, and we wish to know the time lapse as perceived by the observer:

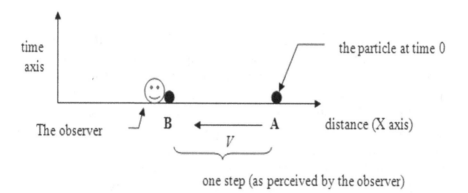

one step (as perceived by the observer)

In the previous section, I derived that there are $c/V - 1$ frames taken during that one step. But I just said that this is the wrong formula, so let's start over. First, let's remove the $- 1$. After we find the correct answer, the -1 will be put back.

The observer won't see the particle move immediately because it takes some time for the light from the particle to reach the observer. By that time, the observer is located at the dotted circle below:

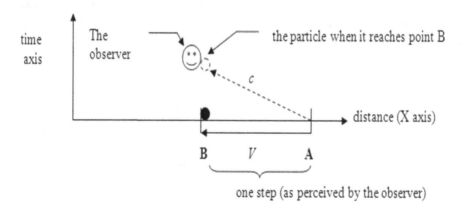

one step (as perceived by the observer)

Note that the dotted diagonal has a speed of c, the speed of light. This speed is involved here because nature can only sense the presence of a particle by the energy it gives off, which occurs at the speed of light. Remember in Chapter 1 where I said that nature is only sensitive to energy. On the X axis, the distance AB has c/V frames taken in what's *perceived* by the observer as one step. But in reality, the particle has gone through many Planck steps P_S, say s steps.

How many steps will be on the diagonal? The diagonal represents what comes at the observer. It's clear that what comes toward the observer is the same thing as what goes along the X axis, except for the delay of time. So there's bound to be the same number of steps *s* on the diagonal. Consequently, there's a mapping of steps from the horizontal to the diagonal, and this explains why the dotted particle is aligned with the real particle at point B (picture enlarged):

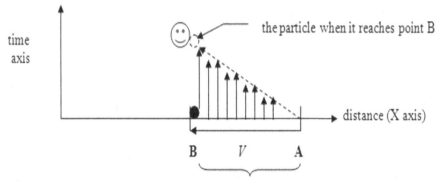

s number of real steps (not as perceived by the observer)

So the same number of frames is mapped onto the diagonal, even though the diagonal is longer! Surely if the diagonal is longer than the horizontal path, then more frames will be taken of the diagonal. How many more? The diagonal is c/V time longer than the horizontal line. So if there are c/V frames on the horizontal line, then there are $c/V \cdot (c/V)$ frames onto the diagonal. But there will be one less frame taken because at the start, no frame has been taken yet. Finally, the actual number of frames as seen by the observer is

$$Number\ of\ frames\ per\ step = (c/V)^2 - 1$$

We answered the question regarding how many frames of time of the object are taken in a single step as perceived by the observer. This formula says that as a particle speeds up, the number of frames taken by nature decreases, so the time "tick" as seen by the observer decreases to: time slows down. Note that when $V = 0$, an infinite number of frames per step are taken of the object. This makes sense because as the object isn't moving, it takes no steps, which isn't an interesting case. So the formula is usable only when the object moves.

The Correct "Tick" Formula

The mathematical developments so far may have helped you have a feel for time dilation at the quantum level. But the formula obtained so far is still not quite correct. This is because the first interpretation used these imaginary frames which are not what happens physically.

Let's now move slowly into the second interpretation of time, the correct one. If you recall in this interpretation, it's the delay between the motion of that particle and the moment that nature detects that motion that determines the "tick" of time of that particle.

In the development of the formula of this previous section, I took into account the time dimension and the time it takes for light to reach the object. We were getting close to the correct formula. However, I still used the analogy of frames taken by nature. This isn't really what happens, as nature doesn't take snapshots of the particle. The reality is the other way around: the observer has a passive role. Energy emitted by the particle is detected by the observer, as we'll see in the next section.

This time, we'll forget about the analogy of movie frames and focus on time. I'll ask the same question again ("how many frames of time of the object are taken in a single step as perceived by the observer?"), except that we'll remove the notion of frames.

The unit of time is of course seconds. So instead we'll ask this question: "How many *seconds* does the object take in a single step as perceived by the observer?" As we no longer are interpreting what's going on in terms of an observer taking snapshots, we'll now concentrate on the motion of the particle during one real quantum step (i.e., of Planck space P_S) rather than one snapshot (during which many real quantum steps were taken by the particle).

So, here it goes: At the moment that the particle starts moving, the observer won't see it immediately because it takes some time t for light from the object to reach the observer. The particle will have moved a distance of tV during that time, landing at point B (but the observer will not know that yet):

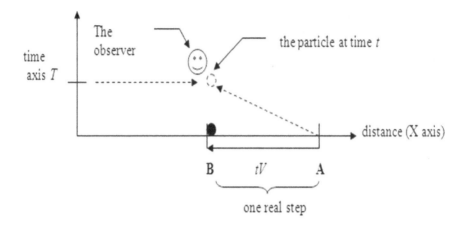

where T is the time according to the observer and t is the time "inside" the particle. In other words, T is the time as seen by nature outside the particle. At point B, the observer has just seen the particle at point A, not at point B. But then why place the observer at point B? This has already been explained in the previous section: there's a mapping between the horizontal and the diagonal.

The second time the observer sees the particle, it will be after some time t again, and the particle will have moved some more in the space–time dimension at the position $(2T, 2tV)$ as shown in the picture at the top of the next page.

Three right-angled triangles are formed from the same diagonal drawn earlier in the context of frames. All three triangles are obviously proportional, so it makes no difference which one we take to derive a time dilation formula. Each diagonal arrow is of length ct, which is the distance traveled by the light in space–time:

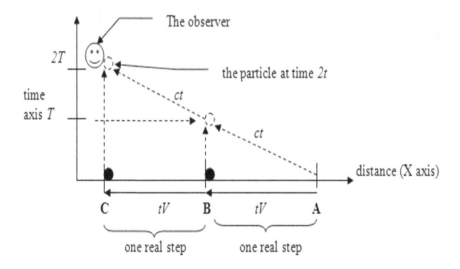

Let's use the Pythagorean equation for a right-angled triangle h^2 $= a^2 + b^2$, where h is the hypotenuse and a and b, are the sides:

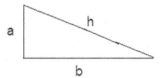

In the space-time diagram, the length ct is the hypotenuse, so it's given by the formula $(ct)^2 = T^2 + (tV)^2$. The energy, or light, emitted by the particle reaches the observer at the speed of light, c, plus V, so $c+V$. But we'll see in the next chapter in the section "Can Anything Go Faster than Light" that nothing can go faster than the speed of light. So $c+V = c$, and therefore, the formula is

$$(ct)^2 = T^2 + (tV)^2$$

or

$$c^2 = (T/t)^2 + V^2$$

The ratio T/t is the rate at which time has changed for the object. It's easy to verify that $(T/t)^2 = c^2 - V^2$

or

$$T^2 = t^2(c^2 - V^2)$$

This formula was derived from the picture at the quantum level of a particle skipping along, so actually the time variables are in units of Planck times. In other words, the formula could be written as

$$P_T{}^2 = P_t{}^2(c^2 - V^2)$$

or equally

$$P_T = P_t\sqrt{c^2 - V^2}$$

In any case, the ratio of change of time, $c^2 - V^2$, applies at our macroscopic level as well as the quantum level because at the macroscopic level; the result is the "average" of what happens at the quantum level.

You may have noticed that there's a unit mismatch in the formula

$$T^2 = t^2(c^2 - V^2)$$

derived a moment ago. Indeed T is a distance, but it represents a time. Let's fix that unit mismatch in the section below.

The Clock of the Universe

At the beginning of this chapter, in my second interpretation of time I explained that "there's an inevitable *delay* between the motion of that particle and the moment that nature detects that very motion. It's that delay that nature will use as the 'tick' of time of that particle." Does the right-angled triangle of the previous section reflect that? The right-angled triangle may be viewed as three vectors—let's say vector C, vector V, and vector T:

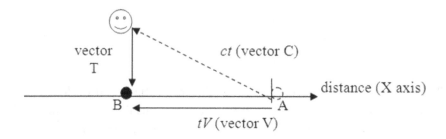

Vector *C* shows when the observer (nature) looks at the particle where it appears to be at *A*, but by then the particle is really at *B*. This is sensed by the observer as a change in direction—hence the difference in angles between vectors *C* and *V*.

If you ever took a college course on vector analysis, you might recall that the difference between two vectors is given by the vector joining their two ends. Here, that vector difference is vector *T*. So the difference between vector *C* and vector *V* is vector *T*. Now we are almost there in our search for the correct time formula. The formula obtained in the previous section is easily manipulated to obtain the formula

$$T = t \sqrt{c^2 - V^2}$$

And guess what? This is the length of the vector *T*. Interesting! Consequently, nature will interpret this difference between vectors *C* and *V* as a *time* difference—even though it's a *distance* difference.

How can a distance be a time? Well, there's nothing mysterious about this interpretation. We do it every day when we read the time on a clock. Our time is measured using hands on a clock.

1. The *longer* a distance taken by the hand, the *longer* the time;

2. The *shorter* a distance taken by the hand, the *shorter* the time.

This one-to-one relationship is possible *only* if the hands go at a *constant* speed. This way the distance traveled by the hand determines the time elapsed. Amazingly, nature does the very same thing. The constant speed required is the speed of light, represented by the letter *c*. So then light—which is everywhere in the universe—acts as the "clock" of the whole universe. Amazing! This goes well with what I said in Chapter 3 that time is everywhere in the universe, and that it's an absolute dimension.

So if it's light that keeps time going, then a consequence is that where there's no light, there's no time. When or where could that be?

1. Before the Big Bang, there was no light, so then time didn't exist. This matches the conclusion that I came to in the section "What Makes Time "Tick?"" of the previous chapter.

2. Inside a black hole, light is sucked and disappears, so then there is no time inside a black hole. The space-time manifold disappears and is reduced to a single point. Note that space still exists though since a black hole has a size.

So, why is time relative depending on the speed of the object? To summarize everything that has been said before, the time clock speed of the object is simply determined by nature to be the difference between the times that nature (this is the observer) "sees" the object at point A in the triangle above and the object reaches nature at point B.

1. The *faster* the object goes, the *sooner* it will reach point B. Hence, it takes it *less* time to reach point B. So the "tick" of the clock will be *shorter*, meaning that its time clock will be *slower*. Therefore, the object will "age" *slower*.

2. The *slower* the object goes, the *later* it will reach point B. Hence, it takes it *more* time to reach point B. So the "tick" of the clock will be *longer*, meaning that its time clock will be *faster*. Therefore, the object will "age" *faster*.

Time relativity is this simple.

As stated in the previous section, there's a unit mismatch in the formula

$$T = t \sqrt{c^2 - V^2}$$

as T represents time, but is expressed in distance. Now that we know that the speed of light is the clock of the universe, to help us resolve this mismatch, imagine a normal clock with hands:

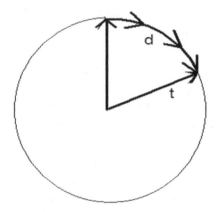

The hand has moved a distance of d in time t. Given that the speed of the clock is V, it's easy to see that the conversion from the distance d to time t is simply $Vt = d$. In our situation, d is T and v is c, the speed of light. So the conversion is $ct = T$. As the variable t is already used in the formula

$$T = t \sqrt{c^2 - V^2}$$

let's use instead the variable t_T. So the conversion from distance to time is $ct_T = T$. Plugging that conversion into the formula above, we obtain

$$ct_T = t \sqrt{c^2 - V^2}$$

Squaring both sides gives us

$$(ct_T)^2 = t^2(c^2 - V^2)$$

then dividing by c^2 on both sides, we obtain

$$t_T{}^2 = t^2(1 - (V/c)^2)$$

and finally applying the square root

$$t_T = t\sqrt{1 - (V/c)^2}$$

You'll notice that this equation is exactly the same as the one derived in section "How Einstein Derived Time Dilation?". The mathematical manipulation of this section was meant to derive this same equation in order to satisfy the units. This formula satisfies us humans in our need for equations with correct units of measure. But do you think that nature cares? The fact remains that in the equation

$$T = t\sqrt{c^2 - V^2}$$

nature interprets this distance T as time. This suggests that it's *not* God that created time but rather nature. How shocking! This interpretation implies that nature has an ability of abstraction, and hence an ability to think (we'll see more evidence of that ability in Chapter 9.). Does this imply that the time dimension really does exist? As far as nature is concerned, yes, and so then in turn, this means that to us humans, the concept of time truly exists, even though time isn't a physical entity.

But let's be careful here. Nature has invented time as a mechanism, just like the invention of the clock. The person that invented the clock didn't invent time; however, as the maker of the clock, he had to have some concept what time is. Does nature's consciousness of time stem from a higher source such as God? If so, then nature did invent the *mechanism* of time, but not the *thought* of time itself. I'll provide a logical argument in Chapters 7 and 10 that this thought may very well indeed have been handed down to nature by the Creator of the universe.

Recall the formula from the previous section

$$P_T = P_t \sqrt{c^2 - V^2}$$

should now instead be

$$P_T = P_t \sqrt{1 - (V/c)^2} \quad (1)$$

There's another way to see that the "tick" of time varies depending on the speed. It's simply by its definition $V = d/t$. In the section "Reality Versus Abstraction" in Chapter 2, I argued that all three of these (velocities, distances, and time) are abstract notions. If they're abstract, then their observation may very well depend on the frame of reference of the observer. This is precisely what the formulas in this section prove, and this is also what Einstein's Special Theory of Relativity states. Wow! I'm getting somewhere with all this.

Note that while I suggested that the time dimension might not actually exist, I still use time in my algebraic manipulations. Isn't that contradictory? No—time is used simply because the speed of light c is expressed in meters per *second*. That's the reason the equation

$$T = t \sqrt{c^2 - V^2}$$

had to be changed to

$$t_T = t \sqrt{1 - (V/c)^2}$$

Finally, the variable T actually represents the difference in *distance* between the distance traveled by emitted light and the motion of the particle. This seems also to contradict the Chapter 2 section, "Reality Versus Abstraction," where I suggested that distances are an abstraction as far as nature is concerned. So how can nature feel distances in order to interpret them as time? This will be further explained later in this chapter in the section "An Absolute Frame of Reference" and the section "The 'God's Light' and the Expansion of the Universe" in Chapter 8.

The above formula based on the *reality* that it's the delay between emitted energy and the detection of the particle that defines the time has been known to be correct by physicists for over a hundred years. Nonetheless, my first interpretation of time that imagines nature as taking "picture frames" of particles is somewhat incorrect, but not by much. If you're mathematically inclined, you may wish to look at Appendix 1 where I demonstrate the two different interpretations of time are very similar because they yield nearly the same equation.

Emitted Energy Versus Internal Energy

As mentioned before, the delay between the motion of the particle and the moment that nature detects that motion is what nature uses to assign the "tick" of time of that particle. Nature detects the particle when it senses the energy emitted by that particle. Then one tick of time later, nature senses the particle's mass. No fewer than four puzzling questions arise from this:

1. How does nature sense the particle? As nature only has senses for energy, it has to be that it senses the energy within the particle, and that internal energy (those tangled- up waves) is its mass (as discussed in Chapter 1). But if the energy is trapped inside the particle, how can nature detect it? We'll answer this question later.

2. How can nature tell the difference between the emitted energy and the internal energy? Energy is energy, isn't it? The answer is already in the question! Emitted energy radiates outward, whereas internal energy stays within the particle. Thus, nature senses them differently.

3. How can internal energy remain within the particle? Is there a wall around the particle that prevents that energy from get-

ting out? Is that energy like a bunch of balls bouncing around within that wall? I doubt it because as balls bounce off the wall, the wall would emit energy. Consequently, the balls inside would gradually lose their energy. Remember that internal energy makes up the matter of the particle. So should this scenario be reality, then all particles would eventually lose their entire matter. Nonsense!

The solution is that the energy within the particle is like a tangled mess of waves. This interpretation was already proposed in Chapter 1. Those waves are closed into themselves, thereby ensuring the energy remains inside. So as an internal wave "wiggles," that energy is absorbed by another internal wave. Consequently the energy never goes outward. It's like a lump of spaghetti. The waves are so tangled up that they can't free themselves and thus can't emit energy. Only when they collide with great force against another particle can that energy be released in the form of light. This effect has been already been confirmed in laboratories for many years. Perhaps a better analogy is Jell-O, a substance that wiggles back and forth. Its energy moves back and forth, but that energy is never dissipated—instead the Jell-O conserves its energy.

4. How does nature know that the emitted energy felt and the internal tangled energy that follows are both from the *same* particle? These energies have no tags, so how does nature pair them? Imagine three particles that travel at different speeds causing their emitted energy to be apart from the particle in different intervals:

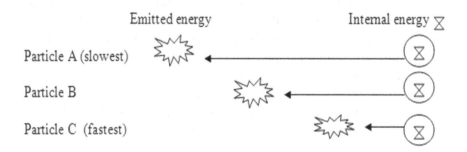

Nature would detect the emitted energy of the fastest particle first, and then it would detect a particle that would *have to be* the particle that emitted that energy (because it's the fastest particle). Then nature would detect the emitted energy of the next fastest particle, and so on. Should nature detect two emitted energies at the same time, it would assign the energy to the first particle that crosses that energy in its path. The problem with this scenario is that nature would always detect the fastest particles, never the slower ones. So there has to be more to the scenario.

It has to be that nature detects all particles of any speed *synchronously*. As the detection of that emitted energy and the internal energy is used by nature to assign the tick of time, then the nature in question has to be the time dimension (or the fabric of space, as will be suggested a few paragraphs below). Thus, the time dimension is very busy. Keep in mind that this dimension is everywhere in the universe so it has no trouble detecting all particles synchronously. It's as if each particle in the universe has a guardian angel that follows it everywhere. So there you have the scenario: there's no confusion as to which emitted energy goes with what particle.

In the previous paragraph, I imply that it's the time dimension that detects the emitted energy. This is most bizarre because the time dimension is nonphysical (as will be seen in the section "The Time Dimension is Nonphysical" coming up). How can a nonphysical entity be able to sense energy—which is physical? This is a paradox. In the section "Reality Versus Abstraction" in Chapter 2, I indicated that most things in nature are in fact abstractions from our interpretations of what we sense. Perhaps the time dimension is our interpretation of another real physical dimension X. It's that dimension X that would detect the emitted energy. What could that dimension be? I suspect that the universe is made of four *physical* dimensions as I suggest in Chapter 6 and the later sections of Chapter 8.

What about the energy inside the particle? Is it that same X dimension that detects it? My hunch is that this X dimension is simply the fabric of *space*—not to be confused with the three dimensions of space. Space would therefore have the ability to sense both emitted and internal energy. So it's not the time dimension that acts as the guardian angel, but rather the makeup of space! It then becomes easy to see that all particles are given a time value synchronously simply from the fact that space is everywhere. We'll come back to this fourth dimension in Chapters 5, 6 and 8 where we'll see that my view seems correct.

Note that there's a distinction between the three dimensions of space and space itself. So it's conceivable that space has an additional dimension, as I'll prove in Chapter 8.

You might recall that in the section "Space Tension" of Chapter 2, I proposed the idea of floats that act as spots in space that a particle can jump onto as it skips in space. I also wondered: how can these floats gather around a particle? How can they know that a particle is around? There has to be some sort of communication between one float to the next such as shown here so that they can get closer to the particle:

It's now clear what that communication is. Based on the scenario of emitted and internal energy of the quantum particle, it becomes clear that it's the *emitted energy* itself that creates the float. There are four reasons this is definitely what happens:

1. This scenario agrees with my arguments about the relationship between the speed of a particle and its time.

    a. The *faster* the speed of the particle, the *shorter* its tick of time (i.e., its Planck time, P_T) and the *shorter* the delay between the detection of its emitted energy and its internal energy. So the float would be created *closer* (i.e., its Planck space, P_S). This agrees with the consequence that a *shorter* tick of time leads to a *shorter* P_S value.

b. The *slower* the speed of the particle, the *longer* its tick of time and the *longer* the delay between the detection of its emitted energy and its internal energy. So the float would be created *farther*. This agrees with the consequence that a *longer* tick of time leads to a *longer* P_S value.

This scenario agrees with the ratio $P_S/\ P_T\ =\ c$. Indeed, the emitted energy travels at the speed of light. P_S is the distance from the particle to the emitted light, and P_T is the time it takes the particle to reach that float (this is again as perceived by the observer, which is nature itself). At the top of the next page is a depiction of a particle through two skips. You'll notice that this is the same particle trajectory that was used earlier in the figure with the three right-angled triangles (see the section "The Correct 'Tick' Formula" to derive the formula)

$$T^2\ =\ t^2(c^2\ -\ V^2)$$

[or $t_T^{\,2} = t^2\ (1 - (V/c)^2)$ that I derived earlier in the previous section to avoid distance–time mismatch units as was derived], but now the emitted energy is depicted:

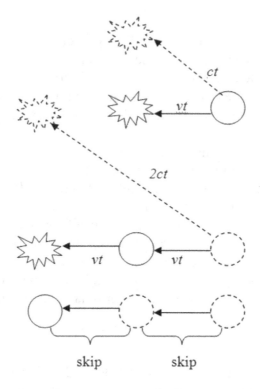

Each skip is of length *vt*, but are perceived by nature to be of length *ct* because this is the length by the time the particle is detected, and it takes *t* time to do so. So then *length/time interval* = *ct/t* = *c*, which is the same ratio as P_S/P_T. This picture also clarifies what I said in the section "Energy at the Speed of Light" that both P_S and P_T change at the same rate as a particle changes speed.

2. This scenario agrees with the fact that quantum particles skip in space.

3. The section "Motion Finally Explained!" of Chapter 8 will prove that the emitted energy is indeed what creates the "float" in space that enables motion!

We're getting a bit closer to a physical explanation as to what makes motion possible. It can be seen now that the explanation given in Chapter 2 for the quantum steps taken by a particle being due to the time dimension *lifting* the particle above the interval is incorrect, even though the resulting equation in the section "The Correct 'Tick" Formula" of this chapter is determined to be correct.

The correct reason now is seen to be that nature simply cannot see the particle instantaneously, but rather only after a delay. As far as nature is concerned, the particle *vanishes* between every P_S step. This all seems weird, and I explained before that nature seems to have a hard time visualizing waves as particles. There's a level of fuzziness about the motion of a particle. More will be said about this topic of quantum steps in the sections "Quantum Steps Taken by Particles" of Chapter 5 and "Motion Finally Explained!" of Chapter 8 where I'll explain why the particle vanishes every P_S step.

If a float is a place where the energy of a quantum particle is emitted in space, then we now have a precise definition of *energy density* in space: it's the number of floats in a given volume. So then based on item (1) above:

1. The *faster* is the speed of particles in a given region, the *closer* (i.e., its Planck space, P_S) together are their floats.

2. The *slower* is the speed of particles in a given region, the *farther* (i.e., its Planck space, P_S) apart are their floats.

The energy density depends not only on the speed of particles, but their frequency. For instance, a higher frequency photon emits more energy than a low frequency photon. In the section "An Absolute Frame of Reference" in Chapter 5, I'll show that:

1. The *higher* is the frequency of particles in a given region, the *closer* (i.e., their Planck space, P_S) together are their floats.

2. The *lower* is the frequency of particles in a given region, the *farther* (i.e., their Planck space, P_S) apart are their floats.

As a consequence, the "thickest" energy density is where there's only light (in outer space) while the "lightest" energy density is where there's very little light emitted in comparison. Where's that? Answer: inside matter!

My definition of space density is very counterintuitive because to us, matter is obviously a lot denser than light. In fact, to us light has no density at all! I'll use my bizarre space density idea in Chapter 6 to develop my own theory of gravitation—and it will work.

What's most important to realize is that there's a *one-to-one relationship* between time and energy. Keep this in mind because my theory of gravitation presented in Chapter 6 is based on this realization.

Going at the Speed of Light

If a particle reaches the speed of light, there are a few implications:

1. Its mass will increase to infinity! I recall reading about that as a teenager, but I couldn't make any sense of it at the time. Now it makes sense considering these two factors:

   a. The particle's mass would increase simply because as it picks up speed, its kinetic energy increases. In Chapter 1, I stated that kinetic energy of all particles inside an object contributes to its mass. In fact, kinetic energy is a kind of mass. So mass will increase.

   b. But this factor in (a) doesn't explain why the mass would reach infinity. As it was discovered a few paragraphs ago, the energy emitted by an object occurs at the speed of light Could it be that as the particle approaches the speed of light, some of that emitted energy might be caught back by the particle, and thereby increase its mass? If the particle reaches the speed of light, the emitted energy will surely accumulate within the particle to no end. Consequently the particle's mass will increase to infinity.

2. At the speed of light, time stops, then P_T becomes zero, and as P_S is tied in with it (as seen the previous paragraph), it also becomes zero. In other words, at the speed of light, the unit of space shrinks to 0 (i.e., the object has no size). Consequently, the object would disappear into the fabric of space.

Those two implications are physically impossible. No objects of mass and size can reach the speed of light. Therefore, we can infer that light "particles" have no mass or size. Indeed, that is what Einstein deduced when he imagined the existence of photons as making up light. These are particles of energy, but of no size. Einstein won the Nobel Prize for Physics in 1921 in part for this discovery.

The Time Dimension is Nonphysical!

The formula $(ct)^2 = T^2 + (tV)^2$ found earlier in this chapter means that the length of the diagonal arrow in the space–time axes is always ct regardless of where the particle is in space–time or at what speed the particle goes in space. In other words, in space–time, a particle "travels" at the speed of light, always! This applies to anything, including humans. So how come we can't feel that speed? Three reasons:

1. Most of the momentum is carried by the time dimension. This will be demonstrated in the section "Imaginary Space Axes and Real-Time Axis" of Chapter 5.

2. We can't "feel" or "see" time simply because to be able to feel it, time would have to be in the space dimensions.

3. The reasons (1) and (2) are a consequence of the fact that motion in space–time isn't really motion in the same sense that we understand it in our daily lives. In other words, a particle does not truly physically travel in space–time. More of the meaning of this "motion" will be given in the section

"Motion in Space–Time: The Plot Thickens!" of this chapter and the section "Motion Finally Explained!" of Chapter 8.

In Chapter 2, we examined the motion of an object through space–time with the help of this graph:

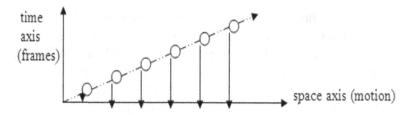

For simplicity, only one dimension of space is shown. We just discovered that at the speed of light, there's no time motion. So the graph is like this:

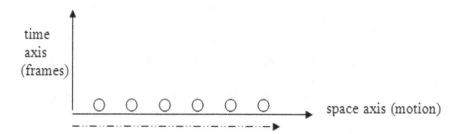

and $(T/t)^2 = c^2 - c^2 = 0$. Note that this path is only possible for a particle of no mass such as a photon, essentially light. If a particle didn't move at all in space (i.e., its speed was $V = 0$), its motion through space–time would be like this:

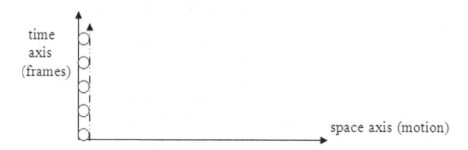

and $(T/t)^2 = c^2 - 0^2 = c^2$. The time arrow would go at the speed of light. So if your body doesn't move at all, you would age at the speed of light! Now, that should convince anyone to get off the couch and get moving!

On a more serious note, there's a small problem about this last graph. It can't happen, at least not to a particle or a wave, because a motionless particle wouldn't emit energy—essentially it would be a dead particle. Remember that we're at the level of quantum particles, much smaller than atoms. There's no such a thing as a dead particle or a dead wave, so neither matter nor energy can take that path above.

If there exists something that can assume that path straight onto the time axis, then an observer (nature) couldn't see it because it would no longer be in the space dimensions—implying that the time dimension isn't the observer after all, contrary to what was assumed up to this point (in the section "Emitted Energy Versus Internal Energy," I proposed that the fabric of space is the observer).

As the arrow in the graph goes at the speed of light, perhaps it is light itself that takes that path straight onto the time axis. But it would be light in a form that we don't know anything about. As light is made of waves, this should mean the space dimensions exist, right? This is because a wave is a form of energy, so it has to reside in the space dimensions. However, the graph above draws a parallel between the speed of light and the time dimension.

Why is the speed of light implicated? I find it very compelling to theorize that *dormant* light exists on the time dimension. This idea will lead me to a startling suggestion in the section "Where Do Souls Reside" in Chapter 10. It would be dimensionless light, a *single point*—the light, but without its energy and space dimensions, and it wouldn't move at all. This is the sort of situation that might have existed at the time of the creation of the universe. This idea has no logic in the realm of physics. However, if there's any entity on the time axis, its properties can't be known and even less studied by physicists simply because the time dimension isn't *physical*! It doesn't belong solely to physicists to claim its existence. Discussion about the time dimension is therefore a *philosophical* issue.

The fact that Einstein's Special Theory of Relativity proves the existence of time as a dimension doesn't imply that this dimension is physical. In the section "Motion Finally Explained!" of Chapter 8, I'll propose a second connection with this dormant light. It turns out that this mysterious light does have energy; it simply doesn't emit it. Can such a light exist? Yes and that light is called *dark energy*, which will be discussed further elsewhere in this book (in particular in the section I just referenced).

If the time dimension isn't physical, does it really exist as far as nature is concerned? Probably, because nature itself imagines that dimension, as this is what I speculated previously in this chapter in the section "The Clock of the Universe." For a moment, let's assume that the time dimension does exist and see later where that leads us. Could there be any other entities existing on the time axis? If so, what can we imagine them to be? Can they escape the time axis into the space dimensions? These questions will lead me to a stunning idea later in Chapter 10.

There's another way to reach the conclusion that the time dimension is nonphysical. In Chapter 3, I suggested that time is a continuous dimension. How can time "tick" in a continuous manner? How does nature manage to make the time tick at intervals when in between intervals, there are infinitely many points? (Remember: this is what continuity implies). This makes no sense unless there is another dimension, a fifth one, that is used by nature to skip over the troublesome continuous intervals of time as needed. This is the same trick nature used by adding the time dimension to allow a particle to skip over intervals of space (see Chapter 2). The problem with this approach of adding yet another dimension is that now the continuity dilemma is with that fifth dimension! So nature would have to create a sixth dimension to get past that continuity dilemma—and this sequence of dimensions would have to go on to no end. This is ridiculous.

So we have a paradox with a continuous time dimension. Either way, we see no *physical* method that allows for a continuous time dimension. Continuity is conceivable only if time is a nonphysical dimension.

Recent experiments in which scientists used the Hubble Space Telescope to look at the most distant galaxies suggest that time is indeed continuous. This is most puzzling to physicists, but as time is a nonphysical dimension, this might be conceivable. However, I suggest in the section "Motion in Space–Time: The Plot Thickens!" in Chapter 5, that time is imagined by nature. How can nature imagine a continuous dimension unless this dimension is elevated *out* of the physical world? As all thoughts are nonphysical and elevated out of the physical world, then nature's thought of the need for time is clearly also nonphysical. It follows that time itself is nonphysical. The only thing that is truly physical is light—the rest are details.

Time Travel and Space Travel

Note that in all space–time graphs in this book, only forward motion in time is shown. Popular belief is that Einstein's Theory of Relativity says that backward motion in time is possible—that is, traveling in the past is possible. But does his theory really say that? No! In the equation found earlier $(ct)^2 = T^2 + (tV)^2$, let $tV = x$ be the distance traveled by the particle. Then the formula becomes

$$(ct)^2 = T^2 + x^2$$

All variables can have negative values. This suggests that the particle can travel backward in space, $x < 0$, and travel backward in time, $T < 0$ or $t < 0$. There appears no doubt that the particle can move backward in space, but is backward motion in *time* possible? Consider this particle before it starts moving:

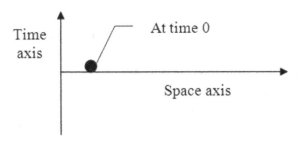

After the particle moved:

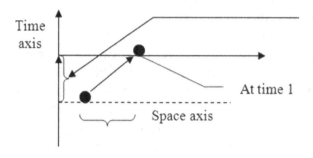

Now for backward time motion, the particle at time 0 would still have to exist so that when I go back to that time, the particle will be there; otherwise, I don't really revisit the past:

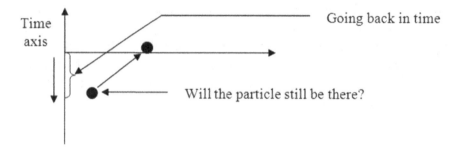

How can a particle leave its mark in the past? Is it by having Nature register it, as in a CD (compact disk)? That doesn't make sense.

Is time travel possible, even in *theory*? No because it simply makes no sense. Traveling into the future makes no sense because by definition the future constitutes of all events that have *not* taken place yet. If I could travel into the future, then the future has to exist already. This is a contradiction. Traveling into the past is equally impossible. Imagine meeting yourself into the past! For instance, you could meet yourself as a child, thereby altering the past and making it no longer the past. This is also a contradiction.

To avoid this contradiction, many physicists have deviced a model of parallel universes whereby I would exist in each one of them, but *in* different times and spaces. This way I would never en-counter myself. In my opinion, this is a crackpot idea. If I were to exist in numerous universes *at* the same time, don't you think that I should be aware of it? If I am not aware of it, then is it really me inside those parallel universes? How can I be in different places at the same time? Physicists have no answers.

Still the formula presented above does allow for negative values of time—it's the interpretation that's crucial. Negative values don't mean that we can go back in time, but simply that our position in time can be negative *relative* to the *origin* of the XYZT axes. As there's no absolute origin of the universe, the origin we choose is always relative.

This is what's meant by relative: the location where the origin is placed is entirely dependent on what we intend to study with the formula. For instance, the origin could be the location of particle 1 when I wish to study the motion of particle 2 in *relation* to particle 1. So, particle 2 might have a negative time value if its time is earlier than particle 1's time. Furthermore, as every variable is squared, the formulas will always work regardless of where the origin is—and that makes sense.

Now it becomes clearer what Einstein's theory says: that time slows down for a person inside a spaceship that goes closer to the speed of light because the person's aging process slows down. But why does it slow down? I am not aware of scientists who ever answered that question. The only way that this slowdown can be possible is that the metabolism slows down. As explained before, as a body speeds up, a shorter delay between the emitted energy and particle detection is experienced.

But how can that translate into a slower metabolism? It might have to do with the shorter distance travel by the particle in space between Planck space intervals (recall that we discovered the dilation of the Planck time and space earlier in this chapter). This translates into a *faster* transfer of energy. Consequently, energy is communicated more *efficiently* around atoms, and so also within cells. Thus, the cells don't have to work as hard and won't age as fast.

There's another way to explain the slow aging process. Each cell within the body requires a constant amount of energy to function. As the body speeds up, it gathers more energy. This includes increase energy within the living cells. Consequently, the cells will demand energy from their surrounding *less often* – the metabolism slows down. An analogy is when I go to the bank to deposit my paycheck. If I am paid once a week, I'll have to make four trips to the bank per month. But If I am paid once a month, I'll have to make a single trip to the bank per month therefore requiring less effort for the same amount of money.

Back to time travel: when that person comes back to Earth, he'll find all his loved ones much older than he. That will give him the impression that he traveled into the future. But he is *not* in the future, and never was during his trip in space even if his loved ones have the impression that a family member from the past is visiting.

There's an easy way to convince you that the traveler never went into the future. Suppose that I was above the galaxy. I take a picture of the galaxy with my digital camera. Assuming that the Earth and the spaceship are at equal distance from the camera, you can be sure that the traveler and the Earth will both show up in the picture (although that they'll appear very tiny). Because only one picture was taken, it's clear that the traveler and the Earth appeared in the picture *at* the same time, although not *in* the same time. If the spaceship was into the future, then it should not even appear into the digital camera photo, right?

Travel in time makes no sense. Let's get real! What Einstein's theory does demonstrate is that objects travel *forward* (only) in the time dimension, but at *different* "speeds," again giving the illusion that some objects are in the future or in the past with respect to other objects. No matter where on the time dimension all objects of the universe are positioned, they are all in the *present*. Nothing is in the past; nothing is in the future.

Here's another way to see this. Consider this diagram of space–time with two particles moving in space. Particle 1 is moving very slowly, and particle 2 is moving very quickly:

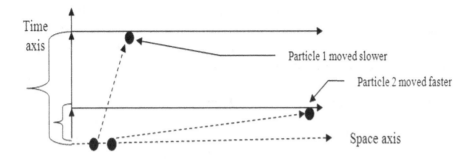

As you can see, at the start both particles are side by side on the same space axis and in the same time. After they moved, their space axis moved into a *different* time frame because the particles moved at different speeds into space. Because they're no longer in the same time, should we conclude that one particle is in the past relative to the other particle? If so, then this should imply that they can't see each other anymore. The obvious reality is that the two particles will still see each other no matter where they are onto the time axis, that is, no matter their speed differential. Although they aren't onto the same place along the time axis, both particles are therefore clearly in the present. As an analogy, on a racetrack, does a car speeding up all of a sudden disappear from space just because it's no longer in the same space-time as the other cars? Of course, it isn't so! It's obvious that both cars still see each other no matter their speed differential.

Recall in Chapter 2 where I suggested that the space axes are imaginary in that there are no absolute space axes. So, it makes no difference if their imaginary space axes are not at the same place: both particles are *in* the same space dimensions, existing *at* the same time, but *in* different locations in the time dimension. I have to admit that the notion of "speed" in the time dimension is a dubious concept (more on that in the section "Motion in Space–Time: The Plot Thickens").

So you see that the word *present* doesn't mean *in* the same time, but rather *at* the same time. In other words, time "ticks" for everything in the universe, just not at the same rate. For an analogy, at my desk it's right now 4:32 p.m., but it's 6:32 a.m. in Japan. I'm not *in* the same time as the Japanese, but I exist *at* the same time as they do.

Most people and physicists say that the difference between the time dimension and the space dimensions is that we have no control in our motion in time—it goes forward—whereas in the space dimensions, we may move in any direction. But this is our point of view.

Does nature share our view? How about backward motion in space? That surely is possible, right? I can walk forward, and then walk backward. I just explained that backward motion in time isn't possible, and it turns out that backward motion in space is *not* possible either! Nature has no notion of backward motion. This seems totally crazy and contrary to everyday experience. To understand my daring declaration, consider a particle that moves along the X axis to its *left* (i.e., *backward* and *negative*):

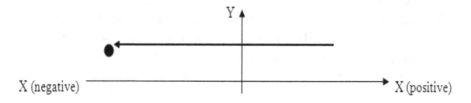

However remember that in the section "Flexible Space" of Chapter 2, I explained that there are no real XYZ axes—they are imaginary and are used only for mathematical reasons. So this particle can just as easily be mapped this way:

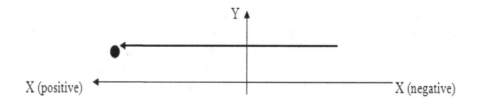

The X axis has been flipped around. And all in a sudden, the particle travels to the *positive (forward)* direction of the X axis! It's the observer (nature) that has flipped to the other side of the graph. We're more used to graphs with the positive X being to the right. So the graph above is just the same as this one:

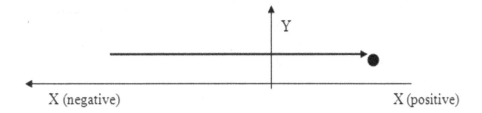

where the negative X and positive X are interchanged. Understand that the particle hasn't changed direction, as it still goes toward the positive X. It is we humans who can view the particle going either backward or forward, but obviously not both at the same time!

So, in the equation $(ct)^2 = T^2 + x^2$, a negative value of the variable x (the same thing with t or T) is interpreted as simply meaning that we choose to look at it the other way around. It's like when I see a car go to the right of me, but if I were on the other side of the road; I would be seeing that same car going to the left. The car hasn't changed direction—it's I who has changed my position *relative* to the car. You see that it's important to properly interpret a mathematical equation. Nature is able to sense a change of direction of a particle, but nature still sees it going forward.

However, we humans are used to giving a qualifier to a direction of motion so we choose the words *backward* and *forward*. As nature can't tell the difference between those two words (after all, nature doesn't know their meaning), we can safely say that all particles of the universe travel in space–time in the *positive* direction of motion. We'll use the term *forward* to express that direction because we can relate to it better. But forget about backward: there's no such concept in nature. Everything goes forward in nature. Backward or negative positioning either in space or time is an abstract concept.

For instance person A is born, grows up (forward), then eventually starts dying, and shrinks a bit. This seems to suggest that the person is going backward. Yes, in health, but the person still goes forward in the process of life. It's just that his/her state changes throughout life. That change of state of health goes forward. Say that person B was born later than person A, and so is younger. We might view person B as having an age negative relative to person A, but that has no significance in biology and nature.

A magnet has both positive and negative tips, but the direction of the magnetic lines is forward from the positive to the negative. We may view it in the opposite direction and say that the direction is negative, but this is just an abstract concept.

One last example: men and women are sexual opposites. Does that mean that one is positive and the other negative? Of course, that is not so.

I hope that this notion of relativity has now become clearer to you.

Some might attempt to refute my argument by saying that if nature sees only forward motions, then two cars approaching each other on the same lane couldn't collide because they are both going forward, and thus in the same direction. My argument still holds because both cars go forward, yes, but *relatively* in *opposite* directions:

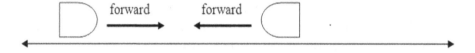

and the imaginary XY axes of both cars show them going in the positive (forward) direction:

Another situation is that one car goes backward, and the other forward, both in the same direction:

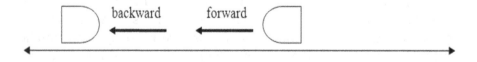

and the imaginary XY axes of both cars shows both of them going positive (forward):

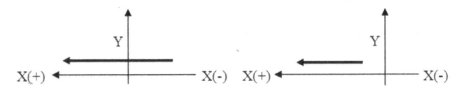

The car going backward is actually going forward as far as nature sees it. Nature doesn't know what a car is so it has no concept of a car going backward. Everything in nature moves forward, believe me! If a car turns around and moves in the opposite direction, it still goes forward, just in a different direction. Imagine yourself at the wheel of a car on the highway. It doesn't matter what direction the highway takes, you always move forward, right? And if you decide to put the car in reverse, the car is still going forward as far as nature is concerned.

There's another reason that backward motion in space is no more possible than in time. Suppose the car moves from point A to point B, then it backs up to point A again. For backward motion to be successful, the car would have to be back at the *exact* position it was before. But the second time the car is at back at point A occurs at a *different* place in space. Indeed, the car actually ends up in a different place because the earth has moved a little bit since the first time the car was at point A. So going backward in space is truly impossible.

Now back to the two cars. It seems ridiculous that I placed the cars on different XY axes. After all, I know that the cars are going to collide, so it's more reasonable to put them onto the same XY axes as follows:

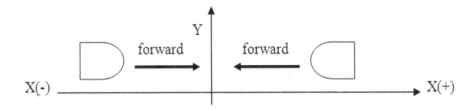

This graph makes more sense, and we can see that one car has a positive direction while the other has a negative direction. But nature doesn't know that these cars are in the same system so it "gives" each car its own XY axes. Nature doesn't understand the *semantics* of a system, especially when imagined by humans. It has no way of knowing that these cars should be on the same XY axes. This is another reason I said in the section "Flexible Space" of Chapter 2 that the XYZ axes are imaginary, not real. This applies not only to systems imagined by humans, but systems from nature too. For instance, imagine the moon circling earth. Here are three of the many ways of placing the XY axes for motion:

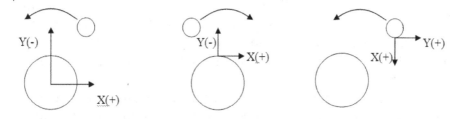

Nature has no trouble about any direction that we see as a system because regardless, the laws are never affected. Nature has no sense of where the origin of the XYZ axis should be.

Past, Present and Future

The previous section made the case clear that space-time travel can only go forward. The past constitutes all events that took place before 'now', and the future constitutes events that will take place after 'now'. However, as I stated before in Chapter 2, it's not clear what the 'now' is. Consequently it's particularly not entirely clear what the past is.

For instance, suppose that I look at the night sky and that I see right 'now' two stars shine brightly at the same time into two super novas (these are stars that explode). Have they really exploded 'now'? No, because we know that light from the stars takes a long time to reach us. So these events happened into the past.

OK—then we should instead be asking: did the two stars explode at the time into the past? Did the two events occur *simultaneously*? It sure appears that way to us on earth. But these events most certainly weren't simultaneous because they occurred most probably at different distances from us, so their light reached us from different times into the past.

Here's another way to see that these two events weren't simultaneous. Suppose that there existed a planet inhabiting intelligent life nearby one of these two stars. Quite obviously, living beings there would *not* have seen the two stars explode at the same time: they would have seen their nearby star explode a long time before the other star did. Yet, we did, here from earth, see them explode at the same time.

So, depending on the frame of reference that you are in, two events may seem to have happened simultateneoulsy when they actually did *not*. The reverse is also true: two events may have happened simultaneously while appearing otherwise to observers. For instance, suppose that I'm observing Jupiter with my telescope and then I see a meteor slam into it. Just 32 minutes before that observation, I was waving good bye to my wife on her way to her night shift at the convenience store. These two events don't appear at all to be simultaneous, yet they are! Here's how. It takes light from Jupiter about 32 minutes to reach the earth (depending on their respective position). So the meteor hit occurred 32 minutes *ago*—just at the precise moment that I waved good bye to my wife.

Time relativity implies that it is *impossible* to tell if two events were simultaneously or not. It doesn't mean that simultaneous events don't occur. Of course, they do. It's just that an observer can't determine one way or the other. Who knows? As I'm writing these lines, little "green men" from a planet from a far away galaxy might be preparing at the same present moment as mine a voyage into the cosmos. But because the light from that planet will take millions of years to reach us, their event appears to come from the future.

All this brings me to another famous one-liner from Einstein: "The distinction between past, present and future is only a stubbornly persistent illusion."

Many physicists believe that his statement reflects well his theory of Relativity. Indeed many physicists believe that the past, present and future all exist at the same time, but in different parallel universes. In my opinion, this is *total* nonsense! I'm fairly certain that Einstein didn't have parallel universes in mind.

What Einstein most likely meant is that the time at which an event happenned cannot be used to determine if that event took place before or after other events. This is due to time dilation whereby time ticks at different rate depending on the speed of the object. When the astronauts from the Appolo mission were speeding toward the moon in 1970, their clock was going slower than my clock here on earth. So we were in different times. Yet my existence and their existence were clearly simultaneous. But if they and I were to look at our watch, we would read different times, and so, we would wrongly conclude that one is in the past compared to the other.

It should become clearer to you that the past, present and future are notions distinct from time. In other words their meaning have nothing to do with time! Am I crazy? No. Time belongs to the world of consciousness. The past belongs to the physical world. Indeed the past constitutes events that took place in *space*. They didn't take place in time. The future never actually exists in the physical world. Just like time, the future belongs to the world of consciousness. There's no physical reality to the future therefore it's surely impossible to visit it.

Recall the figure from Chapter 3 where I introduced the idea of the absolute time axis:

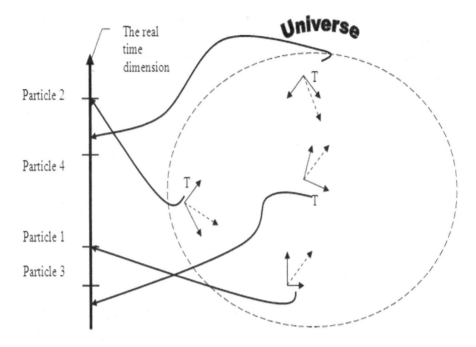

Particle 2's time is ahead of particle 4's time. This does *not* mean that particle 2 is in the future as far as particle 4 is concerned. It simply means that particle 2 is in a different time dilation than particle 4 is. Similarly particle 3 isn't necessarily in the past with respect to particle 1. So, it's impossible to know if an object is in the past or future simply by observing its time clock! This is what Einstien meant from his quotation mentioned earlier. Just by looking at the clock of at which an event took place, it is impossible to tell whether it happened into the past or right now.

My claim seems ridiculous because on the Earth, everything moves more or less at the same speed. So, for all pratical purposes, the time ticks at the same speed for everything and everybody. So, an event that took place at 2 o'clock is clearly into the past compared to an event that takes place at 5 o'clock. Time dilation (and the confusion that it implies on the past and the future) becomes an issue only when an object speeds up much faster than anything we've ever experienced.

What about the present? Well, all four particles in the picture above are in the present somewhere in the universe. Also because the present is neither in the past, nor in the future, it's obligatorily timeless and eventless. Nothing whatsoever takes place in the present. There's no time, no energy transfer. Everything is still. The present is like a single dot in space-time.

This description looks a lot like the conditions that existed... just the moment before the Big Bang when the universe was just a dot (this will be discussed in more detail in Chapter 8). Also I suggest in Chapter 7 that the Big Bang was the result of thoughts. So could it be that the present is also a result of thoughts? Indeed very much so. When you're engaged in thoughts, you're fully conscious. You are in the present. This makes sense because if you were unconscious, your mind surely wouldn't be present! The same thing applies to nature for in Chapter 9, I dare to propose that nature is capable of thoughts.

Does the present exist in a physical sense? Yes, but somewhat just like the very moment of the Big Bang when the universe was a dimensionless singularity point. Perhaps the universe is constantly in a state of trillions of singularity points (dots) that bursts as little 'Big Bangs', so to speak.

Another way to put this, the present—the 'now'— is a universe of its own, but timeless, spaceless and still. Guess what? Using some extremely simple mathematical manipulations, I discover in Chapters 5 (section "Quantum Steps Taken by Particles") and 8 (section "Motion Finally Explained") that, at the quantum level, particles sort of behave like little Big Bangs. They are still a mere instant, then explode into motion, then are still a mere instant, then explode into motion again, etc. This scenario accounts for the effect of skipping of quantum particles that I deduced in Chapter 2.

However the present isn't quite a universe of its own: it depends on the past, obviously. It's like when someone runs. At any present moment, the person is out of balance and should therefore fall. But she does not fall because of past motions. Because the present is timeless, does time really exist then? Is it just an illusion?

Chapter 5: What Makes Motion Possible?

If you ever took an introductory course in physics, you might remember Newton's Three Laws of Motion:

1. An object at rest will remain at rest unless acted on by an unbalanced force. An object in motion continues in motion with the same speed and in the same direction unless acted upon by an unbalanced force. This law is often called "the law of inertia."

2. Acceleration is produced when a force acts on a mass. The greater the mass (of the object being accelerated), the greater the amount of force needed (to accelerate the object). This law is formulated as $F = ma$.

3. For every action there's an equal and opposite reaction. This law is often called "the law of action-reaction."

These laws describe properties of motion, but fall short of explaining the mechanism that drives motion.

Newton didn't know quantum physics, so these laws obviously are meant to apply to our macroscopic level; however, the Second and Third Laws sort of apply to the quantum level as well. In fact, I already used Newton's Second Law in Chapter 1 to derive Einstein's equation $E = mc^2$ that formules an effect that takes place at the quantum level. But the first law doesn't apply at the quantum level because motion of quantum particles isn't continuous. Indeed, in Chapter 2 we saw that motion is executed in quantum steps, not in a continuous fashion. Moreover a quantum particle doesn't move in a straight line as this will be demonstrated in section "Motion Finally Explained" of Chapter 8. Finally, in Chapter 4 we discovered that the length of those quantum steps depends on the speed of the particle. So, a lot was discovered about motion up to this point. But very little was said about the *mechanism* that's responsible for quantum motion.

The purpose of this chapter is to shed some light on the answer to this question: what mechanism makes motion possible? This turns out to be an inquiry that is so complex that it's only partially answered in this chapter. We'll pursue this issue some more in Chapter 6 with the final complete answer being revealed in Chapter 8.

Motion in Space–Time: The Plot Thickens!

This section presents dilemmas about motion in space–time. I hinted a few times in Chapters 2 and 3 that the notion of motion in the time dimension is dubious and that motion in space leads to a dilemma.

Particles move through space in quantum steps as we learned in Chapter 2. But that leads us to a dilemma! Let P_S be the distance skipped by a quantum particle and P_T the time that takes. As was mentioned in the section "From the Planck Units to a Time "Tick" Formula" of the previous chapter, $P_S/P_T = c$, the speed of light. Now this raises a dilemma: suppose a particle goes from point A to point B, where the distance between these points is D. Suppose that distance is traveled in time T, so then the particle's speed is $D/T = V$. Between points A and B, the particle performed perhaps many billions of quantum steps, say q times. This means that $D = q \cdot P_S$. It follows that $T = q \cdot P_T$. So then the speed of the particle is

$$V = D/T = (q \cdot P_S)/(q \cdot P_T) = P_S/P_T = c$$

So the speed of the particle is the speed of light! This would make sense if the particle was weightless like a photon. But as we discovered earlier in this chapter, a particle with a mass, however small, can't reach the speed of light. I can think of four explanations for this dilemma:

1. An object is made of billions of quantum particles; even an atom is made of numerous quantum particles. Did you notice that the smaller the object, the faster is goes, and the larger the object, the slower it goes? At the macroscopic level, this

is explained by the inertia of the object, Newton's First Law (it's easier to change the speed of a small rock than of a large one). At the atomic level, the atom moves a lot more slowly than an electron. Note that the atom is a lot heavier than the electron. The behavior of atoms can be very complex affair, so I'll focus my discussion solely on explaining the dilemma stated above. An atom moves more slowly than the speed of light because its quantum particles:

a. Don't move at the speed of light,

b. Don't move all in the same direction,

c. Don't move from one float to the next at the same time. As we'll see in the section "Quantum Steps Taken by Particles" later in this chapter, quantum particles inside an atom don't behave in a synchronous manner. As a result, the atom can't move as fast as the fastest one of its quantum particles. Therefore, the atom certainly can't move as fast as the speed of light. Consequently, any object made of atoms definitively moves *a lot* more slowly than the speed of light.

As an analogy, have you ever seen a group of kindergarten students tied together walking just behind their teacher? A child moving slower than the others will inevitably slow them down as well. The only way that they could move faster would be by running (a) in the same direction, and (b) at the same speed.

2. Notions at the quantum level such as the intervals P_S and P_T can't be extended to the macroscopic level. These two intervals have been discussed earlier in this chapter to explain time "tick," but it's a tick at the quantum level. The speed of an object can't be determined from the behavior of *each* of its particles. One reason that you might remember is that there's a little fuzziness about the intervals Δs and Δv (recall

Chapter 2) related to the behavior of a quantum particle. Note that $\Delta s < P_S$ because the fuzziness occurs within one "skip"—one quantum step. In addition, these quantum intervals should include kinetic energy within the object, but this doesn't contribute to the motion of the object. If we're dealing with a single atom, this formula still doesn't work because (as was demonstrated in the section "Emitted Energy Versus Internal Energy" of the previous chapter) nature determines the ratio P_S/P_T from the *emitted energy* of the particle, not solely from its speed. In fact this is the main reason behind the dilemma. The algebraic manipulation

$$D/T = (q \cdot P_S)/(q \cdot P_T) = P_S/P_T = c$$

does not take into account the particle's speed! You'll grasp better what I mean when you reach section "Motion Finally Explained" of Chapter 8.

3. According to the ratio $P_S/P_T = c$, nature perceives the particle as skipping at the speed of light from one float to the next, then resting there a mere moment before skipping to the next float. This implies that the particle has to accelerate when leaving a float, and then decelerate when landing on the next float, and so on. This is like a person running whose foot touches the ground and obviously stops moving for a moment before springing back into motion. But this view seems at first glance an unlikely scenario for particles for three reasons:

   a. As the particle can't accelerate beyond the speed of light, the acceleration and deceleration would result in a speed P_S/P_T of much less than c. This seems to contradict quantum theory (actually there's no contradiction because we'll see in the section "Motion Finally Explained!" of Chapter 8 that in the span of a single quantum step, the particle becomes a wave that goes at the speed of light, then returns as a particle. This accounts for the particle going more slowly than the speed of light).

b. What fabric of space could make the particle accelerate and decelerate? The floats where the particle lands are just empty space. There's nothing there to interact with the particle—but perhaps there is! If space could interact with a particle, then space is made of something. In other words, empty space would itself contain energy. Is that possible? I think so, because there's no void in space to start with (as will be seen in Chapter 6 and later sections of Chapter 8). Moreover, in Chapter 2, I suggested that the constitution of space is continuous.

Space would act like a springboard, decelerating the particle, and then accelerating it in the *same* direction. Space would also act like a crowd of people tossing a ball from one person to the next. Does this mean that space applies friction against a particle? If that were so, then a particle in free motion would slow down gradually. This is obviously not the case, and it would violate Newton's First Law of Motion: "A body continues to maintain its state of rest or of uniform motion unless acted upon by an external unbalanced force."

An illustration of this law is the constant motion of the earth: space doesn't apply friction against it. We'll discuss a bit further the fabric of space in the sections "The Energy Dimension" and "Motion Finally Explained!" of Chapter 8.

c. Let's suppose that explanations (a) and (b) are correct. Suppose that the particle could accelerate to the speed c at halfway the interval, and then decelerate to 0 onto the next float. Let's contemplate the energy this would require.

As I remember from my high school physics, the relationship between the acceleration and the force required is given by Newton's formula $F = ma$. You might recall that we already used this formula in Chapter 1. The acceleration is the change of speed over time, or *change of speed/time*. Here the particle would reach the speed c at the halfway point, that is, at time $P_T/2$. So the acceleration is $a = c/(P_T/2)$, and

$$F = ma = m\left(\frac{c}{P_T/2}\right) = \frac{mc}{P_T/2}$$

As $P_S/P_T = c$ then $P_T = P_S/c$. If that P_S/c is plugged into the force formula, that yields

$$F = \frac{mc}{(\frac{P_S}{c})/2} = \frac{mc}{P_S/(2c)} = \frac{mc^2}{P_S/2} = \frac{2mc^2}{P_S}$$

This formula doesn't seem to make sense because it's based on the assumption that the particle reaches the speed of light, which is known to be impossible unless the particle is a photon. However, notice that Einstein's formula $E = mc^2$ is implicated in the force. This formula gives the energy contained in the mass and matter of the moving particle. The term $2mc^2$ is the amount of energy that passes through the interval P_S. The factor 2 implies that the particle spends its entire mass energy in the first half of the interval P_S (suggesting that it vanishes from space!), then it regains *all* that energy in the second half, hence reappearing in space as a particle. Because the energy is regained, the particle does not slow down, and Newton's First Law applies.

What a strange scenario, but we'll see in the section "Motion Finally Explained!" of Chapter 8 that this scenario is *correct*!! The particle does disappear from space, and then reappears in the interval P_S. So, as far as space is concerned, the amount of energy it felt is indeed $2mc^2$. Let's denote that energy $\Delta E = 2mc^2$. The formula then becomes

$$F = \frac{2mc^2}{P_S} = \frac{\Delta E}{P_S}$$

Keep this result in mind for the next section. The scenario of the particle being pushed around in and out of space seems bizarre to say the least. I present in paragraph (4) below a more complete scenario based on this formula.

4. The space interval P_S represents the shortest distance a quantum particle can move, and the time interval P_T represents the time it takes to discharge energy. In the case of light, that energy is purely speed, so it's entirely kinetic; therefore, it's no wonder that the ratio P_S/P_T is the speed of light. But in the case of a quantum particle, the discharge of energy is a mixture of kinetic and other forms of energy that the particle gives off. As shown in paragraph 1(c), the total energy is $\Delta E = 2mc^2$. So only a fraction of that goes to kinetic energy. Consequently, the particle moves *slower* than the speed of light.

This is a more plausible and complete scenario. So does the particle really accelerate, and then decelerate from one float to another? It does, and this will be proven in section "Motion Finally Explained!" of Chapter 8. So then this proves that space interacts with a mass as the particles that it's made of travel through space. However, there's no friction because the energy absorbed by each quantum particle of the mass is entirely given back to the particle one Planck time later. Space is *not* empty of energy and *not* empty of structure. This will become even more obvious in the next two chapters.

Motion in space presents a dilemma because our vision is that *particles* move in space. This view is incorrect, as hard as it may be to believe. As the section "Motion Finally Explained!" of Chapter 8 will show, only *waves* (that the particles are made of) move in space. The interaction with the fabric of space mentioned above is with the wave of the particle. To view particles as waves eliminates the dilemma encountered in Chapter 2 in which space could neither be discrete nor continuous.

Therefore, we can safely say that space is continuous because waves have no shape. They are just intangible energy that occupies a region in space, but can't be localized to a specific point. Mathematically, a continuous space implies precisely this: the inability to locate a point in space. Therefore, space may very well be continuous after all, as unbelievable as this may seem. Although most physicists believe that space is discrete, the jury is still out on that sticky question. Based on this particle–wave duality, in the section "Quantum Steps Taken by Particles" in this chapter I present a more complete view regarding the motion of quantum particles.

Motion also poses a dilemma in the time dimension. In this chapter and the previous one, I used the word *motion* when talking of a particle moving in time. We know what motion in space is: distance over time. If a particle has a speed, it's in motion and vice versa. So then *motion = distance/time = D/T*. This also says that the particle moved a *distance* of *T* on the time axis. I never fussed about these terms before, but the notions of *motion* and *distance* on the time axis lead to the following dilemma.

To us, time isn't a distance, and so we can't talk of motion in it. Nonetheless, let's assume a different meaning of the word *distance* in the time dimension. Suppose that distance in time means the time span *T* of a particle traveling in space with a distance *D*. Speed in the space dimension is *D/T*, that is, *speed = distance/time*. In the time axis, that division becomes *speed = time/time* because on that axis, distance is time. So on the time axis, speed has no unit! If it has no unit, then speed in the time axis is a *ratio*.

What's that ratio? As the particle traveled in *T* time, the distance on the time axis is *T*, and the time traveled in space is *T* as well. That time is mapped onto the time axis according to the space–time diagrams shown many times in this and previous chapters. But then the speed on the time axis is *T/T = 1*, so the ratio is *1*. Consequently, the speed of a particle along the *time* axis is always *1* regardless of the speed of the particle in the *space* dimensions! This implies that although particles move in time at *different* intervals (as discussed earlier in this chapter), their speeds on the time axis are always all the *same*—hence a paradox (this also implies that there's no acceleration in the time axis).

Another way to see this ratio of 1 is that time moves at one second per second, at 2 seconds per 2 seconds, and so on. So what should we think of all this?

1. If the notion of distance is meaningful in the time dimension, then that dimension is self-referential, as expressed by the ratio T/T above: time is distance and distance is time. Some would suggest that this is illogical, and therefore the time dimension means nothing and thus doesn't exist.

    If the time dimension were a physical one, then its self-referential property would indeed present an impasse. In physics, self-referential systems are impossible. It's like the elusive perpetual-motion machines that some people claim to have invented. These are machines that supposedly draw new energy from within themselves without any energy input from the outside. Any serious physicist will tell you that such machines are impossible.

    But as we discovered earlier in the section "The Time Dimension is Nonphysical!" of the previous chapter, if time is a nonphysical dimension, then perhaps a self-referential system is possible with time. The term *distance* here has a slightly different meaning—it's not a real distance because the time dimension isn't real in a physical sense; rather, it's a *conceptual* dimension. Conceptually, a self-referential system is just fine in a nonphysical world. An example of this is the notion of the factorial used in probability theory. The factorial of the number n is note $n!$ and it equals $n \cdot (n-1) \cdot (n-2) \cdot ... \cdot 1$. The factorial can be defined in a self-referential manner as $n! = n \cdot (n-1)!$.

There's another possible interpretation of the ratio $T/T = 1$. It says that distance and time are the same. Doesn't this sound familiar? Indeed, recall what we discovered in the section "The Clock of the Universe" in the previous chapter: "Nature will interpret this difference between vectors C and V as a *time* difference—even though it's a *distance* difference." To nature, time is distance and distance is time. Also in the section "An Absolute Frame of Reference" later in this chapter, it's said that "dark energy allows nature to assess *both* distance and time."

2. As you saw earlier in this section, motion in space occurs because the particle interacts with the fabric of space: it accelerates and decelerates repeatedly. As an analogy, it's like leaves that blow in the wind. The wind allows the leaves to move. Another example is the water flowing down a river. The bed of the river allows the water to move.

    If motion in time is possible, then what is time interacting with to allow that motion? It certainly cannot be anything physical because time is nonphysical. To my knowledge, physicists have never provided an answer to this question. That dilemma doesn't stop them from believing that motion in space-time is possible. But let's be serious. *The plot thickens* in my opinion. Space-time theory seems to be based on a notion that has no reality. I will further discuss the trouble with space-time in section "A Problem with the Space-Time Model" of the next chapter on gravitation.

3. Perhaps the notion of motion in the time dimension does *not* exist! But then how does a particle move in space if it can't move in time? I think the resolution of this paradox is simply that the time dimension is—again—*not* a physical one. This is in agreement with the section "The Time Dimension is Nonphysical" of the previous chapter. So by definition, a particle—which is physical—*cannot* move along the time dimension as essentially there's nothing physical in the time

dimension. It's an abstract dimension very much like the ratio of women to men in some colleges. Both women and men are physical, but the ratio associated with them is abstract. The ratio of physical distance tV traveled by a particle over physical distance T representing time is also abstract (see diagram below).

There are countless abstractions around us that escape our consciousness. Another example is in the game of hockey. The player with the best score is determined by a calculation: the addition of the number of goals the player scored and the number of assists which gives the total number of points. The player with the most points has the best score. Goals and assists are physical and concrete concepts, but the number of points, the addition of goals and assists, is an abstraction without any physical meaning.

4. In the diagram presented in the section "The Clock of the Universe" of the previous chapter

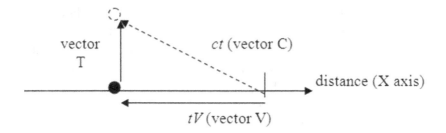

vector T is said to represent the time delay between the time the emitted energy is detected and the detection of the particle itself. With vectors, the difference between two vectors going in the same direction is the vector joining their two ends. So the difference between vector C and vector V is vector T, and this is the time for which we wish to find the formula. The length of T is determined by the formula

$$T = t\sqrt{c^2 - V^2}$$

Now this doesn't appear to make sense, T is a distance, but we use it as a time value! Is it trickery? No! The *distance* on the time axis represents *time* simply because of the nature of the axis. For instance, suppose there's a graph depicting a correlation between the height and weight growth of a person:

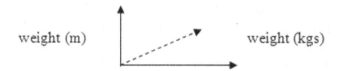

weight (m)   weight (kgs)

The intervals on both axes are measured using a ruler (thereby in distances), and the distance represents height in meters or weight in kilograms depending on the axis. Yet height and weight aren't distances.

Because the time dimension is an abstract dimension, and therefore motion in it isn't physically possible, some people don't consider time a true dimension, and so also believe the space–time motion to be a myth and that the entire space–time theory of Einstein should be thrown out. I have issues with that theory too, but I won't throw it out because of the view below that I thought of. Consider again this space–time diagram shown earlier:

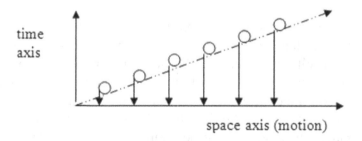

time
axis

space axis (motion)

The particle is mapped onto the time axis, but this does *not* mean the particle *moves* parallel to the time axis. Again nothing physical can move along the time axis. It's the *time inter-*

vals associated with that particle that "move" along the time axis. You can think of the particles along the time axis as being like mirror images. The image in the mirror is some distance inside the mirror, but the object itself is obviously not inside the mirror! A mirror image—called a virtual object—is an abstraction of the real object.

So, if time is a dimension at all, it's a *virtual* dimension, and it is *nature* that has to interpret it that way, not God. This suggests that nature has to have a *consciousness* at the quantum level for the need for such a complex interpretation (more about consciousness in the section "On Consciousness" of Chapter 9). All physicists believe that time is a dimension of the universe, but I'm not convinced, as to me it's more like a dimension imagined by *nature* and more precisely imagined by the fabric of space.

So, I draw a distinction between nature and the universe. Nature is a part of the universe. There will be more on this distinction in the section "The Beginning of Nature" in Chapter 8.

But what about the paradox mentioned earlier that particles are associated with *different* time intervals, yet their speeds on the time axis are always all the *same*? This simply implies that the particle can't tell whether its time ticks faster or slower. Only an outside observer would detect the dilation of time—and that is precisely what Einstein's Special Theory of Relativity says. So in my opinion, no paradox really exists so long as we view the time dimension as a *virtual* one. We as humans may wish to believe in its existence, but God doesn't make use of it, although nature is conscious of its existence (because nature invented it) as discussed in the section "The Clock of the Universe" in the previous chapter.

Another Formula for Fuzziness

Let's go back to the energy formula from the previous section, $\Delta E = 2mc^2$. Recall that Chapter 2 provided the reason that a particle seems to skip around: it's because nature can't keep track of the location or the speed of the particle. There's a little fuzziness. At the time, I expressed that fuzziness as being the multiplication $\Delta s \Delta v$. This fuzziness occurs within a Planck space, P_S. The fuzziness implies that the value of P_S is itself fuzzy. So in the formula

$$F = \frac{\Delta E}{P_S}$$

found in the previous section, the term P_S should be replaced by its fuzzy value Δs:

$$F = \frac{\Delta E}{\Delta s}$$

It seems that nature is just as interested or sensitive to the overall energy of a particle as it is to its speed. That makes sense because speed is sensed by nature as a transfer of *energy*. Everything is about energy in nature. Note that from this fuzzy force formula $F = \Delta E/\Delta s$, it follows easily that $F \cdot \Delta s = \Delta E$. This agrees with the definition of energy, which is a force spent over a distance: *energy = force × distance*. (You might recall that we had used that definition in Chapter 1)

The force is given by Newton's Second Law's formula, $F = ma$. Because $a = \Delta v/\Delta t$ by definition then $F = m \cdot \Delta v/\Delta t$. This yields

$$F \cdot \Delta t = (m \cdot \Delta v/\Delta t) \cdot \Delta t = m \cdot \Delta v$$

Finally multiplying both sides by Δs, we get

$$(F \cdot \Delta t) \cdot \Delta s = (m \cdot \Delta v) \cdot \Delta s$$

or just the same, due to the commutativity and associativity of multiplication:

$$(F \cdot \Delta s) \cdot \Delta t = m \cdot (\Delta v \cdot \Delta s)$$

As $F \cdot \Delta s = \Delta E$, this equality comes to

$$\Delta E \cdot \Delta t = m \cdot (\Delta v \cdot \Delta s)$$

or in a shorter form

$$\Delta E \Delta t = m \cdot (\Delta v \Delta s)$$

This is interesting! In the section "Space–Time Fuzziness" of Chapter 2, I proposed that

$$fuzziness = \Delta s \Delta v$$

As nature is sensitive to energy, it seems fitting to define *fuzziness* via a formula that involves energy, which is just what we derived just now:

$$fuzziness = m \cdot (\Delta v \Delta s) = m \Delta v \Delta s = \Delta t \Delta E$$

Note that, if we ignore the mass m in $\varDelta E = mc^2$, the fuzziness $\varDelta t \varDelta E$ in units m^2/s is just the same as the units of the fuzziness $\varDelta v \varDelta s$ of Chapter 2. The term $m \varDelta v$ is a measure of the *quantity of motion* of a particle that nature is unsure about. The $\varDelta E$ has a range of values $0 < \varDelta E \le 2mc^2$. The term $\varDelta t$ is the fuzziness about the amount of time that it takes nature to detect the particle. The $\varDelta t$ has a range of values $0 < \varDelta t \le P_T$. You might recall the equation (1):

$$P_T = P_t \sqrt{1 - (V/c)^2}$$

derived in the section "The Clock of the Universe" of the previous chapter.

Wow! I just made a link between the Special Theory of Relativity (this formula above) and quantum physics (the fuzziness). According to physicists, these two domains of physics have no such link. But this formula seems correct based on my derivation of it earlier in this chapter. Also Appendix 5 provides a mathematical proof whereby relativity theory and quantum theory are used together—and it works. What a revelation!

The equations I've derived here, $m\Delta v\Delta s$ and $\Delta t\Delta E$, are named the *Heisenberg uncertainty principle* after the physicist who discovered them in 1925. Both equations have a common minimal value that I'll derive later in Chapter 8.

Depending on the formula, Heisenberg's principle states two things:

1. Formula $m\Delta v\Delta s$: the position and momentum of a particle (identified by $m\Delta v$) can't be simultaneously measured with arbitrarily high precision. There's a minimum for the product of the uncertainties of these two measurements.

2. Formula $\Delta t\Delta E$: the time and energy of a particle can't be simultaneously measured with arbitrarily high precision. There's a minimum for the product of the uncertainties of these two measurements.

These aren't statements about the inaccuracy of measurement instruments or a reflection on the quality of experimental methods. They arise from the fact that at the quantum level a particle behaves *both* as a *particle* with a position and speed (i.e., the $m\Delta v\Delta s$) and as a *wave* with energy and time (i.e., the $\Delta t\Delta E$).

As explained in Chapter 1, quantum particles don't have a well-defined contour because they're made of wobbly, tangled-up waves. This is the origin of the uncertainty principle. Moreover, in section "Motion Finally Explained" of Chapter 8, I'll show that quantum particles move as waves. Because a wave isn't a particle, in such a state, the position of the particle is fuzzy whereas the speed is easy to determine because a wave moves at the speed of light. When the wave tangles itself up into a particle, its position is less fuzzy, but now its speed is unclear because a particle does not move—it wiggles. This explains the fuzziness factors $m\Delta v\Delta s$. As for the fuzziness factors $\Delta t\Delta E$, while the particle moves, nature has an easier time to detect its energy because waves give off their energy in well-defined lumps. When the wave tangles itself up into a particle, its position in time is less fuzzy, but now it doesn't emit energy, so its energetic state is fuzzy.

Even with perfect instruments and technique, the uncertainty re-mains—it's inherent in nature's inability to feel with *perfection* its quantum particles in space or time! We must conclude then that na-ture isn't perfect. Is it conceivable that God created a universe that's not perfect? Shockingly, the answer is *yes* as will be proven in Chap-ter 7.

Note that this uncertainty can be felt mostly at the level of quan-tum particles. This is because, at levels above that, as the size of particles increase, the effect becomes negligible. There's very little fuzziness about an atom as a whole. Later in the sections "The Quantum World" and "Quantum Steps Taken by Particles" I'll dis-cuss among other things the requirement for a particle to be consid-ered quantum and "fuzzy."

So nature is uncertain about its own makeup! This has a bizarre implication that will lead us to a startling idea in Chapter 8. A con-sequence of this is that nature can't see exactly where electrons are around an atom because (a) electrons move too fast, and (b) elec-trons behave or move as waves. As a result, nature sees the electrons as a foggy patch. Not surprisingly, this is also how atoms are seen by scientists through an extremely high magnification microscope. An analogy is water on the surface on swimming pool. You can't see well the bottom of the pool because the water's waves blur the light coming from the bottom.

Some books give an incorrect description of this uncertainty principle. The worst descriptions are those that say that the uncer-tainty derives from the interaction of the observer with the quantum particle in that the observer inevitably affects the behavior of the particle. If that observer is nature, then there's no experimental in-teraction. If that observer is an experimenter, then this principle has nothing to do with experiments.

This uncertainty principle leads the experimenter to use prob-abilities to guess the behavior of a quantum particle. This led to Ein-stein's famous rebuttal, *"God does not play with dice."* Many scien-tists disagree with this statement. Yet Einstein was correct: God does not play with dice because He isn't even aware that this principle exists! It's nature that's aware of its inability to precisely "feel" quantum particles. Who knows? Perhaps nature isn't aware of its own limitation in this regard.

Having said this, there's nothing probabilistic about the behavior of a particle. Nature does have a hard time keeping track of the behavior of particles because the particles move as waves, but nature does *not* use probabilities to locate the particles because nature simply doesn't give a darn—just we humans do!

Again to us, a particle is more than the sum of its parts. So a particle is a concept to us, and we can't bring ourselves to let an uncertainty principle make that concept "uncertain," so hence the introduction of probabilities. Probabilities are for the *experimenter*, not nature. If nature used probabilities, then it would have to perform calculations. But it's clear that nature isn't a calculator.

What about larger objects, such as a golf ball? It certainly doesn't move anywhere near the speed of light. Yet all the particles inside it do move at some speed or discharge energy at the speed of light. What's happening is that inside the ball, there are *more* particles that move in the direction of the ball than in any opposite direction:

It's that simple. In the picture above, there are six particles going to the right, with only three to the left. Now do the particles going to the left keep going in that direction? They certainly don't! If they did, the ball would split apart. The molecular forces keep particles going to the left to change direction. Of course, the opposite also happens: some particles going to the right may turn around and go to the left. The important result is that there are *always* more particles moving to the right than particles moving to the left. Essentially the speed V of the ball is the result of the *average* of the sum of the speed of all particles inside the ball. But does this imply that nature does a bit of statistical calculations? No—nature doesn't perform any calculations.

Consider this analogy: when a crowd of people enters the doors of a church and chooses a seat in preparation for the religious service, there's some crowd dynamics involved. A researcher could use statistical analysis to study people's behavior such as to where they choose to sit and make an inference as to the reason for that choice. But do you think that people use calculations to decide where they'll sit? Of course, they don't. The idea is the same with nature.

Can Anything Go Faster than Light?

Suppose that there's some kind of entity that can go faster than the speed of light. As that entity necessarily contains energy, that energy goes faster than the speed of light. But energy is a kind of light as was proven by physicist Maxwell in the 1850s. So then the question to ask becomes: what makes that light (energy) go faster than the speed of light? By definition, light is the energy that's emitted from a body that enables nature to sense it (and for humans to see it). Consequently, if anything can go faster than the speed of light, it is light itself!

This begs a new question: can the speed of light increase (or vary)? If the speed of light (c) varied in speed for whatever reason, then the ratio $P_T/P_S = c$ would change, implying that either P_T changes or that P_S changes or both change (for whatever reason) at a *different* rate. But we know from the section "Emitted Energy Versus Internal Energy" that this rate never changes according to Quantum Theory.

Also according to the formula

$$T = t\sqrt{c^2 - V^2}$$

if something went faster than the speed of light, then $c^2 - V^2 < 0$. The square root would be negative, which is impossible. So nothing can go faster than the speed of light.

But this equation doesn't really answer the question because it was developed assuming that the value of c is *constant*. Is it really constant? If it's constant, what prevents it from changing? Physicists don't yet have an answer. My hunch is that the energy's speed is limited by "God's Light" that I discuss in Chapter 8. We will see in that chapter why I use the word God. Then I'll rename "God's Light" to *dark energy*—a term familiar to physicists, as it's the fabric of space. That kind of light (it's not really a light in the sense that we are familiar with) is all over the universe, and the floats it creates as it travels around are the shortest, closest possible:

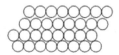

Therefore, no particle or energy can have or squeeze floats shorter than that. As the speed of a particle is limited by the intervals between the floats, it follows that nothing can go faster than the speed of light.

Another way to come to the same conclusion is that the "speed" at which God's Light interacts with energy (light) never varies. This is the only way that light can't change speed. Let me expand on this explanation. Dark energy (God's Light) is at the source of the reason that the speed of light can't vary. Recall a couple of sections ago, I suggested that space interacts with quantum particles, thus implying that space isn't void of structure. It contains energy because it takes energy to make the particle move from one float to another. What's that energy of space? As noted above, it's what physicists call *dark energy*. Toward the end of Appendix 6 I prove that this is so.

Now think about this: because there's interaction between the quantum particle and that dark energy, some time must be involved. The longer the time, the slower the particle moves. No matter how fast the particle moves, its speed is limited by that interaction. It follows that in the case of light, that interaction is *minimal*. Consequently the speed of light is the limit. I theorize that the reason that this interaction is minimal in the case of light is that it interacts with dark energy, which is also a kind of light (that's another reason why I call it God's Light).

Could dark energy be responsible for that limit? Could dark energy explain why light travels at 3×10^8 meters per second rather than some other speed? Physicists have no definite answer. My hunch is in the affirmative.

Dark energy and its relationship to motion are further discussed in the sections "The Energy Dimension" and "Motion Finally Explained!" of Chapter 8. Also we'll continue a discussion about God's Light in the section "An Absolute Frame of Reference" of this chapter and the section "'God's Light' and the Expansion of the Universe" in Chapter 8.

Here are two notes are worth making about dark energy:

1. Based on a search on the Internet, I found some confusion about the type of energy that this is. Most references say that it is energy *in* space that cannot be removed, no matter how hard nature tried. My understanding is that dark energy is the energy *of* space, that is, it's the energy that makes up the constitution of space. This is the definition that I will employ throughout this book.

2. I theorize that dark energy is energy that enables motion. This will be explained in detail in the mentioned section "Motion Finally Explained!" of Chapter 8. Early modern physics of the late 19th century proposed the existence of a medium of the *ether*, a space-filling substance or field, thought to be necessary as a transmission medium. Nowadays, physicists have long abandoned this theory because it is believed that if such ether exists, it would cause friction against objects thereby slowing them down toward a motionless state. Yet, I dare to bring back this ether theory because it makes no sense that motion would take place in nothingness. Space has to be constituted of something—energy.

Imaginary Space Axes and Real-Time Axis

In Chapter 2, I suggested that the space axes are imaginary in that there are no absolute axes, whereas in Chapter 3, I suggested that the time axis is absolute (i.e., it's the same axis used by everything in the universe). The algebraic development below seems to bear that out.

The position of a particle in space is given by the formula

$$d^2 = x^2 + y^2 + z^2$$

The variable d tells how far from the origin the particle is at as shown in the diagram below:

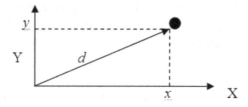

For simplicity, the Z axis is ignored. Also the location of the origin is irrelevant. Without the Z axis, the formula is reduced to $d^2 = x^2 + y^2$. That particle has traveled that distance d in time t or T (as seen by the observer). The formula $(ct)^2 = T^2 + (tV)^2$ used before in section "The Correct 'Tick' Formula" of the previous chapter that ties space and time together is used again here. However, here the particle travels a distance d in time t, so $d = tV$, and the formula is expressed as

$$(ct)^2 = T^2 + d^2$$

or

$$(ct)^2 = T^2 + x^2 + y^2$$

The way to interpret this equation is that the result of the particle moving at the point (x,y) in time t is that it traveled the time "distance" of T. Recall that we saw that equation for T

$$T^2 = t^2(c^2 - V^2)$$

in the section "The Correct 'Tick' Formula" in the previous chapter. It's the same idea as in the graph above where the result of the particle moving at the point (x,y) is that it traveled the distance d. So T is the result of the motion, and we isolate it:

$$T^2 = (ct)^2 - x^2 - y^2$$

But this formula doesn't follow the same pattern as the one $d^2 = x^2 + y^2$ as it involves subtraction. This can be resolved by using the imaginary number $i^2 = -1$:

$$T^2 = (ct)^2 + (ix)^2 + (iy)^2 \quad (1)$$

This formula says that from the origin to the point (x,y,t), the particle has traveled the "time–distance" of T. As was pointed out in the section "The Clock of the Universe" of the previous chapter, T isn't really a time but rather a distance. In that section, we found that the time t_T associated with the distance T is simply $T/c = t_T$, which led us to the formula

$$t_T = t \sqrt{1 - (V/c)^2}$$

But we won't have to worry about that detail in this section as it's unnecessary for the discussion.

Note that the terms ix and iy are imaginary. An imaginary number is a mathematical invention that has no reality in physics. Consequently, formula (1) says that nature isn't able to map the (x,y) position of a particle onto space XY axes. Now why is that so? This is because there are no real X and Y axes as I explained in the section "Flexible Space" of Chapter 2. Nature simply doesn't know what the origin of reference is for the position (x,y).

Moreover in that section, I explained that there's an absolute time axis, which is reflected in formula (1): the *ct* value isn't imaginary. The term *ct* in the formula shows also that most of the motion of the particle occurs in the time dimension. It also reflects that light is everywhere, and that it's the clock of the universe. This is in agreement with the content of the section "The Clock of the Universe" in the previous chapter.

Formula (1) also says that the distance traveled by a particle is always measured in *time* not space units. This further suggests that the time dimension is the frame of reference in the universe (later, we'll see that it isn't quite time that's the frame of reference but rather light that's used by nature to measure time—this is the term *ct* in the formula above).

Formula (1) implying an imaginary XYZ axis doesn't suggest that nature doesn't "feel" a particle moving in space. It does feel its energy, but it only needs to keep track of its position in the time dimension. So nature is just like someone's boss! Often the boss doesn't care how you do your work or where but just wants to keep tabs of the time it takes you to do the job.

Despite all this, should the time dimension not really exist, the formula is still correct for two reasons:

1. There's nothing wrong with accepting time—even as an abstract concept. We can associate a time value to every particle in the universe just as we associate a mass to every one.

2. Should time not exist as a dimension, it's just the interpretation of the formula that changes. Formula (1) also says that nature keeps tabs on delays of *emitted energy* of particles (recall that delays are the second interpretation of time as introduced at the very beginning of this chapter). This goes also well with the section "Nature's Sensitivity to Energy" in Chapter 1 where I argued that energy is what nature really cares about. So nature doesn't care where in space a body is—it just cares about its energy use. This is just like my utility company, as it doesn't care how I use my electricity at home or where within the house. Rather, it just cares how much energy my house consumes. As a corrolary, nature

doesn't really care at what location a body is in space or even whether it's going forward or backward. Recall this agrees with the arguments advanced in section "Time Travel and Space Travel" of the previous chapter.

The debate is open: is time a real dimension? Or is it just in our imagination (or nature's imagination)? I argued in Chapter 2 that the universe has to contain at least four dimensions to allow for motion, including the skipping of particles in space to jump over intervals of "empty" space. In the section "The Energy Dimension" in Chapter 8, I'll propose a fourth dimension—a truly physical one—to explain motion.

Notice that time is the reverse of speed. The faster that a particle moves, the less time it requires going from state A to state B. But due to the formula $F = \Delta E/\Delta s$ found earlier, the relationship is more general than that. A change of state occurs when a particle loses or gains energy, so the more energy that a particle loses or gains, the less time it requires going from state A to state B. Again time is *dependent* on something. By definition, a dimension is supposed to be independent of other dimensions or parameters. That doesn't seem to be the case with time. This is another reason for my doubt that time to be a true dimension.

Energy seems to be the physical realization of time. Could then energy be the real fourth dimension? This is perhaps the third time so far that I've suggested this possibility. Unlike time, energy is physical. But energy and space are related: where there's energy, there's space, and vice versa. So they're not independent—or are they? Whatever one chooses as the fourth dimension:

1. It has to be a dimension that is sensitive to what goes on in the universe.

2. It has to be sensitive to the exchange of energy no matter where in the universe.

3. Most importantly: it has to account for motion.

Remember, energy travels in space, and so does the internal energy of matter. It seems that space, or more precisely its *fabric*, is the best candidate to detect energy and to perform the necessary "calculations" for the "tick" of time. This is precisely the conclusion we'll come to in the section "Motion Finally Explained!" of Chapter 8.

But space isn't a calculator! What actually makes the calculations? Answer: *nothing*. Motion takes place all by itself, just like the orbit of the moon around the earth. The moon's path can be calculated by us, but nature doesn't bother with calculations—we humans do. Could the fabric of space be itself a dimension? If so, then it has to be independent of the other space dimensions. We'll come back to this is the next two chapters and find that indeed the fabric of space is an issue separate from the XYZ axis.

It seems that the universe can do without the time dimension although humans cannot. But what about the motion of particles that requires them to skip in space? In Chapter 3, I said that the time dimension is necessary to allow for those quantum steps. This is one way to look at it. But now that we know from the previous chapter that time is just a difference in distance between emitted energy and the arrival of the particle, then that delay might account for the skipping. Essentially the "fuzziness" felt by the fabric of space is what makes quantum steps possible. We should thank Heisenberg's uncertainty principleas otherwise the motion of space and energy wouldn't be possible. Without the uncertainty principle (i.e., the imperfection of nature), the universe would never have begun.

Despite the confusion surrounding the existence of the time dimension, I'll continue to assume its existence for the remainder of this book, but for philosophical and spiritual reasons only. Moreover, that dimension will allow me to make a connection with the nonphysical entities in Chapter 10 as there's something "supernatural" to time. For that reason, I wish to continue using the time dimension for nonphysical reasons.

Who Is the Observer?

I wish to clarify a few things. I described in the section "The Correct 'Tick' Formula" of the previous chapter that as a body picks up speed (as perceived by the observer) it starts to shrink and its "tick" time slows down. In the section "Another Formula for Fuzziness," I derived the Heisenberg uncertainty principle that says that the observer either can't tell the exact speed or the exact position, or the exact time or the exact energy amount of a particle.

I recall a book relating a debate regarding this that took place between physicist Niels Bohr (a pioneer in quantum physics) and Einstein. Einstein said to Bohr that he couldn't accept the uncertainty principle (because "God does not play with dice") or that the uncertainty depends on the observer. Bohr replied that Einstein's Special Theory of Relativity isn't any better because the phenomenon of the body shrinking and its "tick" as time slows down also depends on the observer.

In my opinion, neither physicist clarified what's meant by the *observer*, but I intend to do so now.

1. In the case of the uncertainty principle, the observer isn't a human because the principle states an uncertainty that is inherent in nature itself. The observer *is* nature itself! The uncertainty has nothing to do with an experiment made in a laboratory where the experiment itself may in fact alter the state of the quantum particles observed by the experimenter. There's an uncertainty principle when an experiment is conducted, but it's another principle called *Schrödinger's uncertainty principle*.

2. In the case of a speeding body under the Special Theory of Relativity, the observer can be either a human or nature. But what makes the human see the body shrink and its "tick" time slow down? Answer: nature, for nature itself "sees" the body shrink and its "tick" time slow down!!

Why do I implicate nature? Because if nature weren't involved, the shrinking would be no more than an *optical illusion* as seen by our eyes. This isn't at all what the Special Theory of Relativity says. In the case of a body speeding up toward the speed of light, the body shrinks because nature senses its time shrink. This implies that nature feels the dimensions of the body faster. Consequently, the body's length has shrunk. As we humans are part of nature, clearly we are limited to what nature sees or interprets. Nature doesn't have eyes, so we actually see a portion of what nature "feels." I did say *a portion* because there are things of nature that we can't perceive such as infrared light. Our physical reality is dictated by what nature interprets itself of what it feels! Therefore we see the body shrink.

So in both principles (items 1 and 2 above), the observer is nature.

I feel compelled to suggest that the time dilation as described by the Special Theory of Relativity has a link with the fuzziness of motion, the uncertainty. So could Einstein and Bohr have been debating the same thing? There's not a single physicist who would suggest such a connection between the Special Theory of Relativity and quantum theory as they aren't supposed to have any connections. Yet I seem to have found one, and in Appendix 5, I demonstrate via a mathematical development that the two theories can work together!

I wish to add the following to the item (1). I've read in various books and on the Internet the claim that quantum physics states that the observer alters the outcome of an experiment by simply observing it. For instance by simply observing electrons shot at a target, I, the observer, alter the behavior of the electrons. The way that this result is phrased leads people to believe that thoughts from the observer may alter events at the quantum level. This claim is false. Quantum physics implies *nothing* of the sort. The notion of *observer* is misunderstood. The observer isn't meant to be a human looking down on the experiment. The observer is some device that detects the electrons or whatever else that the experiment uses. It's the event of detection of the electrons that may affect them and the result of the experiment.

Discrete Energy and Frequencies

In earlier chapters, I found that motion occurs in quantum steps as well as time. Unsurprisingly, it turns out that energy transfers occur in quantum lumps as well, as the next few sections demonstrate.

What's energy? Energy applied to an object makes it move. To move, an object requires a force applied to it over the distance through which it was moved. The result is energy spent. So then by definition $E = F \cdot d$, where F is the force, and d the distance. The force is defined as Newton's formula $F = ma$, where m is the mass, and a is the acceleration. So $E = ma \cdot d$. Imagine a rope that is wiggled up and down at a regular rate f. This is the frequency of the wave. It forms a wave as follows:

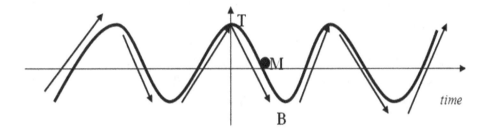

Now we wish to devise a formula for the energy of one of those waves. Three different approaches will be used, starting with a simple approach and finishing with the most complex but most accurate one.

This first approach uses simple mathematics without trigonometry or calculus. Imagine you wiggle a rope up and down:

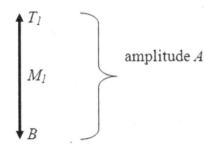

You start at the top T_1, at which time the speed is 0, and then accelerate to the halfway pont at M_1, and at the bottom B, the speed is back to 0. Note that in T_1, T doesn't mean time—it means the top of the wave, the crest. Halfway, the speed reaches a peak of v, and then starts to decelerate. So then spread over time, the situation below forms (a wave):

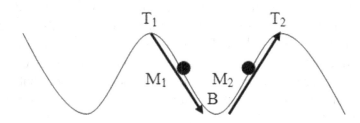

At the point M_1, the distance traveled by the *wave* (and not the hand) is greater than $A/2$.

But in the calculations here, we concentrate only on the motion of the hand up and down. What's that distance in relation to the peak speed v and time t it took to reach the distance to M_1? The graph below shows the progress of the speed within that distance:

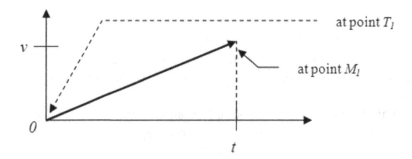

As we assume that the acceleration is constant, the line is straight up. The slope is the acceleration $a = v/t$, and the peak speed is $v = at$. The area within that rectangle is the distance d traveled from T_1 to M_1. So the distance traveled is $d = (vt)/2 = ((at)t)/2 = \frac{1}{2}at^2$. The force necessary is $F = ma$. The energy is defined to be the force applied over a distance. So, the energy is

$$E = F \cdot d = = ma \cdot \left(\tfrac{1}{2}at^2\right) = \tfrac{1}{2}m(at)^2 = \tfrac{1}{2}mv^2$$

But this is the energy for only the first half of the wave. For a full swing of the wave, it's *four* times that amount (from T_1 to B and back up to T_2). So, the energy is:

$$E = 2mv^2$$

Note that while waves give out their energy in discrete lumps, the waves themselves move along in a continuous manner. This seems contradictory, but it isn't. Consider as an analogy a bicycle. The force applied as the pedals occurs in waves. When the foot is at the top, it presses down on the pedal while the other pedal is at the bottom and no force is applied to it. Energy is applied in discrete bursts. Meanwhile, the bicycle moves along continueously in a constant speed. This is also the sort of thing that happens to pure energy, light essentially. Because energy moves along continuously, we have to conclude that space is continuous, not discrete! Yet nearly all physicists believe that space is discrete. I dare to propose that they are wrong.

An Absolute Frame of Reference!

This result in the section above, $E= 2mv^2$, is quite interesting because it's of the same form as the formula $\Delta E= 2mc^2$ found in the section "Motion in Space–Time: The Plot Thickens" earlier in this chapter. Recall in that section, the formula was derived from assuming that the *particle* skips from one float to the next by accelerating and decelerating (or vanishing from space and reappearing). The latest formula of energy found above, $E = 2mv^2$, is for a *wave*.

Interesting—this suggests strongly that, as it moves, a quantum particle behaves like a wave. Wow! Just like waves from the sea hit the shore in succession, a quantum particle also seems to give off its energy in succession rather than in a continuous fashion. So the quantum particle gives off its energy in quantum lumps. Later in this chapter, we'll search for an equation that expresses this quantum energy given off by particles.

Also, as the short trigonometric manipulation in Appendix 2 shows, nature detects the quantum particle (more precisely, its quantum energy) only when its wave is at its top amplitude (at its crest):

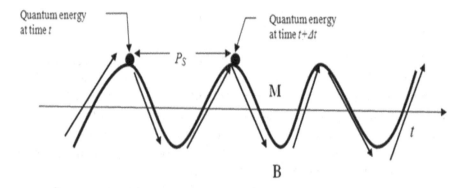

That is, this picture shows that $P_S = \lambda$, where λ is the wavelength. The difference in time between two waves (expressed by Δt in this graph) is by definition the Planck time, P_T. The variable P_T derived in the section "The Correct 'Tick' Formula" of the previous chapter

$$P_T = P_t \sqrt{c^2 - V^2}$$

actually represents the difference in distance between the distance traveled by emitted light and the motion of the particle. OK, to prevent unit mismatch, we should instead use the formula from the section "The Clock of the Universe" in the previous chapter:

$$P_T = P_t \sqrt{1 - (V/c)^2}$$

Indeed as $P_S/P_T = c$, the formula above may then be expressed as:

$$P_S/c = (P_S/c) \sqrt{1 - (V/c)^2}$$

The factor c, the speed of light, cancels out:

$$P_S = P_s \sqrt{1 - (V/c)^2}$$

But hold it! You might have objections here:

1. A few paragraphs ago, we determined that P_S is simply the length of the wave of energy of a quantum particle. Now the equation above says that P_S depends on the speed of the particle. How can P_S represent two different things?

2. The section "Reality Versus Abstraction" of Chapter 2 suggests that distances are an abstraction as far as nature is concerned. If so, how can nature "feel" that distance P_S?

Let's tackle the first objection (in fact, we've already dealt with this one). Do you recall the graph depicting the right-angled triangle from the section "Motion in Space–Time: The Plot Thickens!"?

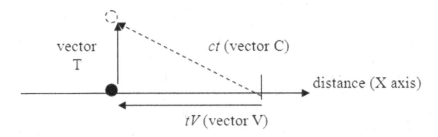

It shows the difference in distance between the times that the energy is detected and that the particle arrives. You might recall that nature uses this to determine the time "tick" of the particle. Because $P_S = \lambda$, the equation

$$P_S = P_s \sqrt{1 - (V/c)^2}$$

may be expressed as

$$\lambda_S = \lambda_s \sqrt{1 - (V/c)^2} \quad (1)$$

where λ_s would be the wavelength of the particle at rest. So the right-angled triangle above may also be expressed differently as

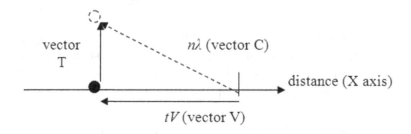

The diagram assumes that n waves are emitted, but the value of n isn't important for the argument. Because the particle speeds toward that emitted light, the actual wavelength perceived by the observer (nature) will be *lower* than λ as per formula (1) above. The *faster* the particle moves, the *shorter* the wavelength as detected by nature. The *slower* the particle moves, the *longer* the wavelength as detected by nature. This is the same as the Doppler Effect that you might have heard about. Suppose a train comes toward you. The faster the train moves, the sharper its sound will be as heard by you; the slower the train moves, the lower its sound will be heard. This analogy also applies to the case of the particle and its emitted light.

Let's deal with the second objection now. Although nature may not associate a physical meaning to time (because it's abstract and nonphysical), it certainly can associate a wave as something physical, which includes its frequency. As waves of energy (light) are the only things that nature really feels, it feels its frequency f in units of *waves/second*, and wavelength λ in *meters/wave*. Recall in the section "Motion in Space–Time: The Plot Thickens!" that I said that quantum particles may be viewed as carried by space itself via successive deceleration and acceleration. I've mentioned a few times already that the fabric of space is made of dark energy. In the section "A Formula for Discrete Energy" a little further in this chapter, I'll prove that energy is carried by waves. This means that dark energy is made of waves. In Chapter 8, we'll see that dark energy contains a massive amount of energy.

In the section "A Formula for Discrete Energy" of this chapter we'll derive an equation that says the higher the energy, the smaller its wavelength; the lower the energy, the longer its wavelength. Therefore dark energy has a very short wavelength. I have a hunch that the Planck interval P_S of a quantum particle is a multiple m of the wavelength of dark energy. Here's a picture that depicts the concept as a particle travels through dark energy in quantum steps marked by the circles:

Here is the multiple $m = 3$, so the particle's Planck length of the particle is three times the wavelength of the dark energy. Dark energy is like a perfectly tuned ruler so its frequency and wavelength never change. There you have it: the *absolute frame of reference* of the universe is the dark energy of space that everything travels through!

Furthermore in the section "The Time Dimension is Nonphysical" of the previous chapter, I drew a parallel between the time dimension and dark energy. Therefore, dark energy allows nature to assess *both* distance and time. Isn't that amazing? So then, space–time theory is really all about space and its constitution. Time is just an abstraction.

As this dark energy light serves as the clock of the universe, it must satisfy these three conditions:

1. It never changes frequency.

2. It never changes speed.

3. Nothing can alter it. This is a consequence of point 1, for if there were interaction, its frequency could change, and so the clock of the universe would be altered! The universe's clock would be broken.

As far as I could tell from my research in the Internet, dark energy fits those three conditions. Moreover, it's energy that can't be detected because it doesn't radiate. I'll reflect some more on dark energy in Chapter 8.

However, there's a problem with choosing dark energy as the frame of reference for motion. Because it acts as the absolute frame of reference, its wavelength and frequency can't change. Yet when a particle travels through it, some sort of interaction has to take place between that particle and that dark energy so that the Planck length of the particle can be "calculated." Hmm! I'll let physicists debate this point.

Light with a Spin

The realization above allows us to deduce that a quantum particle is actually a wave when it moves. In the picture at the beginning of the previous paragraph, the wave is depicted in two-dimensional, but in reality, it flows in three-dimensional. This implies that the particle that is carried by the wave spins around the crests of the waves with a motion very much like a corkscrew:

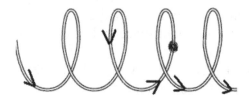

This also has to be the motion that light takes: light moves in a straight line, but while spinning, it moves like an airplane propeller! This seems unbelievable but it's true. Note that only extremely tiny quantum particles would adopt such a path as an atom would go straight. There's a minimal mass after which a particle wouldn't adopt this wavy path. I don't know what that mass is, but in the section "The Quantum World" of this chapter I'll make a suggestion.

In Chapter 1, I said that waves have four properties: frequency, amplitude, ability to carry energy, and spin. What's a spin? This refers to a kind of angular momentum intrinsic to a body (i.e., the rotation of the body around its own center of mass). For instance light (photons) has a spin of 1 because, as the picture above shows, it spins a full 360 degrees onto itself. An electron has a spin of ½ so the wave of which it's composed of spins less than does a photon. The electron is made of a twisted wave such as those discussed in Chapter 1:

This isn't an exact picture of the spin of the electron. The proton, neutron, and neutrino all have a ½ spin as well.

Light is a bit of a mystery. Because light travels—well, at the speed of light, its own Planck space is

$$P_S = P_s\sqrt{1 - (V/c)^2} = P_s\sqrt{1 - (c/c)^2} = 0$$

This implies that light doesn't skip at all; hence, its trajectory is *continuous*. Recall that in Chapter 2, I stated that a continuous space or trajectory is a physical impossibility because it leads to an infinite regression. But back then, I was thinking about the trajectory of *particles*. So I have a hunch that in nature, light (i.e., *waves*) travels in a continuous trajectory, and particles travel in discrete quantum steps. Essentially free-flowing energy travels in a continuous fashion, but not particles. However in the section "Motion Finally Explained!" of Chapter 8, I'll prove that unbelievably, quantum particles move as waves too!

Note that I already indicated in Chapter 3 that I suspected that the time dimension is continuous, and in the section "The Time Dimension is Nonphysical" of the previous chapter, we found a link between light and time. So perhaps it shouldn't be a surprise that the motion of light should also be continuous because ime is continuous.

As a result, this suggests that the constitution of space is continuous. Yet physicists believe that space is discrete. I beg to differ.

However, you might try to refute my reasoning on the account that light is made of photons, and photons are quantum (hence not continuous) as was discovered and proven by Einstein. This is true, but photons are a manifestation of the way that light delivers its *energy*. It says nothing about the way that light *moves*. The same argument applies to quantum particles.

A Formula for Discrete Energy

Let's get back now to our search for a quantum energy formula. In the section "Discrete Energy and Frequencies" above, I found the energy of a single wave to be:

$$E = 2mv^2$$

The complex mathematical manipulations shown in Appendix 3 transform this formula into this new one:

$$E = 8\pi^2 E_p f_s$$

where E_p is the lowest energy possible in a photon, so $8\pi^2 E_p$ is a constant. This formula shows that energy is emitted in quantum discrete amounts (such as photons), based on the frequency f_s of the wave. Therefore, the higher the frequency, the more intense is the energy; the lower the frequency, the less intense is the energy. This is fairly intuitive (if you wish to read all the mathematical manipulations behind that equation, consult the mentioned appendix).

It turns out that the formula relating the frequency and energy of a wave was first discovered by Planck in 1902 and is formulated as $E = h\nu$, where $h = 6{,}6 \times 10^{-34}$ J•s, and ν represents the frequency. The energy unit is the joule. For instance with the Planck formula $E = h\nu$, the units are

$$E = J{•}s \cdot s^{-1} = \text{Joule}.$$

This is correct because energy is measured in joules. If you consult the appendix, You'll notice that my frequency f_s has units m/(m•wave^{-1}) = wave, and so my formula yields the units

$$E = kg• (m^2/s^2) • (wave) = J•wave.$$

This "wave" unit isn't a real unit, but if the equation $E = hv$ made use of it, its units would become J•s • (wave•s^{-1}) = Joule•wave. This comes to the same units as my formula. Therefore, my formula $E = 8\pi^2 E_p f_s$ is correct. Wow!

I learned from the Internet that the lowest frequency of a wave to be known is about $v = 5$ x 10^{-3}/s. That gives a lowest energy wave of (6.6 x 10^{-34}J.s) • (5 x 10^{-3}/s) = 3.3 x 10^{-36} J according to Planck's formula $E = hv$. That is also the lowest energy of a photon. Thus, the lowest photon energy is $E_p = 3.3$ x 10^{-36} J. So then my constant in

$$E = 8\pi^2 E_p f_s$$

comes to $8\pi^2 E_p \approx 78.9$ • (3.3 x 10^{-36}J•s) = 2.6 x 10^{-34} J•s.

This is *very* close to the actual Planck's constant $h = 6.6$ x 10^{-34} J•s as discovered by Planck himself! Let $g = 8\pi^2 E_p$ be my constant. Then my energy formula is:

$$E = gf_s$$

For convenience, I'll drop the subscript s from now on:

$$E = gf \quad (1)$$

Planck's formula is $E = hv$. Keep my formula in mind (and Planck's formula too) because they'll be needed in the section below and in section "The Beginning of Nature" in Chapter 8.

The Quantum World

I've brought up the topic of quantum particles many times already. I wish now to give you a better idea of the extremely small scale of the quantum world than what I've provided so far. The word *quantum* means the smallest discrete quantity of some physical property that a system can possess (according to quantum theory). So far in this book, I've come across three such physical properties:

1. The smallest unit of time of 5.4×10^{-43} second.

2. The smallest unit of length of 1.6×10^{-35} meter.

3. The smallest unit of energy/momentum of 6.6×10^{-34} J•s.

These are *quantum* units, all discovered by the father of quantum physics, Planck, in the early twentieth century. The quantum world is small truly beyond imagination! To give you an idea of how incredibly small the quantum world is, let me compare it to our macroscopic human world:

1. The smallest time that we can "feel" is probably no less than 10^{-2} second, the time unit used in the Olympic 100-meter sprints. This is roughly 10^{41} times longer than the time unit in the quantum world.

2. The smallest length or size that we can "feel" is perhaps around $\frac{1}{16}$ mm, which is the size of a tiny grain of sand, or a little bigger than 10^{-5} meter. This is about 10^{30} times bigger than the Planck length unit of the quantum world.

3. The smallest energy that we can "feel" is about 20Hz. This is the weakest sound heard by our ears or about 10^{-32} Joule according to the Planck $E = h\nu$ formula. This is about 100 times bigger than the energy unit of the quantum world.

Suppose that I could magnify the quantum world to our human dimensions so 10^{-43} second becomes 10^{-2} second (a factor of 10^{41}), 10^{-35} meter becomes 10^{-5} meter (a factor of 10^{30}), etc. From this, the following comparisons can be made:

1. The average size of an atom is 10^{-11} meter. So to me, this atom would be about $10^{-11} \times 10^{30} = 10^{19}$ meters in size, nearly the size of our own galaxy! Therefore, to the quantum world, an atom is of an *astronomical* size. The nucleus of an atom is about 10^{-15} meters. So to me, this would be about 10^{15} meters, many millions of time larger than our solar system.

2. A bacterium is about 2×10^{-6} meter in diameter, so a bacterium would appear to me as a ball $(2 \times 10^{-6}) \cdot 10^{30} = 2 \times 10^{24}$ meters wide! According to the Web site www.space.com, the universe is 156 billion light-years wide, which is about 10^{27} meters. So a bacterium would look to me nearly as large as the universe itself. Absolutely incredible!

3. The average time for molecule binding is about 10^{-10} seconds per atom. So to me, I would have to wait about $10^{-10} \cdot 10^{41} = 10^{31}$ seconds for two atoms to bind, or 10^{18} billion years. That is more than the entire age of the universe! Because in the comparison at point (a), a galaxy is calculated at somewhat the size of an atom, then it would take 10^{18} billion years before two galaxies to "bind" or merge. Therefore, to the quantum world, 10^{-10} second is an *extremely* long time to wait for a single molecular event to happen. This suggests that the binding of atoms is a very complex affair, much more so than the interaction between galaxies.

If it can't be seen or felt then it doesn't exist, right? From these comparisons above, we can see this reasoning is clearly false. It appears that the quantum world is *not* a vast empty space where next-to-nothing happens. Considering the comparisons calculated above, you can be sure that the quantum world doesn't wait that long (10^{-10} second) for something to happen. In other words, there are plenty of events happening millions of times faster than the time for atom bounding.

The molecular binding itself is most likely the result of thousands of quantum events unknown to me. Also, as an atom in the quantum world is the size of a galaxy, you can be sure that the atom contains numerous quantum particles yet to be discovered that buzz around at an incredibly fast speed—hence the uncertainty principle.

Finally, as a bacterium is nearly as wide as the universe in this comparison, it's as if every single bacterium or cell is truly a universe of its own. The human body is made of many billions of individual universes. Also this comparison makes it abundantly clear that a bacterium is an *extremely* complex organism so it's impossible that it appeared on the earth "out of the blue" from some magical lightning strike—Creationists failed the test again! The appearance of bacteria was an intelligent evolutionary process, and in the section "Codes and Algorithms for Evolution" of Chapter 9, I show that despite the incredible complexity, nature probably had time to create bacteria and even humans through evolution alone.

I theorize that the quantum world is a lot busier than the activity of the *entire* universe of our macroscopic level. There exist most certainly physical phenomena in the quantum world that don't exist in our macroscopic one. Perhaps additional physical dimensions exist in the quantum world (although I remain unconvinced).

Based on this realization, a question comes to mind: what other thresholds or units exist between these quantum units and the units of our macroscopic human world? Also what's the smallest possible particle that carries a mass in the form of matter?

I can't help but suggest that the quantum world is a miniature version of the universe, except that the laws governing the quantum world are surely much more complex. We just need to study a little bit of chemistry to realize that the extremely small world of molecules, atoms, and subatomic particles abounds with very complex interactions.

The previous paragraphs made comparisons about size and time, but what about energy? I mentioned that in our human macroscopic world, about 10^{-32} joule is the lowest that we can hear according to the Planck $E = hv$ formula. A quick search on the Internet revealed that 5×10^{-3} Hz is the lowest frequency wave known in nature. That translates to $(6.6 \times 10^{-34} \text{J.s}) \cdot (5 \times 10^{-3}/\text{s}) = 3.3 \times 10^{-36}$ J. This implies that no amount of matter can contain less than that amount of energy. So using the formula $m = E/c^2$, that gives the smallest possible mass of a matter to 3.3×10^{-36} J/$(3 \times 10^8$ m/s$)^2$.

As 1 joule equals 1 kg•m^2/s^2 the calculation becomes

$$(3.3 \times 10^{-36} \ kg \cdot m^2/s^2)/(3 \times 10^8 \ m/s)^2 =$$
$$(3.3 \times 10^{-36} \ kg \cdot m^2/s^2)/(9 \times 10^{16} \ m^2/s^2) = 3.7 \times 10^{-53} \ kg$$

Considering that the electron weighs only 9.1×10^{-31} kg, this is a very tiny mass. So it appears that in the quantum world, numerous particles might indeed exist that are smaller than an electron. Sure enough, a quick search on the Internet revealed to me a list of sub-atomic particles that fit this description, but they all belong to the family of neutrinos.

The smallest and lightest of those neutrinos is the one with a mass about ten million times lighter than the electron or about 10^{-38} kg. This is billions of billions of times heavier than the theoretical smallest mass of 3.7×10^{-53} kg calculated above. So clearly the lowest wave energy of 3.3×10^{-36} J can't physically convert to any matter because it does not carry enough energy to shatter waves into tangled-up waves—which are the makeup of matter. There's another reason. That wave energy has a wave frequency of only 5×10^{-3} Hz, and the wavelength corresponding to this is 6×10^{10} meters. This is about half the distance from the earth to the sun! It doesn't take a genius to realize that such low frequency waves can't possibly create matter for two obvious reasons:

1. That wave simply doesn't have enough energy to curl up into a particle (as mentioned above).

2. Remember that in Chapter 1, I suggested that a quantum particle is made of "balloons" of tangled waves:

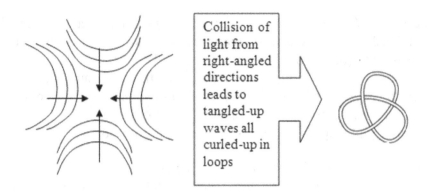

Collision of light from right-angled directions leads to tangled-up waves all curled-up in loops

Can that wave of 6 x 10^{10} meters long fit inside such a tiny particle? Of course, it can't.

From Einstein's equation $E = mc^2$ and Planck's equation $h\nu$, we can easily deduce that only very high frequency waves are good candidates for creating matter, and that the matter created is extremely small. So creation of matter occurs only at the subatomic levels. An example of that are chemical reactions that change a substance into another, at times releasing energy.

We've made some calculations implicating masses and energy. Physicists are more interested in the mass of quantum particles than their size. The reason might be that the size isn't well-defined—a quantum particle is wabbly like Jell-O. Also it's not the size that best determines the behavior of a quantum particle, but rather its mass, frequency and energy.

You may have noticed that the mass expressed in kilograms of a quantum particle is rather awkward to work with. For this reason, physicists have invented a quantum unit called the *electron-volt* symbolized by *eV*. One eV equals to the amount of kinetic energy gained by a single unbound electron when it accelerates through an electrostatic potential difference of one volt. Physicists have calculated one eV to equal roughly 1.6×10^{-19} Joule. From the equation $E = mc^2$, another unit was invented symbolized by eV/c^2. It expresses a measure of quantum mass. When c is set to 1, eV becomes a unit of quatum mass. So the electron has a mass of 1.6×10^{-19} J $/(3 \times 10^8 \text{m/s})^2$

$\approx 1.8 \times 10^{-36}$ kg. This is one eV. Since the mass of the electron is 9×10^{-31} kg, then the mass of the electron expressed in units of eV is

$$(9 \times 10^{-31} \text{ kg})/(1.8 \times 10^{-36} \text{ kg/eV}) = 5 \times 10^5 \text{ eV}$$

or 0.5 MeV. The proton has a mass of about 94 MeV or 0.94 Giga eV.

When an energy wave can transform (somehow) into matter, then we necessary have the equality

$$h\nu = mc^2$$

The left-hand side gives the energy as the wave (the way that matter moves); the right-hand side gives the same energy as a particle (the way that the matter gives its energy). This formula says that the *lower* the mass, the lower the frequency associated with the wave, consequently the *higher* the wavelength. Let's change the formula to $h\nu = h(c/\lambda) = mc^2$, where λ is the wavelength, remembering that $c = \nu\lambda$. The formula becomes $h/\lambda = mc$ or $\lambda = (h/c)(1/m)$.
The graph of this formula looks something like this:

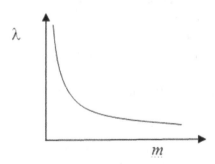

This graph seems to contradict my two points above (i.e., that low frequency waves can't possibly create matter). This graph clearly allows for low frequency waves (i.e., high wavelength λ). The graph says that the wavelength is inversely proportional to the mass (i.e., a *smaller* tangled wave making up the mass implies a *longer* wave-length of that wave, as per the formula). How can a longer wavelength fit in a smaller "ball" of tangled waves? There seems to be a contradiction, but is there? No—there's none.

The formula $\lambda = (h/c)(1/m)$ is applicable only for *waves that can convert directly into matter and vice versa*. It doesn't apply to the whole range of waves and masses—a beach ball can't convert itself into a wave or vice versa! So then what are the ranges of wavelengths and masses to which this graph applies? Remember the analogy used in Chapter 1 in which a rock is repeatedly cut in half until there's nothing left to cut. Just the moment before the last cut, the matter present is what the variable m represents in the formula. Thus, it represents the *smallest possible matter that isn't pure energy*.

As I said in an earlier paragraph, the smallest matter (at least as far as the physicists can tell) was discovered in recent years to be the *neutrino*. The implication is that regardless what material is repeatedly cut in half until the last cut (it could be a rock, copper, or what have you), the last piece left of matter cannot be any smaller than a neutrino.

Note that the last piece doesn't have to be a neutrino. In fact it *cannot* be a neutrino because this quantum particle isn't bound to atoms making up matter. Indeed neutrinos behave like photons—they are free-flowing particles showering the universe. In reality, when you cut an object repeatedly until the last piece, that piece is simply a quantum particle that isn't made of other particles. Matter is made of atoms which are made of protons, neutrons and electrons. Protons and neutrons are made of *quarks*, and these are not divisible into smaller particles. So the last piece after the last cut would likely be a quark.

Consequently, the formula $\lambda = (h/c)(1/m)$ is applicable only to the neutrino, quarks, electrons and other quantum particles not divisible. The most massive of such particle known is the *boson* which is about one million more massive than the electron, and the smallest is the neutrino.

It turns out that the neutrino constitutes actually a family of quantum particles of various masses. So the formula implies that the wavelength (and frequency) that results from "cutting" the neutrino isn't necessarily always the same. Nevertheless, there exists only *a narrow range of type of waves* that can create matter, and of always the same narrow range of wavelengths! Physicists have known for a hundred years already that the electron is capable of behaving like a wave so it belongs into that range as well as the neutrino and others. In the graph above, only a tiny section of it applies:

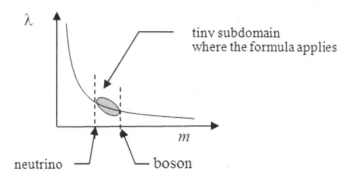

As the graph above shows, that narrow range of type of waves resides in the lowest wavelengths, and low mass, but not the lowest mass mathematically possible. A search on Wikipedia revealed wavelengths for various emissions of energy:

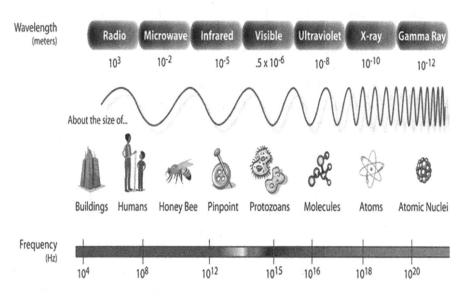

This picture clearly shows that tiny particles such as atoms and quantum particles have the lowest wavelengths. Matter is created at the quantum level. In the graph above, we can easily see that the wavelengths where matter is created are around 10^{-10} to 10^{-12} meters (actually the range is known to reach 10^{-29} meters). These are in the range of the smallest waves possible in nature, and that is the range shown in the graph above.

In Chapter 1, I gave a couple of reasons to try to persuade you that matter is indeed made of a "balloon" of tangled waves. The formula $\lambda = (h/c)(1/m)$ can be used to verify that it is so. Consider for instance the electron. Its mass $m = 9.1 \times 10^{-31}$ kg, giving a wavelength of $\lambda = (h/c)(1/m) = (6.6 \times 10^{-34} \text{J.s} / 3 \times 10^{8} \text{ m/s}) \cdot (1 / 9.1 \times 10^{-31} \text{ kg}) = 2.2 \times 10^{-12}$ meters.

Note that an electron is about 10^{-18} meters across, so the electron's wavelength is about one million time longer than its size. Thus, the tangled-up wave will be a very tight squeeze into the "balloon" making up the particle of the electron.

In the picture above, we can see that the wavelength calculated for the electron falls in the correct range: the atomic nuclei. Of course the electron isn't in the nuclei, but it's of the same domain of particles. In light of this discovery, my conjecture in Chapter 1 whereby I suggested that the creation of a "balloon" of tangled waves is only possible with light of very *high* frequency possible is correct. As to how that "balloon" is created, I have no answer. It's clear that the waves have to somehow bend and curl up in order to create matter, but I have no idea what condition could bring that about. I know that photons colliding can create quantum particles, but how it all happens is unclear to me.

But think about this: if matter is created from light coming into contact at right angles, then it should be very easy to create mass by simply projecting light from a lamp directly toward the light of another lamp. Of course that doesn't produce matter. As it has been argued in the earlier paragraphs, matter is produced from a special kind of light of very high frequency that can't be produced by our day-to-day technology.

A final note: the formula $\lambda = (h/c)(1/m)$ may be manipulated to be expressed as $\lambda = h/(mc)$. This is interesting because the term mc is called the quantity of movement of the particle, which is a notion that involves the mass and the speed of the particle when it is in motion. Physicists use the symbol p for the quantity of movement. So my equation $\lambda = (h/c)(1/m)$ is the same one as $\lambda = h/p$. Interesting! This is the very formula discovered by the famous French physicist Louis De Broglie in 1924 that was the basis for his particle–wave duality principle. This is the principle that is behind the explanation for the motion of matter. It turns out that matter moves as waves, but gives off energy as particles. We'll look at this in more detail in the section "Motion Finally Explained!" in Chapter 8.

Quantum Steps Taken by Particles

In the section "An Absolute Frame of Reference!" I deduced that a quantum particle behaves or moves like a wave because the equation for the energy of a quantum particle, $\Delta E = 2mc^2$, is similar to the equation for the energy of a wave, $E = 2mv^2$ (which becomes $E = 8\pi^2 E_p f_s$ or equivalently $E = gf$ after a lengthy mathematical manipulation shown in Appendix 3).

But surely not all matter behaves like a wave. In that same section, I wondered about when a particle no longer behaves like a wave. What's the maximum mass m at which a particle no longer behaves like a wave?

The answer is given by the formula $\lambda = (h/c)(1/m)$ because this formula contains the range of masses applicable for when a wave can be transformed into a mass and vice versa. So the small grey region I depicted in the graph earlier in the previous section represents the range of mass of particles that can behave as waves.

In the section "Another Formula for Fuzziness" in this chapter, I said that a particle that behaves *both* as a *particle* with a position and speed (that is the $m\Delta v\Delta s$ incertainty) and as a *wave* with energy (that is the $\Delta t\Delta E$ uncertainty) is impacted by the uncertainty principle. These particles are the ones for which the formula $\lambda = (h/c)(1/m)$ applies because this formula is for particles that can become pure energy, then go back to being particles again, possibly in a mere instant. Equivalently this is the formula for entities that are particles and waves apparently at the *same* time! But it's not quite at the same time. A quantum particle is a wave when it moves then rests as a particle, and so on. This change of states happens many billions of times a second. It can't be noticed even in the most sophisticated laboratories.

Every once in a while (the Planck time P_T) nature sees the particle, then sees it as a wave, then sees it as a particle again, and so on. This is one reason for the fuzziness principle mentioned in an earlier section. This is the phenomenon that enables motion of bodies and gives the impression that a quantum particle skips in space as it moves along.

Let us thank God for quantum particles that are particles and waves, back and forth! Indeed if nature always saw things as particles, then motion wouldn't be possible at all; if nature always saw things as waves, then particles wouldn't exist at all—the universe would be void of matter. Luckily this particle–wave effect takes place at the quantum level. (At the atomic level and above, it's not noticeable.)

In the section "Motion in Space–Time: The Plot Thickens!" I explained that the particle skips to where its emitted energy is (i.e., where its next float is):

Internal energy   Emitted energy   Internal energy

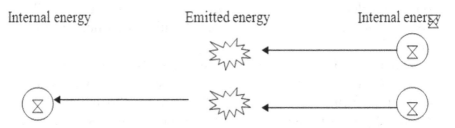

But what if the quantum particle doesn't emit energy? There's no problem because it always emits energy. The correct explanation for the motion of the particle has been given in the previous paragraph. The emitted energy is simply the particle as a wave with its energy, which allows it to go from one float to the next. Then it vanishes as a particle where the two floats are. These two adjacent floats are simply the particle as a tangled-up wave. When a particle vanishes, it doesn't move because its entire energy is concentrated into that tangled-up wave, so it emits nothing.

Based on this new understanding, how does an atom move? First, an arbitrary set of its quantum entities (say Q_1) are seen by nature as particles, and hence don't move. Meanwhile, another arbritary set of quantum entities Q_2 are seen as waves and move. There are numerous such sets Q_i, some being in the state of particles and others in the state of waves. The picture below gives the idea. Suppose for simplicity that the atom is made of six quantum particles. At Planck time 1,

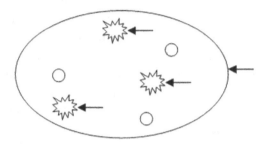

three quantum particles are seen as waves, and so have moved (the energy signs) while the three other particles stay because they aren't seen as energy. Then at Planck time 2,

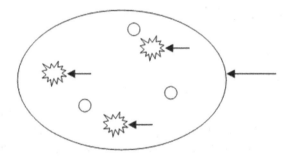

the three quantum particles that were seen as particles are now seen as waves, and so have moved (the energy signs) while the three particles that were seen as waves are seen now as particles and so don't move. So as a whole, the atom moves a bit. Because the atom has millions of such sets, the effect is that the atom appears to move continuously, even though it doesn't. Thus, as I move a pen across to draw a line, the motion appears continuous. Note that the space dimensions themselves are most likely continuous as we concluded in Chapter 2.

In Chapter 8 in the section "Motion Finally Explained!" I'll prove that quantum motion does take place this way. To summarize, the particle moves only when it's being emitted as energy and stays put when it returns in form to that of a particle.

Note that because the quantum particles don't move synchronously, the atom can't move as fast as the fastest of its quantum particle. If the fastest quantum particle (such as a photon) moves at the speed of light, the atom is guaranteed to move more slowly than the speed of light. The speed of the moving atom at a given time would be the average speed of all its quantum particles at that time.

Remember the analogy of the kindergarten students tied together walking behind their teacher? We noted that a child moving more slowly than the other students will inevitably slow down the other students as well, and that they could move the fastest by running in the same direction at the same speed.

Why Does Energy Move as Waves?

It seems so many things in nature move in waves: sounds, vibrations, water shattering onto the shoreline in succession, heat, radio transmission, light and so on. This chapter and section "Motion Finally Explained" of Chapter 8 make the case that quantum particles move as waves while giving off their energy discretely. Also I've thrown hints a couple of times that energy might be traveling in a continuous fashion. There are two puzzling questions:

1. *How can quantum energy travel in a continuous way?* I've dealt with this question numerous times so far. To help explain continuity:

    a. I've used the analogy of the bicycle in section "Discrete Energy and Frequencies" of Chapter 5 and

    b. I've used the analogy of the waves from the sea in "An Absolute Frame of Reference" of the same chapter.

c. I discovered in section "Light with a spin" that light has a continuous trajectory. Then in section "Motion Finally Explained" of Chapter 8, I discover that quantum particles move along as waves, light essentially. Therefore a quantum particle actually has a continuous motion!

d. In section "The Singularity Theorem" of Chapter 8, Stephen Hawking's Singularity theorem states that there was an infinite amount of energy at the point of singularity when the Big Bang occurred. If that's true, then there's an infinite amount of energy in the universe, and consequently perhaps every point in space has energy. This would certainly enable energy to travel in a continuous motion.

2. *Why does energy travel in waves?* If you took an elementary course in physics, you know that sounds travel through the air in form of waves. What we call the pitch of the sound is simply the frequency of the wave of that sound. High pitch sounds are caused by a high frequency, and the reverse applies for low pitch sounds. The energy from the sound needs to gather enough "strength" (i.e. force) in order to pass through the air. When there's enough energy gathered, it's strong enough and the air resistance gives in and lets the sound pass through for a short distance, and then the sound needs to gather some more energy and strength to penetrate the air resistance again, and so on. This phenomenom is expressed with the equation *Energy = force x distance* that we used in section "Discrete Energy and Frequencies" to prove that energy is given off in discrete amounts.

The real cause behind the formation of waves is Newton's principle of *action-reaction*. This principle states that for every action (a force associated with energy traveling); there is an equal reaction (a force of resistance to that action). The same principle of action-reaction occurs with waves on the sea or any other sources of waves.

In section "An Absolute Frame of Reference" of this chapter, it was stated that, as it moves, a quantum particle behaves like a wave. This will be further demonstrated in section "Motion Finally Explained" of Chapter 8. So, since a quantum particle behaves like a wave, it is subject to the action-reaction principle. The action's origin is simply the energy of the particle. The reaction's origin can only be space, more precisely, the fabric of space. This proves that quantum particles—including light—interact with space very much like a swimmer interacts with the surrounding water.

This inference corroborates well the statements made in section "Motion in Space-Time: The Plot Thickens" of Chapter 5, and again in the section "Motion Finally Explained" of Chapter 8, where it is said that quantum particles interact with the fabric of space. It's this interaction that creates the waves in light, for instance.

Energy is given off in "chunks" called waves. When comes to think of it, energy, no matter its form, is always given off in chunks. Think of the athlete on her regular jogging routine. She enhales air containing the energy, oxygen. Then she exhales. The energy is taken and given off in discrete chunks. But the running motion of the athlete is continuous.

Based on what I read on the Internet, physicists say that quantum particles move in discrete steps, jumping from one "float" to the next. This is also what I deduced in Chapter 2. However I've come to realize that this scenario is incorrect. It's the energy of the particles that's emitted in such discrete chunks. The waves themselves associated with the particle moves in a continuous fashion. You'll learn more about these ideas in section "Motion Finally Explained" of Chapter 8.

A final note is worth making about the emition of energy. This isn't always the case, but often, it's the energy that travels along, not the matter. For instance the waves approaching the sea shore don't bring along the water from the far distance. It's the energy that's captured by water molecules and then carried forward to other nearby molecules, and eventually that energy makes it to the shore. The same applies with electricity. Electricy travels at the speed of light. However the electrons inside the wires don't travel forward at the speed of light. It's the *energy* emitted from one electron to the next one that travels at the speed of light. Because electricity is carried in such chuncks, not surprisingly, waves are associated with electricity. But the amount of electricity—amperage (current)—is discrete.

Chapter 6: What Is Gravitation?

Apparently physicists have figured out what mechanism makes all forces of nature work such as electromagnetic, nuclear, and other forces. However, the most obviously present force that we feel every moment of our life—gravitation—remains unexplained.

Physicists don't know for sure what *causes* gravitation despite Einstein's General Theory of Relativity that was supposed to clarify this. This debate is still open, and I'll shortly propose my own humble theory that (I dare suggest) explains it! But be ready for a different view from that of university physicists. My theory is so different that your immediate reaction might be to reject it. It's necessary for you to read the whole chapter before correctly grasping my theory.

From Newton to Einstein

Numerous months went by before I could come up with my theory. I first had to reflect on some basic facts and discoveries about gravitation:

1. Everywhere we look in the cosmos, gravitation manifests itself when a celestial body made of matter makes satellites orbit it. So it's clear that matter is at the center of the question of gravitation. Note that matter may not necessarily be a hard rock as it can also be a cloud of gasses. For instance, Saturn has a small rocky core and most of its volume is just gas. Nonetheless, the gasses contribute to its gravitation.

2. In the seventeenth century, Newton was the first to publish the idea that the force that makes an apple fall to the ground is the same force that makes the moon orbit the earth. What an amazing insight! Then he used Galileo's discovery assuming no friction from the air, that two objects (say a feather and a hammer) that fall from the same height will reach the ground at the same time regardless of their mass.

Newton reasoned that if this were so, then it has to be that a more massive object is "pulled" to the ground with a greater force than a lighter object. This led him to formulate his famous equation $F = ma$ indicating that all objects accelerate toward the earth at the same rate. But he didn't stop there.

He reasoned that if the force is proportional to the mass, then the same applies to the earth: the object also applies a force onto the earth! Therefore, the gravitational force is proportional to the multiplication of the two factors: the mass of the earth and the mass of the moon. So $F = m_1 \cdot m_2$ where m_1 and m_2 are the masses of the earth and the object respectively. Next Newton reasoned that the farther apart the earth and the object are, the weaker the force. (This is fairly intuitive.) This led him to the final formula

$$F = G \frac{m_1 \cdot m_2}{r^2}$$

that says that the gravitational force is inversely proportional to the distance squared from the body. Why is the distance squared? This is simply a consequence of the geometry of the sphere of radius r. For instance, if I were to travel halfway between the earth and the sun, the intensity of the light coming from the sun would become *four* times stronger. It's the same idea with gravitation. The term G is called the gravitational constant whose value is believed to be the same everywhere in the universe.

3. A consequence of the gravitational formula above is that the closer an object orbits the earth; the faster its orbital speed will be simply because the stronger is the force applied to it. It's a simple task to derive a formula for that orbital speed. Let's take for instance the solar system with the sun and the planets orbiting it. Consider Newton's gravitational formula

$$F = G \frac{m_s \cdot m_p}{r^2}$$

where m_s is the mass of the sun and m_p, the mass of the planet. The centripetal force—the force that pulls the object to from the sun—is given by

$$F = \frac{m_p v^2}{r}$$

(If you're interested in reading the mathematical derivation leading to that formula for the centripetal force, consult Appendix 4.)

To orbit the sun, the planet has to find a stable orbit. As no planet has fallen into the sun, we may safely assume that their orbits are stable. Johannes Kepler was the scientist who in the early seventeenth century first formulated this stability of orbits. Kepler's second law of planetary motion states that angular momentum is preserved all around the orbit of a body. The consequence of this is that an object that orbits the body (the sun, let's say) travels *slower* the farther it is from the planet. This is easily confirmed with a search on the Internet of the speed of the orbits of the planets around the sun:

Planet	Orbital Speed (kms/s)	Escape Speed (kms/s)	Distance from the Sun (kms)
Mercury	47.9	67.7	5.7×10^7
Venus	35.0	49.5	1.1×10^8
Earth	29.8	42.1	1.5×10^8
Mars	24.1	34.1	2.3×10^8
Jupiter	11.9	18.5	7.8×10^8
Saturn	9.6	13.6	1.4×10^9
Uranus	6.8	9.6	2.9×10^9
Neptune	5.4	7.7	4.5×10^9

The stability of the orbit is assured when the gravitational force equals the centripetal force, that is

$$F = G \frac{m_s m_p}{r^2} = \frac{m_p v^2}{r}$$

or

$$F = G \frac{m_s}{r} = v^2$$

So the orbital velocity of the planet is

$$v = \sqrt{Gm_s/r}$$

With the data in the table above, it's easy to see that this formula is correct. Note that $G \approx 6.7 \times 10^{-11}$ m³·kg⁻¹·s⁻² is the gravitational constant, $m_s \approx 2 \times 10^{30}$ kgs is the mass of the sun, and r is the average distance of the planet from the sun. For the earth, $r \approx 1.5 \times 10^8$ kms $= 1.5 \times 10^{11}$ meters. Using the formula gives us a speed revolution $v \approx 29440$ m/s $= 29.4$ kms/s. The table shows 29.8 kms/s, which is quite close.

4. In Chapter 2, I found that moving particles skip from one float to another. In Chapter 4, I discovered that the floats get *closer* as the particle moves *faster*, and that the floats are *farther* apart as the particle moves *slower* (motion of quantum particles is counterintuitive because the faster a particle moves, the closer are its floats, so the shorter are the skips. This is quite unlike animals in which the faster they run, the wider are the steps).

In that same chapter, I also found that the *closer* the floats are, the shorter (*slower*) is the ticking of time; conversely, the *farther* the floats are, the longer (*faster*) is the ticking of time. The formula of orbital speed obtained in item (3) states that the speed of a planet increases if its orbit is closer to the sun, and that its speed decreases if its orbit is farther from the sun. Given the quantum motion results of Chapter 4 that I derived, then

a. The distance between floats of the quantum particles within a planet *decreases* as the planet's orbit is *closer* to the sun. As a consequence, the ticking of time of the planet *decreases*, meaning that time *slows down*.

b. The distance between floats of the quantum particles within a planet *increases* as the planet's orbit is *farther* from the sun. As a consequence, the ticking of time of the planet *increases*, meaning that time *speeds up*.

These two effects are precisely what Einstein's General Theory of Relativity predicted. Moreover, this result of time dilation has been scientifically proven.

A Problem with the Space–Time Model

The result in item (4) of the previous section was used by Einstein to suggest what's called a space–time continuum. My understanding of it is that space and time are coupled somewhat like in a marriage. The curvature of space caused by a celestial body in a particular region can affect the ticking of time there. The ticking of time is more affected near a celestial body than farther away in space. Likewise, the ticking of time can affect the fabric of space. The coupling of these two partners causes the space–time manifold to curve where there's a concentration of mass, a celestial body.

Einstein's space–time theory is very original and elegant to say the least. Nonetheless, allow me to propose my most daring opinion in this book: there's no real space curvature as proposed by Einstein's General Theory of Relativity. Hold it! Don't throw my book away—give me a chance to explain. My objection has to do with the implication of time in his model.

Space–time theory does indeed work as its effects have been proven in numerous scientific observations. My objections are related to the *model* that implicates time in the General Theory of Relativity:

1. Space–time theory implicates time in the explanation of gravitation. Surely time is implicated in the sense that when a body is attracted by a gravitational region, it speeds toward it, and this happens in a time span. But time explains nothing about what causes gravitation. In fact, the effect is the other way around: time is affected by gravitation—it doesn't cause it. The mathematics surrounding the General Theory of Relativity is far too complicated for most of us, so let me use analogies instead to explain. Here are four easy examples to illustrate the fallacy about space–time:

    a. Take for instance the simple equation $d = \frac{1}{2}gt^2$ that describes the distance d covered by a body as it falls straight down due to gravity g. The variable time t isn't there to explain what gravitation is; it's there simply as a consequence of the effect of gravitation. The variable t is there to explain the *behavior* of a body around a gravitational field. It's clear that the time dimension in that formula explains nothing about the nature of gravitation. Time doesn't explain motion; it's *energy* that explains motion.

    b. Another example is the result of a 100-meter Olympic sprint. Suppose that I look at the clocked time of the winner and ask the trainer what made the runner arrive first. He explains that he won because he took the shortest time to cross the line. Yes, but time explains nothing. The fitness and energy use efficiency of the runner are the real reasons.

    c. Yet another example is of a movie in DVD format. Suppose a child seeks an explanation from a parent as to how a movie can be played from a DVD. The parent answers that it's because the DVD player spins the DVD so many revolutions per second. Again, time explains nothing. The real answer is that the movie is stored inside the DVD. It's the fabric of the DVD that explains it.

d. A final example is when we cook a nice roast beef in the oven. Suppose we leave it in the oven too long and it overcooks until it turns into charcoal. What explains the burned-out roast? It's not time in that it spent too long in the oven, but rather that it was exposed to too much heat and too much energy. Energy is the explanation, not time.

I admit that my examples are simplistic, but space–time theory has the same fallacy as these examples show. It's what's inside *space* that explains gravitation—and what's inside space is energy. Sure time is required to observe physical phenomena, but time itself never explains any physical phenomenon. Time is a philosophical concept, not a physical one. Therefore, it doesn't fall to physicists to claim its existence or significance.

2. Most books or science television broadcasts that explain space–time theory do so by using the following model. A big ball is placed on a cloth, and the weight of the ball makes the cloth sink. This represents the curvature of space–time. Then a small ball is rolled close by, and the curve of the cloth makes that small ball orbit the big ball:

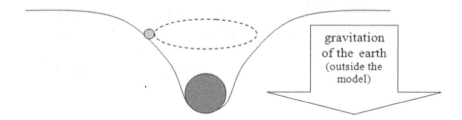

And that's it. Supposedly this model illustrates gravitation. I'm not aware of anyone objecting to this. However, either this picture has been badly interpreted through the past century or there are two important flaws with this model:

a. To make the small ball begin its trajectory around the big ball, we have to give it a push. *Assuming* that the ball is already moving around the hole and that it has enough energy, the model predicts perfectly well what the trajectory will be. However, the model doesn't explain how the small ball can *move* in the first place.

b. After the small ball is given the proper push, it will orbit around the big ball. This is where explanations of the space–time model end. What the model fails to explain is why the small ball keeps its orbit. The answer usually given is that the ball follows the curvature of the cloth representing space–time. Yes, but what gives it the energy to keep that orbit? Answer: the earth's gravitation in which the model is used. Yes, but that source of energy comes from *outside* the model. The model on its own doesn't explain where the energy comes from that makes the small ball move.

3. Space–time curvature of space assumes that time is a physical dimension. In this book I dare to propose that there's no such physical dimension—only a nonphysical dimension, possibly little more than a virtual dimension.

    As we discovered in the section "The Time Dimension is Nonphysical!" of Chapter 4, the time dimension is immaterial. In other words, time has nothing to do with physics. How can a nonphysical dimension be part of the explanation of gravitation, which is a physical phenomenon? All physicists will confirm that space–time theory explains well the *effect* of gravitation. I agree. The fact remains that space–time is a nonphysical model. Therefore, its space curvature has no physical reality. Also if space–time is curved, with respect to what is it curved? The theory is unable to make that clear. The model suffers the same dilemma expressed in section "Motion in Space-Time: The Plot Thickens" of the previous chapter whereby it couldn't be determined with respect to what motion in time takes place.

The model misses in explaining what makes a planet keep moving around the sun. It takes no new energy to keep a body moving at a constant speed in a straight line. But in this case, the body's orbit curves, so new energy has to be fed into it. We know that the source of energy is the pull from the sun, but the model fails to explain how that source makes the planet orbit it. The curvature on its own explains nothing. Therefore, the model misses the point completely.

In order for the small ball to orbit the big ball, it needs to be going fast enough to obtain an orbital speed. Where could that energy come from? The model fails to answer that question. To put it simply, the space–time model with the cloth curvature fails to explain the mechanism that accounts for the *motion* of a planet in the first place! The model explains the *trajectory* of a moving body. Indeed, the trajectory does follow the curve as per the space–time theory. But what's the physical reason?

In the model of the ball on the cloth, the trajectory follows the curve of the cloth due to the gravitation of the earth. That is the source of the small ball's energy, but it's a source outside the model. In the real case of the solar system, where does the energy come from? It's from the gravitation of the sun of course. But that answer fails to explain what gravitation is.

Some would oppose my arguments by saying that the curvature of the cloth explains the revolution of the body around the big body. That's true, but only assuming that the body is already moving at the correct speed and that somehow energy is fed into that body to keep it moving in a circle (an ellipse to be more precise).

Everybody explains that it's the curve of space that gives the body its energy. This is what the model demonstrates, but only from the fact that the energy of the small ball really comes from the gravitation of the earth—a force *external* to the model—where this simulation experiment is conducted. But in the real situation of the solar system, it remains unexplained how such curvature of space translates into energy. To my knowledge, this has never been explained, or not to the general public anyway.

The space–time model is a *mathematical* model that describes correctly the trajectory of a body in a gravitational field. Einstein's gravitation equations explain quite well the trajectories, but they fall short of explaining what gravitation is. In the next few sections, I propose an explanation based on a *physical* phenomenon. Note that my explanation will become complete only after I tackle the subject of gravitons in the section "Gravitons Revisited" of Chapter 8.

Space Density

In my objections set forth in the previous section, I mentioned that it's energy that should be the focus of the coupling with space, not time. Notably Einstein did realize that the constitution of space had something to do with the mechanism of gravitation. As space is constituted of energy, Einstein must have understood that energy had a role to play in gravitation. But I believe that he mixed up his cards by coupling space with time rather than with space energy. I'll suggest a possible reason later in this chapter as to why he chose to implicate the time dimension instead of energy. For now, understand that there are two types of energy:

1. The energy *in* space

2. The energy of the constitution *of* space.

My own theory of gravitation that I'm about to begin presenting will show that gravitation implicates both types of energy.

In this section and the next two, I'll show that energy *in* space is the answer as to *why* gravitation exists. The last section will show that energy *of* space is the answer to *how* gravitation is carried out.

In the section "Emitted Energy Versus Internal Energy" of Chapter 4, I introduced my own definition of space density. We learned there that a float is a place where the energy of a quantum particle is emitted in space, and also that the *energy density* in space is the number of floats in a given volume:

- The *faster* the speed of particles is in a given region, the *closer* are their floats.

- The *slower* the speed of particles is in a given region, the *farther* apart are their floats.

Let me call that density the *kinetic* energy density (or KED) of space. Based on this definition, the highest KED possible is when floats are adjacent to one another without any space in between them. This happens with photons (light essentially), as was proven in the section "Light with a Spin" in the previous chapter.

As light has the highest KED, then it must be that nature senses light better than it senses matter—and my gravitation model is based on this realization. You might recall that as early as Chapter 2, I'd already suspected that nature was more sensitive to energy (light) than matter. This will be proven in the section "Motion Finally Explained!" of Chapter 8.

My model has also a philosophical angle to it. You may remember the Heisenberg uncertainty formulas $m\Delta v\Delta s$ and $\Delta t\Delta E$ I derived in the section "Another Formula for Fuzziness" of Chapter 5. Philosophically, these formulas imply that nature is uncertain about the motion and even the existence of matter while being very certain of the existence of light (this is because $\Delta t=0$ for light, so then $\Delta t\Delta E = 0$).

So, where does this philosophy lead me? Rather than space–time, I propose a space–*density* theory in which space isn't curved but rather has a "thickness" very much like air in the atmosphere has a different density depending on the temperature. Let's define the space *density* of a region of space to be based on the amount of *free-flowing* energy compared to the amount of matter.

What do I mean by free-flowing energy? It's pure energy, light essentially that isn't disturbed such as by an atom. It might be light given off by an electron for instance. Also, as we will see in section "Motion Finally Explained!" of Chapter 8, matter moves along by giving off pure energy.

More precisely, the energy density of a region is a measure of the ratio of the amount of light present in a given region over the amount of matter energy there. Let me call that type of energy density the potential energy density (or PED) of space. So my space density theory is based on this PED of space. Space density is 100 percent where there's light only, and 0 percent where there's only matter (is there such a thing?). Two questions arise:

1. What's the space density of a celestial body? As a body contains mainly matter, its space density is certainly much less than that 100 percent. Why is that? As we saw in Chapter 1, the energy inside matter is "trapped" and isn't emitted. Accordingly, it isn't free-flowing energy.

2. What's the space density in outer space? Out there, there's very little matter. Therefore the energy density is high, although not quite 100 percent because there's a little bit of matter such as dust particles. Based on my development of this idea in Chapter 8, I suspect that the only place where the space-energy density is 100 percent would be near the edge of the universe.

Note that my definition here of space density, denoted PED, differs slightly from the one in the section "Emitted Energy Versus Internal Energy" of Chapter 4, which is noted KED. That definition KED is for energy density that comes as a result of the *effect* of gravitation. The other energy density, PED, is based on a ratio of free-flowing energy. This is the kind of density that *causes* gravitation. This is my theory. To help you understand the difference, consider water flowing down a river. The behavior of the flow of water is to the KED of space what the slope of the river is to the PED of space.

The consequence of my definition of space density is that when nature senses a celestial body such as the sun, it tends to feel it as *low level* energy while it senses the light all around it in the dark outer space as *high level* energy. However, this is very counterintuitive because the celestial body sure feels to us as being much denser than the emptiness of outer space, right?

You might remark that in Newton's gravitational formula, it is mass that's a factor, not *density* of matter or energy. Indeed—keep in mind that the term *density* used in the context here refers to free-flowing energy density as compared to matter density.

My new density concept is quite counterintuitive because it's exactly the opposite of the density we understand in our daily life: mass divided by volume. The denser the matter is, the *less* dense the free energy inside that matter. The denser the free energy, the *less* dense matter is. So the densest space is in outer space, and the less dense is right here on earth, or rather *inside* the earth. But this defies our senses! So the equations I propose are simple:

high free-flowing energy density = *high* PED =
denser space = *low* gravity

and

low free-flowing energy density = *low* PED =
lighter space = *high* gravity

Now that I've explained enough of my idea, here's the picture of space density:

denser space (less gravity)

denser space (less gravity)   lighter space (more gravity)   denser space (less gravity)

denser space (less gravity)

Consider two extreme cases that agree with the two equations above:

1. *A black hole.* It has a very low space density because it has just about zero free-flowing energy. Indeed scientists know that light is "trapped" in a black hole—it can hardly move at all. So the PED of space of a black hole is extremely low be-

cause energy is no longer free. And guess what? A black hole has an extremely high gravity.

2. *A region of free-flowing energy* (a sort of matterless "cloud"). This is where there's no matter at all (i.e., high PED of space):

dense space (less gravity)

dense space (less gravity)   Free flowing energy (less gravity)   dense space (less gravity)

dense space (less gravity)

This isn't a cloud of gas, but rather just a cloud of pure energy, assuming that such a thing is possible. What do you think would happen next? The cloud would just dissipate. In other words, the cloud produces no gravitation—actually it would produce the opposite: antigravitation. Consequently pure energy on its own doesn't create gravity. It absolutely requires matter. Consequently, the gravitation of a celestial object does the opposite of antigravity: it *pulls* energy and matter toward its center.

Wait a minute! What did I just propose? Pure energy produces *antigravity*. In Chapter 8, I show that the universe started with the Big Bang and that it's *pure energy* that came out of that source of origin. Consequently it's antigravity that caused the energy to shot outward with great force to create the Big Bang. Much later once matter started to form, gravitation slowly neutralized antigravity. Note that gravitation didn't "kill" antigravitation. So, antigravitation is still around, hidden. Antigravitation exists only where there's pure energy, and no matter at all. Where's that? Is it in outer-space? No, because there's matter there although very little of it. I explain in more detail in section "The Expansion of the Universe and Its Fate" of Chapter 8 that the edge of the universe contains energy only—there's no matter there. Guess what? Physicists found evidence of acceleration at that edge. They speculate that some antigravity is the source of that acceleration. This confirms my humble view on anti-gravity.

My model of a space density dimension provides a physical explana-tion for the motion of a body in a gravitational field. In contrast, space–time has the time dimension to offer. But time isn't even a physical dimension, so why should we believe that it may have an effect on motion? As I explain in the section "Motion in Space–Time: The Plot Thickens!" of Chapter 5, motion in the time dimen-sion isn't even possible! It's true that there's a relationship between space and time. Einstein's theory is quite correct in the sense that it does predict the path of bodies in a gravitational field. However, it fails to provide a physical explanation of gravitation.

I'm convinced that Einstein understood that space density was the real reason behind the cause of gravitation. Why then did he choose the time dimension? I can think of three reasons:

1. His General Theory of Relativity is based on his Special Theory of Relativity that showed time is relative. Time doesn't tick at the same rate as it depends on the frame of reference. So it was natural that Einstein carried that dimen-sion into his theory of gravitation.

2. As will be shown later in this chapter, quantum physics is implicated in what causes gravitation. Let's be fair to Ein-stein: his theory was developed before 1916 at a time quan-tum physics was barely in its infancy. Although he knew that the "shape" of space had something to do with gravitation, he had nowhere to go with that idea other than linking it with the time dimension.

3. Humans are very used to time in our everyday lives. Throughout history, humans have devised ways to measure time. That isn't the case with space density that is based on quantum physics. Humans in general have never even thought of such a concept, and even if people knew of it, it would have no meaning in their lives (You'll never hear anyone ask, "So what's the space density around the earth today?"). Even physicists are to some extent not immune to human perceptions of the nature that surrounds us all.

Here is something worth noting: some say that space is made of nothing, but I disagree. Space is made of energy of various forms. When we look up in the sky, we see darkness—in other words next to nothing. Yet there's more energy in the dark space than in planets and stars! A search on the Internet revealed to me that the universe is 73 percent dark energy that can't be detected. The reason we don't see that immense amount of energy is the energy doesn't radiate due to what I call *cosmic pressure*. Note that this energy is everywhere, such as inside stars and planets. It's called dark not because it's in the dark of space but because it can't be detected. More will be said about dark energy in the section "An Estimation of the Minimal Space Destiny" and following sections of Chapter 8.

Fighting Off the "Void"

The intensity of gravity is determined simply by the difference in space density (remember, this is the PED of space) between the inside of the celestial body and that of the outer space around it. The view of gravity according to my model is that the sun *pulls* a body toward it.

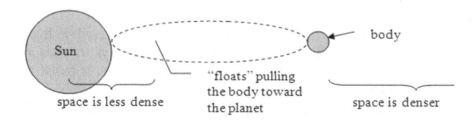

Why would the floats pull the body toward the sun? It's because the space density in the sun is thinner than that of outer space. The atoms inside the sun are like small sponges that soak up the energy. Recall the discussion in Chapter 1 in which I said that matter is energy condensed and "trapped." Well, the "sponge" traps energy, so there's less free-flowing energy, and hence space becomes less dense overall. The less dense space becomes, the thinner it becomes. However, nature fights off *thin space* so to avoid a "void" spot in space—essentially a place where free-flowing energy (light) would be absent. The thinner space is the faster nature will try to fill the "void," and consequently the stronger the gravitational force will be. This is precisely what's observed.

Notice the quotations around the word *"void."* This word suggests emptiness in space, yet although indeed space may be empty of energy, it's never totally empty. As it will be demonstrated in the section "An Estimation of the Minimal Space Density" of Chapter 8, even when space is empty of energy, there's always some energy left. That energy is called *dark energy* or *zero-point energy* depending on the type of energy. The PED of space has a connection with that dark energy as we'll see in the section "The Graviton" later in this chapter.

To receive any credibility:

1. My theory has to account for the *acceleration* of bodies in a gravitational field (e.g., the acceleration of a falling body). The urge to fill thin space inside the sun will produce a motion of objects toward the sun. But will that produce acceleration? Yes—I provide a proof in two ways:

   a. In Appendix 6, while studying the expansion of the universe, I provide an indirect proof in that the use of energy density leads me to none other than Newton's gravitation equation!

b. Here is a more direct proof: consider energy (matter and light) as it bends toward the sun. Imagine matter falling down that proceeds to cross the circle *C1* as shown below:

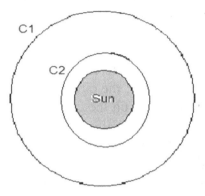

In this proof, energy density is to mean the KED of space (i.e., the density resulting from the effect of gravitation). Why do I use the KED rather than the PED of space? Because I wish to prove that the PED will cause an acceleration that will in turn have an effect on the KED of space. This proof is about the consequence of my theory of PED of space.

It's fair to assume that the energy density on the edge of the circle C_1 is uniform, and that the energy density on the edge of the circle C_2 is uniform as well. As matter moves using floats, let's say that the circle C_1 crosses a number of *f* floats. My model says that the PED of space will cause matter to move toward the sun, and this is the gravitational force. Will acceleration result from that?

As all matter is subjected by the same gravitational force, it's fair to assume that all the matter on the circle C_1 will reach the circle C_2 at the same time. Let's assume that my model is unable to account for acceleration from gravitation, implying that we have to assume that all matter will reach the edge of the circle C_2 with the *same speed* that it had when it crossed the circle C_1. This implies that the number of floats didn't increase while going from circle C_1 to C_2 (the number of floats could increase only if the speed increased too, which we assume that it doesn't). So, an equal number of f floats cross the circle C_2 as on the circle C_1.

Note that the circumference of the circle C_2 is obviously smaller than that of the circle C_1. Consequently the skips between the floats on the edge of circle C_2 are *closer* together than those on the circle C_1. As we discovered in Chapter 4, as a particle speeds up, its skips shorten. Consequently, we have to conclude that matter on the circle C2 gained speed—that is, *acceleration* occurred. This proves that acceleration will result from my model of gravitation. The effect is like a body going down a hill. This is also what Einstein's model of space–time suggests, except that his model doesn't explain the motion in a purely physical way. Mine does!

Essentially the matter that travels from circle C_1 to C_2 has to squeeze in. This causes it to pick up speed. We experience the same effect when we make a hole in a box of juice and squeeze. The smaller the hole, the faster the juice comes out with the converse also being true: the bigger the hole, the slower the juice comes out. Of course this is just an analogy, as light and matter aren't liquids! Nonetheless, we'll see later in this chapter that there may be more than that meets the eye with this thermodynamics analogy.

If you're mathematically minded, you may wish to consult Appendix 5 where I develop a mathematical proof that my model leads to the acceleration of a body around a centripetal force, $a = v^2/r$. That proof also shows that gravitation takes place at the quantum level. So gravitation doesn't just apply to large celestial bodies but to quantum particles too!

Moreover, the proof uses the General Theory of Relativity and quantum physics *together*, and it works! The current belief among physicists is that these two theories don't go together because the first one is for the very large scale while the second is for the extremely small. However, according to my proof, they can work together. In the section "Who is the Observer?" of Chapter 5, I already voiced my hunch that these two theories might go together.

2. My theory has to account for light that is bent as it travels near a celestial body (as predicted by Einstein's theory of curvature of space–time and proven by physicists through observations). With my energy-density model, this result is immediate because light is of higher free-flowing energy density than matter so it will surely move toward the body. In fact, my whole theory is based on this effect with light.

There's a remark worth making here: the reason that gravitation has remained so much a mystery over the last 400 years since Galileo's first experiments is that it seemed so obvious that gravitation is all about matter that attracts matter. I believe this to be the wrong way to look at it. In my theory, I view things the opposite way: based on my concept of PED of space, it's obvious that gravitation attracts energy. What's not at all obvious to me is that *matter* is also attracted! In the section below, I'll show that the consequence of my view is that matter is (indirectly) attracted as well.

3. My theory has to agree with Einstein's space–time theory of gravitation. Yes, it does. Recall in the first section of this chapter, it was said that:

a. The distance between floats of the quantum particles within a planet *decreases* as the planet's orbit is *closer* to the sun. As a consequence, the ticking of time of the planet *decreases*, meaning that time *slows down*.

b. The distance between floats of the quantum particles within a planet *increases* as the planet's orbit is *farther* to the sun. As a consequence, the ticking of time of the planet *increases*, meaning that time *speeds up*.

These two effects of time dilation in a gravitational field are predicted by Einstein's General Theory of Relativity, and have been proven scientifically. My proof in item 1(b) above agrees with this time dilation.

Moreover, in the Chapter 4 section "Emitted Energy Versus Internal Energy" I showed that *time dilation is proportional to energy density*. So the closer the floats are, the denser is space; the farther the floats are, the thinner is space. This is the KED of space—and the space–time model of Einstein is about the KED of space. So Einstein's Theory is a statement about the effect of gravitation, not about gravitation itself.

Based on my theory, the PED of space causes gravitation— the KED of space being the *manifestation* of that cause. So my proof in item 1(b) deals with the consequence of my theory; its result (clustering of floats as a body nears the sun due to acceleration) is a statement about the KED of space as well. The question is: is the KED of space from my theory the same as the KED of space from Einstein's Theory? It sure is because time dilation is equivalent to the KED. My theory does predict time dilation in a gravitational field as predicted by Einstein's General Theory of Relativity.

Therefore, my theory of gravitation is deserving of some merit—and I'm not done with it as there's more in the rest of this chapter.

Solar Systems in Equilibrium

I hope I've persuaded you that my model of gravitation does explain *why* objects fall to the earth with acceleration. In the last section of this chapter, I'll explain the *how*.

Essentially my gravitation model suggests that *gravitation* is a manifestation of nature's attempt to *neutralize* space density differentials. Indeed the PED of space is lighter inside a celestial body than around it, so inevitably free-flowing energy falls into the body. An analogy would be water in a tub. Once the plug is removed, the hole is less dense, so a differential in density is created, and water falls into the hole.

However, a paradox occurs. Ever since Galileo's experiments, and probably ever since the appearance of humans, we've always taken for granted that gravitation attracts *matter* to the ground. It's obvious, isn't it? Nobody would question. I do!

Let's use the sun as an example. Recall that I suggested in the previous section that nature is much more interested in free-flowing energy than matter. By attempting to neutralize space density differentials, gravitation attracts free-flowing energy into the sun. Once in a while, comets slam into the sun, and planets have been orbiting the sun for a few billion years, so it's very obvious that gravitation attracts matter as well. Consequently, both light and matter are attracted toward the sun. So why did I say that it doesn't attract matter? Consider this picture of a rock falling to the sun:

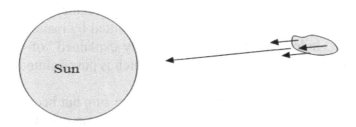

The rock contains some free-flowing energy. It's the light within the rock that's attracted by the sun in an attempt to add free-flowing energy to the sun and therefore make the PED of space inside the sun denser. You might argue that a rock doesn't contain light. Yes, it does because it contains a lot of free-flowing energy as we'll see in section "Motion Finally Explained" of Chapter 8. Free-flowing energy is light.

The small arrows in the picture show the free-flowing energy inside the rock being attracted toward the planet. The actual matter of the rock doesn't contribute to an increase of the PED of the sun. So gravitation isn't interested in matter! Matter doesn't help neutralize space density differentials. It does the contrary: it increases them simply because matter has a thinner space density than free-flowing energy (matter has a lower PED than light). So why then is the matter of the rock attracted by gravitation? It's not—it just appears to be. Consider these two factors:

1. The rock contains some free-flowing energy. Thus, that free-flowing energy definitely will be attracted toward the sun in an attempt to increase the PED of space inside the sun.

2. When the free-flowing energy within the rock is attracted, its matter (made of atoms) will also follow the attraction due to the fact that these atoms are bonded by electromagnetic forces that are a million times stronger than gravitation. So gravitation can't suck the free-flowing energy out of the rock—it inevitably sucks the whole rock. As I said before, as incredible as this may seem, gravitation isn't interested in matter! It's only interested in free-flowing energy.

The resulting effect is that gravitation attracts matter as well as energy. Free-flowing energy is the energy emitted by matter while it's in motion. Indeed section "Motion finally explained" of Chapter 8 will show that matter moves as waves which is pure emitted energy.

Energy thickens space in and around the sun, but because matter traps most of the energy, the space inside the sun continues to thin out, thereby increasing the space density differentials. So the PED of space within the sun remains thin despite all efforts from gravitation. Here is this system in execution:

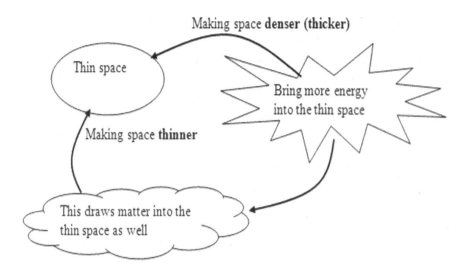

The two effects (thinning and thickening of space density) don't cancel each other out. This is because matter contains an immense amount of energy, but it's more condensed so the resulting effect is a thinning of space density in the sun, thereby further increasing space density as a differential from the rest of the solar system. The result is that the gravitational pull of the sun increases. This in turn will attract more energy and matter into the sun for the same reason given a moment ago.

However, clearly this system isn't in equilibrium, and thus nature can never achieve its goal of neutralizing space density differentials. Hence the space inside the planets (and the sun) can never increase their PED and a vicious cycle is created with the consequence being that the gravitational pull of the planet increases. This is precisely what's observed: the gravitational pull of the sun and planets increases very slowly.

As a keen reader, you most likely consider totally ludicrous my idea of gravitation not being interested in matter. It's been known since Newton's time that the more massive a body, the more gravitational pull it has. Yes, I agree. But the body has more gravitation because the more matter it contains, the *less* free-flowing energy it has. As a consequence, the greater the differential of PED of space of the body becomes with its surrounding in outer space. This causes the gravitation to increase with it being the PED differential that dictates the magnitude of gravitation.

You might ask: what about the free-flowing energy (light) that exists everywhere in the solar system (of course most of it coming from the sun)? It should be attracted by the gravitation of all the planets (and even the sun itself) in order to thicken the space density of the inside of the planets. Indeed, light is attracted, and then the planets should suck that energy so as to attempt to neutralize space density differentials and thereby diminish the planets' gravity.

This effect is obviously not occuring as gravitation never diminishes. Consequently it has to be that light is never sucked into the planets or even the sun. What's preventing the planets from sucking in the light coming from the sun? It's the electromagnetic forces from the rocks and gasses of which planets are composed that keep most of the light out. The light simply doesn't penetrate the inside of the planets.

So, for the most part, the only thing that falls onto the planets or the sun and stays there is matter. Consequently, the PED of space in the planets and sun doesn't increase and the more out of equilibrium the system depicted above becomes. We might remark that just because the sun is full of light, so its PED should be high, not low. Indeed, but that light is generated from matter—namely hydrogen and helium. The sun is still full of matter, so its PED is still low. Remember that the PED is a ratio of free-flowing energy versus matter.

But the (solar) system "fights back" evidently, because not all bodies spiral downward into the sun. Many bodies successfully resist being absorbed by the sun: the planets, satellites, and asteroids. These bodies keep the system in equilibrium. We'll see in the sections "On Consciousness" and "Entropy Versus Organisms" of Chapter 9 that a system not in equilibrium increases entropy and thus leads to its own destruction. To counteract this, nature seems to have a mechanism that tries to place order: it's Kepler's Second Law of planetary motion. Nature seems to have figured out a way to use feedback systems such that the orbit of a body remains stable. However, in the end entropy will win: planets very slowly slip closer to the sun, and eventually will be swallowed by it.

Note that the equilibrium of planetary motion has a narrow margin. A few paragraphs below, I'll demonstrate that the escape velocity of the planets is only 1.4 times higher (squared root of 2 to be exact) than the orbital velocity. This means that if a planet went just a little faster, it would escape the sun's gravitational pull and wander out of the solar system forever. So the planets perform a delicate balancing act as they orbit the sun, very much like an acrobat walking on a tightrope. As the acrobat's walk isn't perfect, a planet's orbit isn't a perfect one either. The orbit doesn't follow precisely Einstein's gravitation equations for the simple reason that the planet knows nothing of these equations! Instead the planet uses a simple feedback system as follows. If the planet gets closer to the sun, it increases speed. Due to centrifugal force, this increase eventually pulls the planet away from the sun, so there's a constant balancing act going on between the gravitational force and the centrifugal force.

I theorize that this balancing act results in the orbit of the planet that is wavy and not smooth. It would be more like this:

Of course, this picture depicts a wavy motion that's *extremely* exaggerated. In reality, the wavy motion of the planet along its orbit would be negligible, and possibly impossible to detect even using sophisticated instruments. As proven in Appendix 5, the reason is that the wavy motion takes place at the quantum level.

In addition, this sort of balancing act results in an orbital shape that's not totally centric. It's well-known since Kepler's discovery that planets adopt an elliptic orbit. The stability of the planets translates into a formula: gravitational force = centripetal force = centrifugal force. What's the centripetal force? The planets are in fact constantly accelerating because of their constant change in direction (a change in direction requires a force applied to the body). It's the gravitational force of the sun that produces the centripetal force, causing a circular motion. I mentioned in the first section of this chapter that the balancing act may be expressed by the formula

$$v = \sqrt{Gm_s/r}$$

What's the escape velocity? It represents the speed necessary for an object to escape the gravitational pull of the body (such as the sun symbolized by the term m_s) that keeps it into orbit. Let's take the case of a planet orbiting the sun. A planet will escape the grasp of its gravitational pull when the planet's kinetic energy (this is the energy due to its speed) becomes equal to the energy necessary to keep it into orbit—this is the energy from the gravitational pull called the gravitational energy.

The equation (4) in section "Derivation of the Famous Equation $E = mc^2$" of Chapter 1 gave us the formula for kinetic energy:

$$E = \frac{mv^2}{2}$$

where m is the mass of the planet in this case.

The gravitational energy may be viewed as the energy necessary to lift the planet from the center of the sun all the way to its orbit. How much energy does this require? The energy is defined as a force applied over a distance. In this case, the force is:

$$F = G\frac{m_s \cdot m}{r^2}$$

and the distance is the radius r of the orbit of the planet. So, the gravitational energy is:

$$F \cdot r = G\frac{m_s \cdot m}{r^2} \cdot r = \frac{G \cdot m_s \cdot m}{r}$$

This energy equals the kinetic energy at the time that the object is about to escape the grip of the gravitational pull:

$$\frac{mv^2}{2} = \frac{G \cdot m_s \cdot m}{r}$$

or

$$\frac{v^2}{2} = \frac{G \cdot m_s}{r}$$

From this, the formula for the escape velocity is easily obtained:

$$v = \sqrt{2Gm_s/r} \quad (1)$$

Let's go back to my theory of gravitation. If you perhaps didn't quite understand my idea of energy differentials as an explanation of gravitation, consider this: I made an analogy in the previous section that matter is like a sponge that traps water. So another way to view the space density differential explained earlier is to regard the body as a sponge full of holes and view the surrounding space as water. If we put the sponge into the water, the differential in density will cause water to seep into the sponge rapidly creating a subtle whirlpool effect in the surrounding water, very much like gravity.

You may respond that once the sponge is full of water, the differential of pressure will be neutralized, so the same should happen with the gravitational pull of a body, which is of course not the case. This is because the analogy is incomplete. There are two other factors to consider:

1. The energy around the body has a hard time seeping into the body due to the electromagnetic forces preventing it. So my analogy with a sponge isn't entirely correct as it should be a sponge that doesn't take in water very well.

2. Even if these electromagnetic forces were absent, it would take an *immense* amount of energy to seep into the body to increase the space density to any significance. Remember that $m = E/c^2$, and mass is what determines the force of

gravitation. For instance as the sun travels through space it absorbs energy from space, but due to this formula, the mass increase remains small. In fact the mass of the sun increases billions of times faster due to the various small celestial bodies (hence matter) that slam into it. But because these bodies are matter, they lower the space density and add to the "spongy" fabric of the sun. As water keeps on being sucked in (referring to the analogy), the sponge can never be full of water, and the differential of pressure remains. However as the sun ages, it'll eventually have given off most of its free-flowing energy, and so its gravitation pull will increase even more. It might become a massive dwarf star or perhaps even a black hole.

Note that the masses of two bodies orbiting each other include the kinetic energy from particles inside their mass, not just the matter. This implies that kinetic energy adds to the gravitational force, not just the matter (or the PED to be more precise).

You might remark that this kinetic effect seems to contradict what I theorized in the "Space Density" section earlier in that free-flowing energy (light) lowers gravity. But no, there's no contradiction: kinetic energy isn't energy from light. It's energy from moving *matter*. Let me put it this way: kinetic energy doesn't emit light, so it isn't free-flowing energy.

To summarize, there's no curvature of space as proposed by the space–time theory. The theory of curvature of space is just an abstraction, a product of imagination. Einstein should have used the phrase space *density*, not space–time. Space isn't curved, but rather has an energy thickness. Einstein's equations are all correct. Space–time does describe the effect of gravitation, but not gravitation itself.

In the section "Gravitation Revisited" of Appendix 6, I'll provide a mathematical proof that:

1. My own theory of gravitation leads to Newton's equation above!

2. The gravitational constant G isn't truly constant! This is what Paul Dirac predicted in 1937, although most physicists today dismiss his result.

3. The effect of gravitation is a manifestation of the Second Law of Thermodynamics! (More on this in the section below).

Gravitation and the Second Law of Thermodynamics

You might have noticed that my space density model seems somewhat similar to some of the rules of thermodynamics such as with gases or water. Indeed I dare to suggest that the *effect* (not to be confused with the *cause* that is indeed a fundamental force) of gravitation might not be a force on the same footing as the other fundamental forces such as electric or nuclear forces. Instead, its effect might be caused by a *difference* of space density, very much like a mass of dense air causes a mass of lighter air to go upward and the cold air downward. Yet nobody has ever theorized wind force as being a fundamental force.

How dare I draw a parallel between the effect of gravitation and the Second Law of Thermodynamics? This seems totally ridiculous: gravitation isn't air or fluid. Correct, but its effet must obey the Second Law because that law applies to *all* forms of energy. This is what I state in the section "The Second Law of Thermodynamics" of Chapter 9 and in Appendix 6 where I provide mathematical work that proposes this parallel between the Second Law and gravitation.

In Appendix 6, I provide a series of mathematical manipulations that use my view of gravitation to lead to Newton's formula for gravitation. So my view for the cause of gravitation therefore seems correct (i.e., it does seem to describe gravitation). That same appendix also suggests that the effect of gravitation is a *manifestation* of the Second Law of Thermodynamics. What a daring suggestion!

However, you might attempt to refute my suggestion this way. The Second Law states that the entropy of a system will tend to increase over time. Remember that entropy is a measure of energy that's no longer available to do work with, or essentially energy that is *neutralized*. In other words, the Second Law indicates that a system tries to spread matter and energy all over space. This is obviously *not* what gravitation does: it concentrates energy and matter toward a celestial body. That is what gravitation is all about. We all know that! So my attempt to connect the Second Law with gravitation seems to be complete lunacy.

But no, it's not. Keep in mind that the Second Law means that a system attempts to disperse energy and matter if *not hindered* from doing so. In previous sections of this chapter, I presented my theory as to why there's gravitation: it's an attempt to neutralize energy density differentials. Isn't that what the Second Law is all about? Indeed—and gravitation does try to obey that Second Law. But, as I explained in the previous section, electromagnetic forces (as possibly other forces too) from inside matter *hinder* this neutralization that gravitation attempts to accomplish.

Two clarifications worth noting:

1. When the Second Law states that energy tends to dissipate, it doesn't discriminate against the type of energy in question. *Any* type of energy tends to dissipate. If it doesn't, it's because something is hindering it.

2. The Second Law is silent about the mechanism that causes dissipation of the energy. This mechanism depends on the type of energy. For instance, energy dissipation on a hot metal plate will occur differently than heat dissipation in the atmosphere. But all types of energy abide by that all important Second Law of Thermodynamics.

As my view of gravitation says that an object is attracted to a body due to energy dynamics, this connection between gravitation and the Second Law has merit, as unbelievable as it may seem.

The Graviton

The sections "Space Density" through "Solar Systems in Equilibrium" explained *why* gravitation exists. It has been explained why a body is attracted to another body. Now let's consider *how* gravitation is carried out. What mechanism makes a body move toward another body?

In Chapter 2, it was shown that a particle moves in space by skipping from one float to the next. The same applies to a celestial body except that it's made of billions of particles. Let's take the example of the sun. I proved in the section "Fighting Off the 'Void'" that, as a body gets closer to the sun, its floats get closer, which makes the body accelerate. Moreover, there's interaction between the fabric of space and the body (the sun in the present example). This was demonstrated in the section "Motion in Space–Time: The Plot Thickens!" of Chapter 5 where it was suggested that a float helps a quantum particle to go from one float to the next. But that section also demonstrated that there's no friction with space.

To give an analogy, the fabric of space can be viewed as having tiny valves that let particles move around. It's like the veins that are made of tiny valves that help move the blood through the blood circulation system of our bodies. To complete a discussion on motion of quantum particles in space, we have to talk about dark energy as well. This will be done in the sections "The Energy Dimension" and "Motion Finally Explained!" of Chapter 8. Those sections show:

1. That the quantum particles that make up a celestial body tend to "sink" into the fabric of space. For instance, as a planet is matter, it will "sink" into the fabric of space. Nature at the quantum level will interpret this as a "void" *in* space. Why? Because nature has a hard time "feeling" matter (because it sinks). Consequently, where there's a celestial body, nature will "feel" a low level of energy. This is the PED of space referred to in an earlier section.

2. That light doesn't "sink" at all into the fabric of space. Consequently, nature has no difficulty "feeling" light. As a result, the PED of space is dense where there's light.

Here is the picture of the space density manifold that shows a shaded celestial body (here the sun) in the center "sink" into the fabric of space:

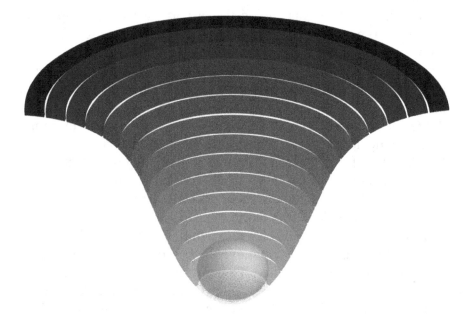

This picture shows the effect of the PED of space, the sinking effect. Of course, this picture greatly exaggerates that effect.

You may note that this space density curvature is *identical* to the space–time curvature of space as proposed by Einstein's General Theory of Relativity. Indeed—so why do I bring it into my own theory of gravitation when I seem to have rejected space–time curvature in the section "A Problem with the Space–Time Model"? Two reasons:

1. I provide a space–density structure to space, which is a physical phenomenon explained at length in previous sections. Einstein's Theory proposes a space–time structure of space based on a nonphysical dimension: time.

2. Although the picture above shows a curvature of space density, there's no real curvature. Instead there's a "thickness" to space in outer space and "thinness" where a celestial body is located. It's impossible to depict this concept of thickness so I had to depict a curve instead.

Although this picture seems to show that a celestial body presses down into the fabric of space, that's not the case. There's no up or down or left or right in the fabric of space. Therefore, for this picture to be exact, it should *envelope* the celestial body from all directions. But that notion of envelope is impossible to put into a picture.

Note the dotted fabric where the sun "sits." These dots are to represent the place on the fabric of space where the sun "sits." Everywhere else, the fabric of space is "bent," but that fabric is located in outer space. The farther away we move from the planet, the less matter there is, so the higher the PED of space. The very dark areas are where there's mostly just light.

Thus, the mechanism that makes gravitation work is simply the PED of the fabric of space, and its effect is manifested by the KED of space. The curvature in the picture above illustrates that the PED of space is high at the top of the curve and decreases toward the sun. Based on the definition of PED of space, a body will slide down toward the sun where the PED is very low. But what *physically* makes a body follow down the curve? We can be sure that it's not the time dimension: it's the PED of space. Two effects:

- At the top of the curve, far in outer space, the floats are dense (close together) so there are *lots* of them; consequently the body will slide down *slowly* because it has to go through *many* floats.

- Then as space density curves down, the floats are less dense so there are *less* of them to go through; consequently, the body will *speed up*. Note that because the PED around and inside the body is thinner than farther out, the space density where gravitation is stronger is thinner. This causes the fabric of space to *stretch* where there's a gravitational field. Keep this effect in mind because we'll come back to it in Chapter 8 when comes the subject of the expansion of the universe.

You might remark that this scenario seems to totally contradict what we discovered in the section "Emitted Energy Versus Internal Energy" of Chapter 4 about the relationship of the speed of a body and the number of floats it goes through:

- The *faster* the speed of particles is in a given region, the *closer* are their floats.

- The *slower* the speed of particles is in a given region, the *farther* apart are their floats.

But there's no contradiction—these floats constitute the KED of space, not the PED of space.

So, then we have to conclude that there exist two different types of floats related to the constitution of space:

1. One type to account for motion. I propose that this first type of floats are shallow and reside at the surface of the fabric of space. They are the floats described in the section "Motion Finally Explained!" of Chapter 8.

2. A second type to account for the PED of space, the type of energy density that generates gravitation. These second type of float are underneath the first type of floats. These deeper floats can't be seen in space. These are the ones that act like valves in a vein. We'll deal with this second type of floats in the section "Gravitons Revisited" in Chapter 8.

Moreover, the KED of space has the exact same space–density shape as the one for the PED of space depicted above except in reverse. In other words, the picture above applies to both the PED and the KED, but in reverse of one another as to their densities. The PED has a high density at the top of the curve but a low density at the bottom of the curve where the celestial body is located. For the KED, it's the other way around—and it's that KED of space that Einstein's space–time curvature of space depicts. Taking again the analogy of the river, the flow of water is higher at the bottom of the valley where the potential energy is low. Then the flow of water is lower at the top of the valley where the potential energy is high. The KED and PED work in reverse, but have the same space density shape.

The first type of float is to the particles of water what the second type of float is to the invisible force pushing the water down. So when the moon orbits the earth, its floats that account for its motion are visible in the sense that their manifestations are "visible" (as in detectable) in space. Keep in mind that the floats themselves are within the fabric of space and so are invisible. Gravitation's cause has at its origin the second type of floats that are also invisible. Physicists gave a name to that second type of floats: *graviton*. Quantum theory predicts their existence, but they haven't yet been found experimentally.

Human experience with gravitation is as a force pulling us downward toward the center of the earth. In reality, gravitation *pushes* down using gravitons; for instance, the fabric of space is what allows me to move my arm as the fabric of space pushes my arm from one float to the next one. The fabric of space plays the same role with gravitation, except that this time I have no control over where my body is pushed. So the fact that a body falls to the earth is a *definite proof* that the fabric of space pushes quantum particles (like valves) as they move around. It's precisely this pushing whose existence I suspected and described in the section "Motion in Space–Time: The Plot Thickens!" of Chapter 5.

Why can't we feel this pushing? Two reasons:

1. It's simply because the pushing occurs at the quantum level where everything is billions of billions of times smaller than our nerve cells, so our nerves can't feel anything.

2. The pushing occurs at the exact same rate of strength throughout our bodies, inside and out. Consequently, our bodies don't stretch at all, so we feel nothing. But if we were

to fall toward a black hole where the "curve" of the PED is extreme, the gravitons wouldn't push the particles of our bodies with the same strength everywhere, so our biological tissues made of those particles would stretch—and that would hurt, believe me!

But wait a minute! Did I say that gravitation *pushes* down using gravitons? If that's the case then gravitation isn't a force originating from the center of the celestial body: it's a force away from the body going toward it. But, don't you think that this contradicts the definition of gravitation which has always been understood to be a force that pulls toward the center of a body? In section "Space Density," I stated that it's antigravity that *pushes* energy and matter away from a center point. So it has to be that gravitons *pull* energy and matter. If not so, that is, if gravitation does push then we would have to conclude that gravitation is… antigravitation in disguise! If you understood the section "Gravitation and the Second Law of Thermodynamics," then in light of what I just explained, it's antigravitation that follows the Second Law of Thermodynamics, not gravitation. It seems therefore that *gravitation is nothing more than antigravitation that's hindered* by electromagnetic forces within celestial bodies. Physicists have been looking in vain for antigravitation for decades while, just perhaps, antigravitation has been around us all along!

Note that physicists are divided about this. Does gravitation pull or push? There's no question that Newtonian physics states that gravitation pulls. Some physicists believe that Einstein's General Relativity Theory model suggests that gravitation pushes because the space-time curvature pushes the orbiting object toward the center. If that is the case, then space-time models antigravitation, not gravitation. I gave my reason in the previous paragraph.

We'll conclude the discussion of gravitons in the section "Gravitons Revisited" in Chapter 8. For now, we should note that physicists believe that gravitons are at the origin of the force of gravitation. I disagree. A graviton is the mechanism that enables gravitation. However a graviton doesn't generate gravitation—it simply carries it out. I've already explained in previous sections the phenomenon that generates gravitation: PED differentials.

By the way, anything between two floats goes no faster than the speed of light. Thus, gravity travels at no more than the speed of light. This is why Pluto feels the sun's gravity five hours later than our earth does. This causes a bit of a wrinkle of space around the solar system. The sun wrinkles space and doesn't curve it in the way Einstein's theory says. A body close to the sun, such as Mercury, feels the wrinkle before a body very far from the sun such as Pluto.

Chapter 7: Did the Universe Have a Beginning?

Surprisingly before the twentieth century, many scientists believed that the universe had always existed and so had no beginning! Yet we don't have to be geniuses to see that this is impossible. Suppose the universe didn't have a beginning—that is, the universe always existed. Then let's say that the line below represents the time that the universe has existed up until today:

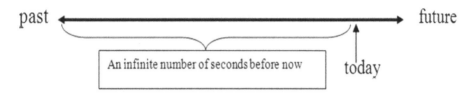

The line demonstrates that if the universe always existed, an *infinite* number of seconds went by before today. Infinite means that the number of seconds before today could never be filled, implying that today can never be reached. Obviously today is here! Consequently, there was *not* an infinite number of seconds before today—and the *current* universe had a beginning. This says nothing about previous universes, if any. Even though there isn't any proof or evidence, many scientists believe that other universes existed before ours or perhaps currently existing parallel to ours.

Note that the same argument used to prove a beginning to the universe may be used to prove that time also had a beginning. However because time is nonphysical, it can't be proven that it started at the same time that the universe started.

The nature of time that I propose in section "The Clock of the Universe" of Chapter 4 suggests that time started at the moment that the universe began. However in section "The Time Dimension is Nonphysical" of that same chapter, the reasoning is that because time is nonphysical, it might have existed before the beginning of the universe. It's therefore conceivable that time may have started some time before the beginning of the universe. However what would have been the purpose? So, it's most probable that time didn't exist before the beginning of the universe.

In the physical scheme of things, most people will agree that all things that have a beginning meet an end eventually. So, as the universe had a beginning, it most likely will do so as well. I'll discuss the fate of the universe in the section "The Expansion of the Universe and Its Fate" in Chapter 8.

Some people have another reason to state that the universe had a beginning: the Second Law of Thermodynamics, which roughly says *"Energy spontaneously disperses from being localized to becoming spread out if it is not hindered from doing so"* (taken with permission from the Web site www.entropysite.com/students_approach.html). In the context here, this law indicates that the universe has the potential to eventually destroy itself. Thus, if the universe always existed, then by now there shouldn't be any galaxy, solar system, or planet left. Obviously that's not the case.

However, this argument is unsatisfactory because the Second Law of Thermodynamics doesn't say that a system will *inevitably* reach *total* disorder. Notice that the quote above says "if it is not hindered." I can think of a few forces that may prevent the universe from decaying such as electromagnetic forces or gravitational forces, or even life. Even if we suppose that at the macroscopic level decay occurs, at the microscopic level there may still be life. For instance, overall the universe might tend to decay, but there may be many regions where galaxies might still be forming.

I don't wish to engage here into an explanation of the Second Law of Thermodynamics because this law isn't at all necessary to answer the question of whether the universe had a beginning. In Chapter 9, I'll come back to the Second Law of Thermodynamics to explain it further in the context of its use in living forms such as cells and in the earth's ecosystem. We'll see that the Second Law isn't necessarily about the doomsday scenarios of destruction that many people imagine.

What Created the Universe?

It immediately follows that as the universe had a beginning, its existence had to have been triggered, hence its creation. Clearly the universe couldn't have triggered its own existence. If it did, then it had to exist in the first place! This is a contradiction. So "something" created the universe. Most people would say that something is God. I even suggested that myself in the preface of this book.

Having said this, this is a philosophical question, not a scientific one. My reasoning given in the previous paragraph isn't entirely scientific and so it might remain debatable. Clearly, the universe was created, but does that imply that there's a Creator? At first glance, it seems obvious that there was a Creator, but this isn't so clear. Here are the reasons.

If there was a Creator, then there was a *before* and an *after* the creation. Let's consider the two possibilities:

1. A *before* creation makes sense only if time already existed.

2. If time didn't exist, then the *before* never happened! Thus, there was no Creator: the universe created itself.

The second possibility seems totally illogical, and the first possibility (the existence of time before the creation of the universe) seems at first glance to be logical. However, it leads to an impasse as follows.

1. If time already existed at the moment of creation of the universe, then the concepts of *before* and *after* existed. In other words, cause and effect was possible. Therefore, it becomes conceivable that the universe had a Creator. That is good news.

2. But because of cause and effect, we have to ask the question of what created the Creator! If the Creator were created at some time, it has to have been created by "something" at that time. But then we have to ask what created that something? The chain of these regressive questions of what created

something goes to infinity. Thus, the question of whether there was a Creator can't be answered.

Here is another way to see this impasse:

1. Suppose that the Creator exists. Could time exist before the creation of the universe? Possibly, because time isn't a physical concept as I proved in the section "The Time Dimension is Nonphysical" in Chapter 4. So it's conceivable that time isn't confined within the universe.

2. As we assume in the present argument that time existed before the appearance of the universe, then the Creator exists, and He was created too. This implies that there was a period of time, perhaps infinite, during which the Creator did not exist:

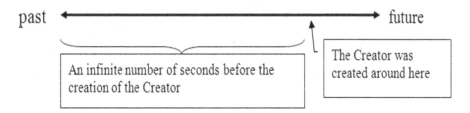

3. Because of the infinite amount of time before the creation of the Creator, it follows that the time of creation of the Creator could never have been reached. Consequently, if time existed before the creation of the universe, then its Creator never existed!

4. If a Creator was necessary to create the universe, then it follows that the universe was never created. But this contradicts the obvious fact that the universe exists.

5. Therefore the universe has no Creator.

6. But then this implies that time didn't exist to create a Creator. This contradicts the assumption in the present argument made that time existed before the creation of the universe.

7. If time existed before the creation of the universe, then there couldn't have been such "infinite number of seconds" before the Creator.

8. If there was no time before the Creator, then cause and effect couldn't occur.

9. It follows then that the Creator could never have come to exist.

The conclusion to the first possibility is that if time existed before the creation of the universe, then the universe couldn't have a Creator. Consequently the universe created itself!

The second possibility leads to the same conclusion as the first possibility. As unbelievable as this may seem, regardless of whether time existed before the creation of the universe, the conclusion is that *the universe created itself*! Many scientists believe in this conclusion, but it simply makes no sense. Something doesn't add up here.

Here's the problem: in my logical reasoning, I assumed that for an entity to be created, it has to have been created at some *time*. But let's suppose for a moment that the Creator was created *before* all times. Is that possible? Can something have been created and exist outside of the confines of time? That is *the big question*.

In my arguments of the previous paragraphs, I assumed that the answer to this question is an obvious *no* because in our physical world, we see cause-and-effect situations from the moment we're born. We associate cause and effect to the passage of time. We look at a clock and see the hands move, and then associate motion to time.

Also cause and effects are driven by exchanges of energy. The universe is made of energy, and it's clear that in the physical world, time exists. But did time exist *before* the appearance of the physical world? In the section "Motion in Space–Time: The Plot Thickens!" of Chapter 5 and other places I suggest that nature "invented" time. Therefore time *did not* exist outside the physical world (God did not create the mechanism of time). However, this realization doesn't help any because my arguments of previous paragraphs concluded that without time, there couldn't have been a Creator anyway.

We're trying to construct a logical argument that leads to a Creator.

To break this impasse, we have to find an answer to the big question: *can something have been created and exist outside of the confines of time?* For the answer to be yes, then that "something" has to be a nonphysical entity. This is because to exist outside the confines of time, this entity had to exist before time was created, and therefore had to exist before the creation of the universe (hence before the physical universe). Yes, we know one nonphysical entity: time, as this was discovered in the section "The Time Dimension is Nonphysical" of Chapter 4.

But this doesn't help because the big question is to find a nonphysical entity that exists *outside time*. Clearly this choice—time—can't exist outside it self. So, we have to look for other nonphysical entities such as *ideas* and *thoughts*. Note that time is also a thought, but it can't be manifested without a physical presence (i.e., nature). That leaves us with ideas and thoughts. Clearly these things have no physical properties.

Ideas and thoughts are generated in our physical world, but as they aren't physical entities themselves, it's conceivable they can exist without any physical structure around. We don't want to assume the same of time for two reasons: (a) this led us to the conclusion that the universe created itself, which is nonsense; and (b) we mentioned earlier that time was created by nature.

To resolve this impasse, we have to conclude that ideas and thoughts already existed *before* the creation of the universe and so before the creation of time. Let's suppose that this is true. It's then plausible that it's a set of thoughts that created the universe! Consequently we've found our Creator: *a system of thoughts*. Humans, no matter the level of intelligence, will never know what those thoughts are—only the Creator and God know.

Despite these logical arguments, it remains unclear whether that Creator is God. But as this system of thoughts was created before the creation of time, it always existed. So nothing created these thoughts. Because by definition, God has always existed, can we conclude that this system of thoughts comes directly from God? I think so. So God created the universe!

Not so fast, though. By definition, God is perfect. Did the Creator create a perfect universe in a logical sense? No, as we'll see in the next section. Note that as thoughts existed before the Big Bang, they existed before time. Consequently, these thoughts are timeless and so will always exist. Could thoughts generated by humans be timeless too? We'll continue this discussion of the nonphysical world in the section "Our Virtual Universe" of Chapter 10.

I've examined supposedly scientific proofs of the existence of God, but those proofs are bogus. Science is about the study of nature. God is beyond that, and is even beyond *everything* as you'll find out in the next section. Science is about the study of how nature works. It seems impossible to combine scientific proof methods and the divine. Some scientists claim that the discovery of the first ripples of the universe, coming from the Big Bang, is proof that God exists, but this proves none of that—it simply proves that humanity is getting closer to knowing how the universe started. The Big Bang and the expansion of the universe will be discussed in Chapter 8.

In this book, I'll assume that God created the universe (most likely indirectly), and also that, most importantly, He created it for a *purpose*. We'll come back to this purpose in Chapter 9.

Our inability to answer this question of what's the ultimate Creator reminds me of one of the greatest mathematicians of all times, Kurt Gödel, who formulated his *incompleteness theorem* in the early 1930s. This theorem simply stated proves that in any system, there are questions for which it's impossible to find the answer using the knowledge *within* the system. The inability to find an answer is *not* due to our limited knowledge or intelligence as humans, but rather due to a limitation of the system itself! To answer all possible questions within a system, let us name it system A, one has to be in a system B that includes the system A. In other words, one has to be outside system A to answer all questions within that system.

This reminds me that when I was in first grade, the teacher taught us to add and subtract numbers. Say for instance, $3 + 8 = 11$. All worked well until it came to my mind to attempt the subtraction $3 - 8$. Oops! I got stuck because I hadn't been taught about negative numbers. As a result, the expression $3 - 8$ had no answer in the system of numbers I'd learned. But once I learned about negative numbers, I was then outside the system of strictly positive numbers and knew that $3 - 8$ had an answer.

What Is God?

I personally believe in the existence of God. So, allow me to ask the question: what is God? This is also a philosophical question, not a scientific one. It's not farfetched to state that humanity will never know what God is, and I'll provide reasons for that later in this section. Nonetheless, we can eliminate a few things that God definitively isn't:

1. God isn't a living thing.

2. God isn't a matter, not a wave, not a source of energy, etc. God is simply not part of the physical universe. God is beyond the universe (although He might be the "observer" of the universe). As shown in the previous section, God, the Creator, was already there before the universe existed. Clearly if God was completely contained within the universe at the time of its creation, He could not have created it!

3. God isn't bound by the laws of nature of the universe. For example, God isn't affected by the forces of nature. This easily follows from point (2) above. This does *not* mean that God can change the laws of nature of the universe at will. In fact, our scientific knowledge points to the evidence that the laws of nature are the same at any location in the universe.

Also, scientific evidence suggests strongly that all living beings are a product of evolution, which is based on the laws of nature. God simply doesn't mess with His own laws!

Note that the fact that God isn't bound by the laws of nature does *not* imply that God is supernatural. The word *supernatural* is really a misnomer, for it suggests things beyond the laws of nature. What humans consider supernatural are things or phenomena not explainable by science.

For instance, a few thousand years ago, humans thought that lightning was a supernatural phenomenon. Another example: suppose for a moment that ghosts do exist. Popular belief is that ghosts are necessarily supernatural because scientists can't confirm or refute their existence. But our limited knowledge about ghosts doesn't imply that, should they exist, they would be beyond the laws of nature. Someday perhaps, humans may obtain sufficient knowledge to answer this question of existence of ghosts, and then perhaps ghosts will become natural phenomena.

Could these two points describe fair characteristics of God?

1. God has a superior reasoning power. Now logical reasoning requires a good working brain, no? As God a not a living thing, God has no brain. Hmm ... to unravel this, we need to ponder the nature of ideas, thoughts, and reasoning. These are definitely not physical or tangible in nature. We'll explore the nonphysical aspects of the universe in various places in this book, especially in the last two chapters.

2. God is totally logical, and beyond any logic that any intelligent being can reach. In fact Gödel, probably the smartest logician to have ever lived, attempted to prove the existence of God using logic, but failed. Of course someone's failure to prove something is no proof that God does or doesn't exist. God's logic ensures that no law of nature contradicts another. And this is the reason that I stated in point (3) above that

God can't change the laws of nature. It would be like lying—and God isn't a liar. So throughout this book, I attempted to remove all possible contradictions in my reasonings.

These well-meaning characteristics are not clear. There's really no means of attaching any characteristics to God. Most religions refer to God as the Creator of all things, and that God is boundless and perfect. You should be able to see that these two characteristics above don't reflect perfection. One reason is simply that it's impossible to describe perfection using logic, as will be seen in the next section.

Does God Exist?

If we can't even determine what God is, is it then even possible to prove that God exists? The implication of Gödel's incompleteness theorem mentioned in a previous section is that *all* logical systems of *any* complexity are, by definition, incomplete (i.e., at any given time, each of them contains true statements that it can't possibly prove according to its own defining set of rules).

Another way to formulate Gödel's theorem is that, given *any* consistent set of arithmetical axioms, there are true mathematical statements that can't be derived from the set. Even if the axioms of arithmetic are augmented by an *indefinite* number of other true ones, there will always be further mathematical truths that aren't formally derivable from the augmented set. Note the word *indefinite* doesn't mean infinite, but rather any finite number of statements, however large.

I like Gödel's theorem so much that I'll phrase its implication another way: rational thought can *never* penetrate to the final *ultimate* truth. Reaching the ultimate truth would lead us to perfection, and perfection can *never* be attained in a system. As people, we kind of know this already: humans aren't perfect. But this isn't quite what Gödel's theorem implies. Humans are not perfect in the sense that we're not perfectly logical beings. Gödel's theorem states that even the *most logical* system will contain statements that it cannot prove true or false. So this theorem is most profound.

You might suggest that God is necessarily imperfect due to Gödel's incompleteness theorem, which says that there are questions that can't be answered in any given system regardless of the logical power within the system. But hold it! Is God part of a system? No, because if He were, then there would exist a system beyond God. However, by definition nothing is higher than God. Therefore God is perfect: He is the ultimate truth.

The only way that God may be perfect then is that He isn't part of any system! This is a consequence of Gödel's incompleteness theorem. I suggested earlier that God isn't part of the universe. Now in order to be perfect, He can't be part of *anything*! This pretty much excludes God from *everything* whether it's in the physical world or otherwise. But does this imply that God doesn't exist? No, it simply means:

1. That God's existence can't be proven in any way at all using logic, no matter how powerful and brilliant the logic may be.

2. That God is beyond any description. *Perfection* is a term that can't be described. There's simply no way of knowing what God is.

So, the question of what God is has no definite answer, which isn't due to the limited intelligence of humans, but to the incompleteness of logic itself (the incompleteness theorem). God is even beyond that theorem!

Will God's existence forever remain in the realm of faith? The good news is that our brains are able to grasp this notion that perfection is beyond description. This ability is perhaps in itself a proof that God does exist. Indeed, I can conceive a thought about something totally beyond the logical system that I'm in, the universe. How can humans think up something completely "outside the box"? I believe this ability is the result of an evolutionary process of the universe to develop living beings intelligent enough to infer the existence of the maker of everything, which is God by definition. You probably have heard the famous René Descartes cogito statement "I think, therefore I am." I propose: "I grasp the notion of perfection, therefore God exists."

In the section "The Thoughts–Energy Dimension" of Chapter 9, I suggest that God is sensitive to thoughts while nature is sensitive to energy. If we can have a thought of God's perfection, then it follows that God is sensitive to that thought. In other words, our brains are complex enough to connect with God.

This suggests that perhaps prayers do have true benefits not only to the person who prays but to another person for whom a prayer is meant. As I suggest in Chapter 9 that thoughts are not confined to our physical dimensions, possibly a prayer may have an effect on a recipient a long distance away.

In 1993, the publishing company HarperCollins published a book by Larry Dorsey, *Healing Words: The Power of Prayer & the Practice of Medicine* in which he details statistics that showed that prayers do have healing powers! He turned to science to prove the efficacy of prayer. In a materialistic age in which only what can be scientifically measured is real, the results of prayer were subjected to the rigors of the scientific method. Dorsey and Joan Borysenko led a team of scientists on an inquiry to measure the effects of prayer on a variety of subjects. More than 140 studies were conducted. They found that even bacteria in a petri dish have a measurable response to prayer, even at great distances. This suggests that bacteria can capture thoughts. This is very interesting, because this is precisely what I suggest in the section "On Consciousness" of Chapter 9. We'll come back to this notion of ideas or prayers having an effect beyond great distances in Chapter 10.

Gödel's theorem is so profound that I took upon myself to attempt to prove it myself. Have you ever heard the phrase, "you're damned if you do, and you're damned if you don't"? The proof is somewhat based on that idea. So here goes.

1. Suppose that you introduce me to a computer called C that is supposed to be able to correctly answer *any* question at all. Note that I specified the verb *supposed to*—I didn't affirm that it does answer all questions.

2. The computer C has to have some sort of program and circuitry inside in order to function. However complicated and sophisticated the program may be, it's clear that it can only be finitely long.

3. I then write the sentence: "The computer C will never say that this sentence is true." Let us call this sentence S. Note that S is equivalent to: "The computer C will never say statement S is true," which is a self-referential statement. Such statements can't be proven true or false as will now be demonstrated.

4. Suppose that I ask the computer C whether S is true or not.

5. If the computer C says that S is true, then "the computer C will never say statement S is true" is false. If the statement "the computer C will never say statement S is true" is false, then statement S is false. This is because statement S is "the computer C will never say statement S is true." So if computer C says that statement S is true, then S is in fact false, and computer C has made a contradictory and therefore false statement. But as computer C makes only true statements, computer C will never say that statement S is true.

6. The reasoning in point (5) proves that computer C will never say that statement S is true. So "the computer C will never say that statement S is true" is in fact a true statement. Thus, statement S is true (as S = "The computer C will never say that statement S is true"), but the computer C will never affirm it.

7. I know that statement S is true. Therefore I know a truth that computer C can never affirm. Consequently, computer C is unable to answer all questions.

Note that the above is a philosophical proof, not a mathematical one. The theorem implies that I can't know everything about myself and that a system can't know everything about itself. As the universe is a system, the universe can't know everything about itself. Let's call U the system that makes up the universe. This suggests that:

1. The universe knows statements. There are entities within it capable of thinking, of formulating thoughts. Humans can think, and so can animals to some extent, and perhaps insects too, although very little. Can nature think? I think so (more about that in Chapter 9).

2. Suppose that there exists a system V greater than the system U that can answer all of system U's questions. Can that system V be God? No, because then according to my own proof above, any system contains statements that can't be answered. But by definition, God is perfect; therefore God has the answer to *all* questions.

3. So then because system V can answer all questions that system U can think of, then perhaps system V is the Creator of the universe. This is because the statement "How was the universe created?" may be answered by a system higher than U such as by the system V.

4. Can system V be God? No, as was deduced earlier, God isn't part of any system. If God is perfect, then Gödel's theorem can't be applied to Him. Therefore, God is *not* the Creator of the universe, at least not directly.

An example of a statement that the system U (universe) can't answer is: what states are quantum particles in at any given moment or place? The fuzziness formula $\Delta s \Delta v$ of the section "Space–Time Fuzziness" in Chapter 2 and also the Heisenberg uncertainty principle discussed in the section "Another Formula for Fuzziness" of Chapter 5 make it clear that nature is uncertain about the state of its quantum particles in time and place. So, there has to be a system V higher than U that can answer that question of uncertainty. Is system V the "Creator"? There's no way to know.

By definition, God is perfect, meaning that God has the answer to all questions. So, then God has the answer to all the universe's Creator's questions. Consequently, God is beyond the Creator. However, I stated earlier, the Creator always existed, so nothing created it. This suggests that the Creator is part of God. Also God has the answer to the entire universe's questions, so God is indirectly involved in the universe.

But my arguments above clarify that God didn't create the universe "by hand." In other words, God didn't fabricate the universe. God is sensitive to thoughts, not to the physical working of the universe. (More will be said about thoughts in Chapter 9 and 10.)

That's fine if you want to believe that the Creator is God, because as there's no way to describe what God is, it's probably impossible to pinpoint the fine line that separates God from the Creator. In this chapter, one thing has been made clear: the universe has a creator.

What Created the Universe?

In the next chapter, we'll discuss what happened after the Big Bang. In this section, let's ponder about what perhaps triggered it. In an earlier section, I concluded that a system of thoughts triggered the creation of the universe. As will be shown in the section "The Thoughts–Energy Dimension" of Chapter 9, in the physical world *all* thoughts manifest themselves at the quantum level. So whatever thoughts triggered the Big Bang had to lead to some sort of manifestation at the quantum level. Indeed, physicists know that the Big Bang occurred at the quantum level. Everything above the quantum level constitutes abstractions as argued in the section "Reality Versus Abstraction" of Chapter 2 and a few other places.

It seems that God only thought of creating energy and it came to exist (that energy will be termed "God's Light" in Chapter 8).

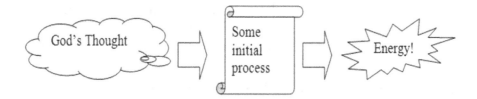

The rest are just details as far as God is concerned. I don't believe that God directly created any physical things other than that initial light. It's His thoughts that led to the existence of physical quantum events.

But what was that initial process that led to the creation of energy? This question puzzled me for many months until I finally came up with the following argument. As will be shown in the section "The Thoughts–Energy Dimensions" of Chapter 9, thoughts do trigger energy. But what was that initial thought that triggered the Big Bang? Do we know of any thought that is associated with energy? Yes, we do. In the section just referred to, I propose that time is actually a thought, and in the section "The Time Dimension is Nonphysical" of Chapter 4, I showed that time is associated with light, which I called *dormant light*. Light is energy. Ah!

So, could it be that the Creator thought of time, and the energy associated with it was then created? That energy would be that of the "God's Light" as is the term I use in the next chapter, or dark energy that is an equivalent term I use elsewhere.

Four notes worth making:

1. God didn't create time and its mechanism. Indeed, I argued earlier in this chapter that time didn't exist before the Big Bang. It's nature that created time as a measurable property and dimension. God (more precisely, the Creator) generated only the *thought of time*. Because that thought was generated before time existed as a dimension, the thought of time is timeless: it always existed! That means that the system of thoughts that existed prior to the Big Bang always existed too.

2. There was no original energy to start with because otherwise, we have to wonder: where did that energy come from? This is a regressive question. There was no energy before the Big Bang. The dormant light, as I call it, was "energyless" light (that is, energy that wasn't doing anything hence the term dormant) that got a boost of energy after God generated the thought of time. This energy led to the creation of space and all other physical things. The next chapter delves into the details such as what might have triggered that boost of energy.

3. As space didn't exist before the Big Bang, it has to be that the energy appeared from a single dimensionless point that physicists call a singularity (more about this in the next chapter). The universe has at its origin a single tiny little dot! The events that led to the Big Bang from that tiny dot remain somewhat of a mystery even to the brightest minds.

4. It's known that when matter meets antimatter, an explosion follows. Some physicists propose that such an event started the universe: antimatter and matter met and caused a massive explosion—the Big Bang. I believe this hypothesis to be incorrect because if the universe had started this way, then we have to ask: where did this antimatter and matter come from? Also this hypothesis doesn't allow for a universe that started from a singularity point which is clearly what happened.

Chapter 8: How Did the Universe Start?

I quickly realized that an attempt to answer this question was impossible without pondering the nature of matter, energy, space, time, and gravitation. This explains why this question is tackled this late.

To get closer to answering the question, we have to go back to the source of the universe— not so much what created it, but *how* it was created. Here are three processes that scientists have imagined over the centuries:

1. *Spontaneous creation.* The universe was created in an instant more or less as we experience it.

2. *Spontaneous creation of space.* Here space is created with all of the universe's energy scattered all over. Eventually matter appears first as dust, then gatherings of gases, eventually making up galaxies etc.

3. *Creation from a single point of origin.* The universe is initially inside that point!

The first process disregards entirely the capacity of evolution of the universe. Scientists now know how stars, planets, and galaxies are formed through billions of years. Why would God have created all those things spontaneously when He already created all the laws to provide for a gradual creation through time? So this first process makes no sense.

The second process makes a little bit more sense, and a few scientists support this sort of creation. But scientific evidence shows that the galaxies are spreading away from one another, suggesting a gradual expansion of the universe and space.

By process of elimination, the third process is therefore the most likely scenario for the beginning of the universe.

Unlike many books that I'm aware of on the subject, my coverage of the creation of the universe will remain nontechnical. The precise explanations surrounding the first moments after the Big

Bang are for too complicated for most of us. My interest isn't so much in what happened the first moment after the Big Bang but in the significance of the events.

In addition, because in the previous chapter I demonstrated that the universe has a Creator, in my upcoming explanation of how the universe started, I'll attempt to see how the Creator (that I'll call God) might have been involved. Most scientists would frown upon such an attempt because the current prevailing trend is that the universe created itself without any divine or supernatural intervention.

It All Started with a Bang!

Physicists say that the universe started from a single point. It doesn't take a genius to realize that, and I provided my simple argument at the end of the previous chapter and I'll provide another argument in the section "The Thoughts–Energy Dimension" of Chapter 9.

But how could the entire universe fit in a single point? Surely the entire mass of the universe couldn't do so! True—so then the only way is that the point contained only energy. Depending on which kind and what's involved, energy might not require any space such as in the case of "dormant" energy that I proposed at the end of the previous chapter. Scientists say that energy comes in many different forms, such as:

1. *Kinetic* (motion)—this one was impossible at the creation of the universe as motion can only take place in space. But space isn't created yet in the single point of origin.

2. *Potential*—this is energy available within a physical system due to an object's position in conjunction with a force (that conserves energy), which acts upon it like gravity. But in the beginning, there was no system.

3. *Electrical*—this one can exist only if electrical particles are present. In the beginning, there were no such particles. However, according to a discovery made by physicist Clark

Maxwell in the middle 1800s, light is made of an electrical field.

4. *Chemical*—there were no molecules in the beginning.

5. *Nuclear*—there were no atoms in the beginning.

6. *Mass energy*—there was no mass in the beginning.

7. *Thermal*—this is heat.

None of those forms of energy could exist in the very beginning because they require matter that didn't exist yet. By process of elimination, that leaves us with the last one, thermal energy. Heat is actually not really a type of energy but rather a *transfer* of energy. So, although it's not matter, it requires space. This implies five things:

1. As the very moment before the Big Bang, there was no heat, no matter, and no forces. Even time probably didn't exist as there was no motion requiring it.

2. This transfer of energy required a source, and that source had to be the light! Indeed, light is the basis of all things in the universe. This is also what I suggest in the section "Entropy Versus Organisms" of Chapter 9. However, this light isn't the one we view every day (more on that later).

3. In the nineteenth century physicist Maxwell proved that light is made of electrical waves. Thus, the electrical field had to be created, perhaps at the same time as light. Where did that electrical field come from? Also, this light isn't the same kind as the one mentioned in item 2 (more on that in the next section).

4. The sequence of events above required time, as a transfer of energy doesn't happen in an instant. This is a classic case of

cause and effect—and that requires an interval of time. Therefore, the time dimension had to be created by nature. Note that God didn't create time as a dimension. (I gave my reason for this in the section "Motion in Space–Time: The Plot Thickens!" of Chapter 5 as well as in Chapter 7).

5. The space dimensions had to be created because a transfer of energy requires space. The space and time dimensions were created at the same time, as well as energy (light). These three phenomena go together, always. It's *not* God that created these dimensions, but rather nature. God belongs in the world of thoughts, not physics.

To sum up, in the beginning of all times, God created the Light—a dormant light—then, through His thoughts (the time dimension was created at the same time), God unleashed the Light, giving it its energy through electromagnetism. As, wherever there's energy, there also has to be space, that dormant light led to the creation of space. The existence of electromagnetism was important as without which energy, matter, and life could never have come about. This short sequence of events must have happened almost *instantly* because energy can't exist without space and space doesn't exist without energy—and all require time.

Remember that in the section "The Time Dimension is Nonphysical" of Chapter 4, I suggested that "dormant" light is contained in the time dimension, and that light is a single point in that dimension. That idea fits well into the scenario above. Note that in the previous chapter, I stated that God generated the *thought* of time—not time itself. So, it's not clear whether the time dimension created the light or the light create the time dimension. This is like the chicken and the egg question: which one came first?

OK—the above sequence of events is a mixture of physical and spiritual events. Let's now concentrate solely onto the physical aspect. At the *very moment* that space was created, the single small dimensionless point had room to expand, and some of the light transformed into other forms as you'll soon find out.

In this section, I'll give the events and the timeline after the Big Bang. I looked up a few web sites to help me. However I found discrepencies across these references. I retained mostly information from Wikipedia because I thought their information more realistic.

What's that "something" that started all this? In the previous chapter, I suggested that a system of thoughts triggered the Big Bang. In Chapter 9, I'll come back to this issue of what started the universe when I discuss the ideas of consciousness and purpose.

Because the universe started from a single point called a *singularity point* (see section "The Singularity Theorem" coming up), we can deduce that the first events that took place after the Big Bang occurred at the quantum level. Based on my deduction in the section "The Quantum World" of Chapter 5, in one single second, there may be up to 10^{43} events that take place at the quantum level! Therefore you can be sure that within the first second of the birth of the universe, many thousands, possibly millions, of events took place.

At the quantum level, a few seconds feel like billions of years! So there were more events that happened in the first few seconds of the universe than the next 14 billions years up until now! In fact, according to Wikipedia, physicists estimate that there exist 2.5×10^{89} elementary particles. All these particles were formed within 3 minutes of the Big Bang! Yet it took some 10 billion years to form the 10^{80} atoms contained in the entire universe. This is how busy the quantum world is.

So, most of the early events led to the creation of quantum particles. I deduce that the things that first came out of the Big Bang were in this order:

1. *Dormant light* (which I call God's Light in the next section) that "woke up" and created the fabric of space (hence its three dimensions) and the time dimension at the same time; then

2. *Pure light* that led to the creation of quantum particles in the way discussed in Chapter 1. This light eventually led to radiation, and much later, the light that we see; then

3. *Quantum particles* that were created from that light a mere 10^{-36} seconds later.

Note that the dormant light was truly... dormant! Indeed it was *not* the vivid kind of light that we see today. Prior to the Big Bang, it was energyless light that was waiting to be lit up. An analogy is an electrical circuit in a house with a light bulb and the switch. There's energy in the circuit, but it is potential energy only. Only when the switch is turned on can the light bulb come on. In this chapter, it'll be difficult to ponder about *what* caused the light to come on, as the cause is probably supernatural. However I'll propose an answer to *how* the light came on.

When the dormant light came on, it was made of virtual photons and gravitons. In section "The 'God's Light' and the Expansion of the Universe" below, I deduce that this dormant light is associated with dark energy. This energy makes up the fabric of space. Since no motion can take place without space, clearly the fabric of space must have been the first thing created after the Big Bang. Moreover in Chapter 6, I deduced that gravitation is a force that occurs at the quantum level and in presence of that dark energy. Consequently, I deduce that gravitation emerged almost immediately after the Big Bang. In fact, gravitation appeared as soon as the fabric of space was created. After consulting Wikipedia, I realized that my deduction is correct from consulting the equation that expresses the value of a Planck length (that I call Planck space, P_S) involved in motion

$$P_S = \sqrt{\frac{hG}{2\pi c^3}}$$

You can see that the Planck length implicates the gravitational constant, *G*. The term *h* is the Planck constant from the Planck energy relation equation $E = h\nu$ that describes energy of electromagnetic waves. Should we conclude that electromagnetism was created at the same time as gravitation? Yes if at least for another reason: light is electromagnetic, and light is what came out of the Big Bang first.

This is so counterintuitive that gravitation came out of the Big Bang so early because (a) gravitation is an extremely weak force, (b) gravitation seems to show up only with large celestial bodies which started to form only many millions of years after the Big Bang!

What about antigravity? Do you recall the argument I developed in section "Space Density" of Chapter 6 where I proposed that antigravity shows up only where this is only pure energy present? This is interesting because this was precisely the condition that prevailed just the moment before the Big Bang. Indeed, as mentioned a few paragraphs ago, the dormant light, full of potential energy, was there just waiting to be woken up. That dormant light was dimensionless, and so had no place to go. What woke it up? The dormant light was pure energy—there was obviously no matter present yet. This is precisely the condition that was favorable to antigravity to spring up. Therefore it is antigravity that woke up the dormant light! The resulting effect was that an immense amount of energy was being pushed out of the singularity point by the antigravity hence generating the Big Bang. The effect is very much like when you burst a balloon full of air. So, it's fair to argue that God created the dormant light (and the singularity point that I talked about in a later section), but not the Big Bang.

So, it's antigravitation—not gravitation—that came out of the Big Bang in its first moment. But gravitation strung up mere moments later even before the first quantum particles appeared.

What was the first quantum particle that came out of the Big Bang? Neutrinos are thought by physicists to be some of the smallest particles possible. So you would think that they were the first particles created because usually things build from the small to the big. This is so counter-intuitive but, in the case the quantum world, the trend is the other way around! Here is how I deduced this.

In chapter 1, I stated that given Einstein's famous equation $E = mc^2$ only very high energy waves can produce matter. Clearly at the moment of the Big Bang, the universe was extremely hot. Physicists believe that a mere 10^{-43} seconds after the Big Bang, the temperature was about 10^{32} celcius! This temperature is called the *Planck temperature* and believed to be the highest temperature possible in nature—any higher, quantum theory breaks down. This means that if the initial temperature had been higher than 10^{32} celcius, the universe would have instantly ceased to exist at its birth.

Given my energy formula $E = gf$ (or equivalently Planck's formula $E = hv$.) that I derived in section "A Formula for Discrete Energy" of Chapter 5, it follows that very hot temperatures implies very high frequencies of waves emitted. At the very instant of the Big Bang, the energy created had a frequency that was far too high for tangled-up waves (i.e. matter as explained in Chapter 1) to form. So the universe had to cool off at an estimated 10^{28} celcius. Again, this happened within the first second of the universe, more precisely, after 10^{-36} seconds of the Big Bang. The frequency of the light was now low enough to allow for the creation of matter.

What was the value of that frequency that allowed for the creation of matter? Physicists still are not entirely sure how or what the conditions had to be to favor the creation of matter. This is still an open question. I looked up the Internet for the highest frequency of light detected by scientists. They are gamma rays. These rays may have a very wide range of frequencies. The highest ones thought to exist are 10^{40} Hz. But, as I explained in Chapter 1, this is far too high for the creation of matter. Astrophycisists have observed gamma rays with a frequency of 10^{27} Hz coming from the Crab Nebula and from some highly active galaxies where matter is known to be created at the moment. Since there's obviously lots of matter in those celestial formations, I have a hunch that this frequency might be the threshold for the creation of matter. So the frequency 10^{27} Hz seems a good bet. Now time to use Planck's energy formula to calculate the energy carried by those gamma rays:

$$E = hv = 6.6 \times 10^{-34} J \bullet s \times 10^{27} Hz$$
$$E = hv = 6.6 \times 10^{-7} J$$

Next let us use the equation $E = mc^2$ to find the mass of the particle that is formed by those rays:

$$m = 6.6 \times 10^{-7} J/(3 \times 10^8 m/s)^2$$
$$m = 6.6 \times 10^{-7} J/9 \times 10^{16} m^2/s^2$$
$$m = 7.3 \times 10^{-24} kg$$

Let us express this mass in units of eV that you learned about in section "The Quantum World" of Chapter 5. Let's divide this mass by the conversion factor 1.8×10^{-36} kg/eV. So the mass is

$$m = 7.3 \times 10^{-24} kg / 1.8 \times 10^{-36} \, kg/eV$$
$$m = 4 \times 10^{12} eV$$
$$m = 4 \times 10^{3} Giga \, eV$$

Is there a quantum particle with such a large mass? A quick look on Wikipedia led me to a very close match: the quantum particle called *boson*. Physicists theorize that this particle was the first one created after the Big Bang. Ah! This is just what I was looking for. They are also believed to be the most massive quantum particle. In comparison, a proton weighs 1.67×10^{-27} kg which is about 5000 times lighter. So you can be sure that protons were created later when the universe had cooled off yet a bit more.

I said in Chapter 1 that many quantum particles are charged, so the electromagnetic force had to emerge before these charged particles. Physicists estimate that electromagnetism showed up almost as soon as the universe's birth, that is, some 10^{-36} seconds after it.

Up until 10^{-6} seconds since the Big Bang, the electromagnetic force was "tangled up" with other forces such as the strong, weak and gravitational forces. Indeed it's believed that at the beginning, all forces were united into a single force. Since all forces are of quantum origin, that makes sense.

Once the electromagnetic force finally separated itself from those other forces, charged particles started to form. Protons and electrons appeared also within a second after the Big Bang. Note that since protons are made of quarks, these had to appear before protons. In fact, quarks appeared some 10^{-6} seconds after the Big Bang while protons and neutrons appeared about one second after that. However I found sources that say these particles appeared 10^{-32} seconds after the Big Bang. By 10^{-6} seconds, the temperature of the universe had dropped to about 10^{10} celcius.

Protons and neutrons were created one full second later! This is like an eternity at the quantum level. Why did it take so long for these particles to form? If you took a physics course in high school, you may recall that the nucleus of an atom is made of protons and neutrons. Yet, it will take three minutes after the Big Bang before atoms appear!

There's a note worth making here about a correlation between the temperatute of the universe and the particles formed—the *hotter* the universe, the *more* massive were the quantum particles formed; the *cooler* the universe, the *less* massive were the quantum particles formed. But you already knew this correlation with the equation

$$h\nu = mc^2$$

expressed in section "The Quantum World" of Chapter 5. The temperature of the wave emissions is proportional to the frequency ν, and the mass produced is proportional to that frequency. For instance, the first quantum particle formed, the boson, has a mass of over one million MeV, while the temperature of the universe at that time was about 10^{28} celcius. After the first second, the universe had cooled to about 0.5 MeV which translates to about 5 billions degrees celcius. This energy matches the mass of the electron. Guess what? This is when electrons were formed. It's clear that electrons could not have appeared until then because the universe was just too hot. The last quantum particle to be formed was the neutrino. It was unleashed when the universe had cooled off some more, and, it has the lowest mass of all quantum particles.

So, 5 billions degrees was just "cold" enough for electrons! This temperature seems like hell to me. In comparison, our sun is a very cold place—its core is only 15 million degrees only, and its surface is a "chilling" 6 thousand degrees celcius.

The neutrino came to exist almost at the same time as electrons. Actually the neutrino existed already before that, but due to the heat, it was unstable and kept coupling with charged particles. Once the heat turned down a little, the neutrino freed itself and started to shower the universe. This happened some ten seconds after the Big Bang.

The photon epoch started about 30 minutes after the Big Bang. This doesn't mean that photons did not exist before that. Photons came to exist as soon as light came out of the Big Bang. But due to the very high heat of the first few seconds of the universe, photons kept clashing and as a result were being transformed into all kinds of quantum particles. There was also a lot of electromagnetic activity.

So, imagine this: light and all kinds of quantum particles went bursting out of the Big Bang in all directions very much like quantum particles in a particle accelerator. The universe was extremely hot: many billions of degree celcius! This chaos went on for the next 20 minutes or so.

What happened next is the emergence of the nuclear force because it's the result of the simplest kind of particles, quarks, which is what protons are made of. These particles gradually formed atoms from electromagnetic forces. By this time, the universe had colled to below one billion degrees. This was cool enough to allow for a stable quantum world hence the creation of nucleus of atoms. Because the core of stars is made of the simpliest nucleuses, we can infer from this that the core of stars cannot be hotter than one billion degrees celcius. Since the core of our sun is only 15 millions degrees, it is considered a lukewarm star.

The expansion of the universe eventually brought it to cool enough (10^4 degrees celcius) to allow electrons to circle the nucleuses. Atoms, starting with the hydrogen, then helium formed. Gases started to form. No more than 20 minutes went by since the Big Bang!

Photons from light had now more room to shoot outward with less electromagnetic interference. So the expansion of the universe started to accelerate. Note that because neutrinos have no electric charge, gravitation is the only thing that could make them merge. However because their mass is incredibly small, gravitation had little effect, and so, just like photons, neutrinos started to shower the entire universe—and continue to do so to this day

For the next 300,000 years, the universe expanded slowly because light or particles could not move outward straight. There was still a lot of quantum force interaction.

After 70 thousands years since the Big Bang, larges masses were being formed, not as a result of gravitation, but of electrical attraction or molecular attraction of atoms. This led to lumps of mass big enough to generate gravity much bigger than what quantum particles could produce. Gravitation (as we experience it today) was the last force to have an effect on the expansion of the universe as otherwise gravity forces would have prevented quantum particles from moving away from the single point of origin—matter would have collapsed back into a black hole, and the universe would have never evolved toward anything.

At this point, the evolution of the universe slowed to a snail pace. Gravitation slowly lumped together gases mainly made of hydrogen and helium atoms. Also molecules started to form, hence chemical energy. At this point, there's a system going, hence the creation of potential energy. At least 50 millions years later, the first stars formed.

It's not well understood at the moment of the writing of this book how galaxies form. So, physicists are not sure when galaxies started to form. My hunch is that they formed around the same time as stars.

By the time that galaxies and stars formed, the univerve had cooled to a bone-chilling -200 celcius. Of course, this is an average temperature. Most of the heat has been absorbed and turned into matter. For the past many billion years, the average temperature has been steady at -270 celcius. This is 3 degrees Kelvin, a mere 3 degress above the absolute 0. This means that there is a background radiation at that temperature everywhere in the cosmos. It is called the *cosmological 3K background light.*

You would think that as celestial bodies were being formed; light could travel more freely outward, hence making the expansion of the universe take off like a rocket. But no, the reverse happened! I provide below a possible reason, and I'll give a second one in the section "The Expansion of the Universe and Its Fate" later in this chapter.

I said in the section "Relation Between Mass and Energy" in Chapter 1 that it's "saturated" high frequency energy in space (energy that comes in at right angles at a single point) that created matter. That energy was going at the speed of light, but as it became matter, some of that energy became locked inside the particles, thereby decreasing the speed of the particles and increasing their mass. Consequently the intervals between quantum steps became larger and space density in that region decreased.

I introduced the concept of space density in the previous chapter. Once space density started to differ from one region to another, gravitation started to manisfest itself more evidently. Due to the fact that the creation of matter concentrated energy within packets of matter, these particles of matter slowed down the kinetic energy. This caused the expansion of the universe to slow down or even stop altogether for a while.

In this short presentation of the history of the universe, you may have notice that antigravitation somehow vanished out of the picture nearly as soon as gravitation showed up. It didn't really disappear. The proper verb should be that it was *neutralized* by gravitation everywhere in the universe except at its edge where antigravity pulls with acceleration the edge thereby accelerating the expansion of the universe. We'll cover in detail this expansion in the section "The Expansion of the Universe and Its Fate" later in this chapter.

Should the Universe Even Exist?

Physicists have discovered that the very existence of the universe, even in its very early stage, was very unlikely! The initial conditions of the universe are based on what physicists call "fundamental constants of nature." These constants include:

1. The speed of light

2. The mass of the electron

3. The gravitational constant

4. The Planck constant that dictates the size of atoms

5. The cosmological constant that dictates the expansion of the universe

6. Numerous other important constants

Physicists tell us that if these constants had been just a fraction of a percent off, the universe would have ceased to exist only seconds after the Big Bang or would never have produced any stars. Physicists have estimated that there was only one chance out of 10^{60} that these constants would have the correct values! Based on this, it's abundantly clear that blind chance had nothing to do with the existence of the universe. There are two schools of thoughts on this:

1. That the universe was created with a conscious purpose such as the purpose of life as expressed in the section "A Purpose Led to Life" of Chapter 9. The universe of consciousness (termed the "hidden universe" that we'll talk about mainly in Chapter 10) created the physical universe only once it had worked out what the proper values of those fundamental constants of nature

2. That there was no purpose, that we just happened to be perhaps God's ten billionths attempt at creating a universe. The other universes were parallel ones that failed to amount to anything.

I favour the first view.

The "God's Light" and the Expansion of the Universe

By the way, in the previous section, the light was eventually the entire spectrum of light, not just the visible light. For instance, light includes all sorts of radiation. Some radiation is also what came out of that single point of origin. However, the first light released was *not* the entire spectrum of light. This is because all the colors of light are composites of frequencies so they had to have been created later through time through a mixture of frequencies. For instance, the color green is the result of blue and yellow. The light that came out of the Big Bang was very high energy waves as suggested in the section "The Quantum World" of Chapter 5. The highest energy waves are known to be gamma rays.

In the section above, I described two different types of lights:

1. The light that's inside a single point, the "dormant" light—God's Light. This is the light that created space and time at the moment that it was "unleashed". In other places in this book, I refer to this light as dark energy. This light doesn't radiate, and so it can't be detected at all.

2. The light that was the immediate result of the Big Bang. This is radiation that eventually developed into the light we see in all its possible forms. All quantum particles were created from this light.

The first type of light created the fabric *of* space while the second light created the energy *in* space. We'll discuss the fabric of space later in this chapter. Both types of light started with a very tiny range of *very high* frequency waves for the following reasons:

1. Intially the temperature of the universe was extremely high therefore producing high frequency waves.

2. In case of God's Light, a high frequency was necessary to create space because high frequencies have high energy and space had to be made of high energy to make motion in space possible. This idea will be further explained in the section "The Energy Dimension" of Chapter 8.

3. In the case of the second type of light, an initial high frequency led over time to the rest of the spectrum of light. Also high frequency waves were necessary for the creation of matter as proposed in Chapter 1 and the section "The Quantum World" of Chapter 5.

As the only things that came out of the singularity were those two types of light, items (2) and (3), it clearly implies that all things in nature—matter, energy, visible light, radiation, and all types of forces and even the fabric of space—are some sort of composite of those two lights! We certainly noticed that the speed of light, c, is involved in *mass* with the formula $E = mc^2$ discovered in Chapter 1, and in *motion* of quantum particles with the formula $\Delta E = 2mc^2$ discovered in Chapter 5. Matter is light, energy is light, and even gravitation is a sort of light. The whole universe contains only one thing: light in various forms. Humans are made of light—nothing else!

God's Light (dark energy) is the light that expands the universe to this day. In fact, if you're mathematically minded, I've given you a proof in Appendix 6. This makes sense as physicists have speculated the existence of that light, the dark energy, which isn't to be confused with the cosmological 3K background light that's a type of radiation that is uniform and found everywhere in the cosmos (dark energy isn't radiation). The 3K stands for 3 degrees Kelvin and is the temperature of that radiation. It's equivalent to 270 celcius *below* zero!

Does the universe have an edge? Absolutely, except that the edge isn't well-defined. The edge is being "pushed" as God's Light (again, that is dark energy) travels outward. We'll never know what that edge is or feels like because the only way to know would be to be present near that edge—which is impossible—and could be the case only if we traveled faster than the speed of light. Even in theory, this scenario is impossible. I devote three entire sections in Appendix 6 to this subject, starting with the section "The Expansion of the Universe" to discuss in mathematical detail the subject of the universe's expansion. A summary of my findings will be provided later in this chapter.

I've heard or read on the internet people and even scientists suggest that the universe is infinite. It's easy to see that's not so:

1. If the universe is infinite, then it isn't expanding, as how is it possible to expand what's already infinite? So it's a contradiction to believe in both states.

2. If the universe is infinite, then it took an infinite amount of time to make it so. This leads to the contradiction set forth in Chapter 7.

3. If the universe is infinite, then it took an infinite amount of energy to fill it (because where there's space, there's energy). This leads to a contradiction similar to the one discussed in Chapter 7.

4. Infinity is a concept elevated *out* of the physical realm. In other words, infinity can't be applied to the physical world. Think for instance of the infinite series:
$$S = 1 + 1/2 + 1/4 + 1/8 + ...$$

Mathematically it can be shown that the sum S is 2. However it's *physically* impossible to get there because the summation never ends.

Essentially if the universe is infinite, then it follows that it doesn't exist!

The Big Bang model and the expansion of the universe are widely accepted views of the evolution of the universe. But the model is not entirely secure. For instance, the Hubble telescope has taken pictures of the most distant galaxies observable. Stunningly, they look as rich and fully evolved as our own, even though they are—according to the Big Bang theory—only 5 % as old as the ones nearby our own galazy.

The Beginning of Nature

I suggested earlier in this chapter and elsewhere too that nature created time. But what created nature? Is nature the same thing as the universe? The two equations I derived in Chapter 5 lead me to a shocking idea. They are:

- The wave energy equation $E = gf$, the equation marked (1) at the end of the section "A Formula for Discrete Energy." Recall that this equation is my way of expressing Planck's equation $E = hv$.

- The formula for "fuzziness," $\Delta E \Delta t$, of the section "Another Formula for Fuzziness" in the same chapter. These two equations can easily be put together as follows.

The energy carried by a wave (emitted energy) has "fuzziness" to its value. This may be expressed as $\Delta E = g\Delta f$ from the equation $E = gf$. If we plug that $\Delta E = g\Delta f$ into the formula of fuzziness $\Delta E \Delta t$, we get $\Delta E \Delta t = g\Delta f \Delta t$. It's obvious that $\Delta f \Delta t \geq 1$ because energy can't be carried by less than one wave! Consequently,

$$\Delta E \Delta t \geq g$$

Remembering that $g = 8\pi^2 E_p \neq 0$, this equation implies that nature is *always* uncertain about the motion of a particle or wave. This has two deep implications:

1. As the above inequality makes no reference to temperature, even at the absolute zero degree Kelvin, this uncertainty applies. But at that temperature, the particle isn't supposed to have any energy left. Actually, that's not true because Kelvin is a measure of heat, and heat is a transfer of energy. So zero Kelvin means that no energy can be *transferred*. Nature can still feel the internal energy, it's just that there's no emitted energy. So the equation says that nature remains uncertain about the energy of the particle even at zero Kelvin.

2. As this inequality is equivalent to $m\Delta v \Delta s \geq g$ (recall from the section "Another Formula for Fuzziness"), nature is uncertain about the quantity of motion of particles, even at zero Kelvin when there's no motion at all. Now if it can't tell that, if all particles of the universe were to vanish, could nature detect that void with any certainty? No—this implies that nature can't be certain of its own existence or makeup. If so,

then nature couldn't possibly have triggered itself. This suggests that nature was created by something external of a higher consciousness!

The consequence of these two points is that there's a distinction to be made between nature and the universe: *the latter created the former*.

In the section "Exponential "Learning"" of Chapter 9, I give a list of different levels of interpretations made by nature, as nature means different things depending on the level of interpretation. For instance, there's nature at the quantum level, nature at the molecular level, nature at the biological level, etc. Clearly, nature at the quantum level was created at the time of the Big Bang. Nature at all levels of interpretation is an observer in the universe. At the quantum level (and possibly other levels too) it senses energy transfers, and from that, it deduces that motion takes place. More was said about nature as observer in the sections "Who is the Observer" and "Quantum Steps Taken by Particles" of Chapter 5.

Note that in the inequation

$$\Delta E \Delta t = m \Delta s \Delta v \geq g$$

$g = 8\pi^2 E_p \approx 8 \cdot (3.14)^2 \cdot (3.3 \times 10^{-36} J \cdot s) = 2.6 \times 10^{-34}$ J•s. So the uncertainty of motion or energy is very low, which is good news at our macroscopic level.

What the Big Bang is Not

There are four popular, yet incorrect pictures of the Big Bang in people's imagination:

1. That it was an immense light, like the most brilliant star we can imagine. This isn't what it looked like because space was being created at the same time that light was being created, and when space was created, much of that light was converted into other type of energies, forces, or matter. In fact,

the light coming out of the Big Bang was invisible. As I mentioned earlier, the light at the very beginning included only (invisible) very high frequency light because only high frequencies could create matter. Light, as we see it was created later. Also for the first few minutes after the Big Bang, light was tangled up with quantum particles. So light could not escape outward very well. As a result, the universe looked like a foggy patch, and not an immense light.

2. That it exploded with a thundering noise beyond imagination. That single point contained all of the energy of the entire universe, so it must have been hot beyond imagination. True—however, an explosion was impossible, at least not a huge one. To explode, pressure is necessary, so there had to be pressure against the Big Bang. But this would mean that there was something *outside* that single point of origin to apply pressure against. This contradicts the fact that everything started with what was inside the Big Bang. Clearly there was no astronomical explosion. As I explained in the previous paragraph, there was *expansion*, thereby creating space as needed, with that space being created with the time dimension and light mixing together as I explain further later on.

3. That its expansion occurred at the speed of light because light came shooting out of the singularity point. It certainly never happened this way as I'll discuss further in the section "The Expansion of the Universe" later in this chapter. The expansion progressed much slower than the speed of light although it has been accelerating for many billions of years.

4. That its expansion was perfectly even (i.e., that the expansion was perfectly spherical). This view is incorrect, if at least for the one reason that nature is imperfect. I stated this in the previous chapter and again in the next chapter: nature is unable to create a mathematically perfect sphere. The expansion occurred unevenly with the consequence being that the space density was never the same everywhere. This is what eventually gave rise to the creation of matter is some places while

other space is empty. This in turn created gravitational pull, thereby creating stars and galaxies.

I might add a fifth misconception:

5. That the entire energy of the universe came out of Big Bang is one "shot".

If you ask them, physicists will admit that they do not know well what happened in the first 10^{-43} second after the Big Bang. And what's known leads to puzzling dilemmas as I will now explain.

In the first section "It All Started with a Bang", it was stated that physicists estimate the temperature of the universe after the first 10^{-43} seconds to have been about 10^{32} celcius. How much energy is that? The conversion factor is that one celcius requires about 1.9×10^3 Joule. So there were $(10^{32}$ celcius$) \times (1.9 \times 10^3$ Joule/celcius$) = 1.9 \times 10^{35}$ Joules of energy that came out of the Big Bang.

Let's now find out how much mass this energy represents. One joule equals one $kg \cdot m^2/s^2$. This conversion comes from the mass-energy equivalence equation $E=mc^2$. So we just have to divide by c^2 to know the mass:

$$(1.9 \times 10^{35} \text{ Joule})/(3 \times 10^8 \text{ m/s})^2 = 2.1 \times 10^{18} \text{ kilograms}$$

According to Wikipedia, the mass of the observable universe is about 3×10^{52} kilograms. This is only for the *observable* universe meaning that the actual mass is certainly much bigger than that. In fact Alan Harvey Guth, a cosmologist, estimated that the actual complete universe is about 10^{26} larger in volume that what can be observed! This means that the universe is actually about 10^8 larger than what's observable.

According to my conversions, there were only 2.1×10^{18} kilograms (a far cry from the actual mass of the universe) that came out of the Big Bang. How to explain the discrepancy? Either

1. The temperature of the Big Bang was much higher than what's reported, or

2. The entire energy of the universe did not come out of the Big Bang in one single shot. That is, energy was being created for "a while" after the Big Bang, or

3. There was already a lot of matter (at least some 10^{50} kilograms) created before the first 10^{-43} seconds.

The first explanation isn't likely because 10^{32} celcius is the Planck Temperature, the highest temperature possible in quantum physics. Only the second explanation is plausible. The third explanation is totally impossible because it would imply that the Big Bang was a huge black hole from which such a Big Bang could never have come to be. This leaves us with the second explanation and the subject of the section "The Singularity Theorem" coming up.

Before going to that section, let's attempt an estimate of the energy density of the universe, 10^{-43} second into its existence. Let's assume that the mass of 3×10^{52} kilograms is correct. This comes to $(3\times10^{52}$ kilo$) \times (3\times10^{8}$ m/s$)^2 = 2.7\times10^{69}$ joules. How big was the universe in the first 10^{-43} seconds? I read web sites that estimate that it was the size of a grapefruit. I doubt that for the following reason: energy travels at the speed of light. Since 10^{-43} second is the Planck time, then the energy that came out had to follow a straight line outward. So, the size of the universe can't have been more than $2 \times (10^{-43}$ s$) \times (3\times10^{8}$ m/s$) = 6\times10^{-35}$ meter or twice the Planck length. Assuming that the volume of the universe was spherical, we use the equation

$$\frac{4\pi R^3}{3} = volume$$

Since $\pi \approx 3$, then this formula may be reduced to $volume = 4R^3$. The volume of the universe was then roughly $4\times(3\times10^{-35}$ m$)^3 \approx 10^{-103}$ m^3 The entire energy of the universe, 2.7×10^{69} joules, had to fit into that tiny spot. Since 1 cm$^3 = 1$ ml, then:

$$10^3 \text{ cm}^3 = 10^3 \text{ ml} = 1 \text{ liter.}$$

Also

$$1 \text{ m}^3 = (100)^3 \text{ cm}^3 = 10^6 \text{ cm}^3.$$

Since 10^3 cm^3 = 1 liter, then 1 m^3 = 10^3 liter.

So the volume of the universe was a mere:

$$10^{-103} \text{ m}^3 = 10^{-103} \, 10^3 \text{ liter} = 10^{-100} \text{ liter.}$$

Using this volume, the energy density expressed in Joules/L must have been:

$$2.7\text{x}10^{69} \text{ joules}/10^{-100} \text{ liter} = 2.7\text{x}10^{169} \text{ J/L}$$

In section "An Estimation of the Minimal Space Density" later in this chapter, I'll estimate the minimal energy density to be $2.31\text{x}10^{22}$ J/L. And this energy density $2.7\text{x}10^{169}$ J/L has got be the maximal value. This is some seriously massive energy density! To give you an idea of its gigantic magnitude, the energy density of our sun is a mere $1.3\text{x}10^{17}$ J/L. Can the fabric of space sustain so much energy in such a tiny spot? If so, then what quantum mechanism can allow for such a massive energy density? If not so, then universe had an initial volume much higher than the one that I calculated. The trouble is that this would imply that the speed of light in the first 10^{-43} second was a *lot* faster than what it is now.

If you look it up on Wikipedia, you will find the *Planck density*. It's the energy density that is believed to have existed 10^{-43} seconds after the Big Bang. The Planck density is $5.1\text{x}10^{96}$ kg/m^3 or $4.6\text{x}10^{112}$ J/L after conversion. My estimate of $2.7\text{x}10^{169}$ J/L is off the mark. It was based on the assumption that the mass of the universe is $3\text{x}10^{52}$ kilograms. The Planck density suggests that the mass of the universe is a tiny 10^{-5} kilogram. Something doesn't add up here.

Actually, it does! This tiny mass is close to what is called the *Planck mass* (about 10^{-8} kilogram) which is the smallest quantum mass possible. How significant is this? Here's the formula for the Planck mass:

$$M = \sqrt{\frac{hc}{2\pi G}}$$

It was said earlier that the universe started with a diameter of twice the Planck length. That length is given by the formula:

$$P_S = \sqrt{\frac{hG}{2\pi c^3}}$$

These equations are easily obtainable for the Wikipedia web site. Let's isolate the Planck constant h from both equations giving:

$$h = \frac{2\pi G M^2}{c}$$

and

$$h = \frac{2\pi P_S c^3}{G}$$

respectively. These two equations are equal, so

$$\frac{2\pi G M^2}{c} = \frac{2\pi P_S c^3}{G}$$

This easily reduces to

$$\frac{G^2 M^2}{c^4} = P_S{}^2$$

and finally the Planck length P_S is

$$P_S = \frac{GM}{c^2} \quad (1)$$

This result is quite a revelation to me because it's half the radius of a black hole! More precisely, it's half the radius of what is called the *event horizon* of a black hole. The event horizon is the region around a black hole where not even light can escape it.

While we're on the subject, let me show you how easy it is to derive the event horizon formula. You may recall that we already discussed this idea of escape in Chapter 5. Indeed in section "Solar Systems in Equilibrium" of that chapter, we had derived the escape velocity, the equation marked (1), of a planet around the sun

$$v = \sqrt{2Gm_s/r}$$

In the case of a black hole, the escape velocity is c, the speed of light. So, the formula becomes

$$c = \sqrt{2GM/r}$$

where M is the mass of the black hole. From this can be derived the event horizon radius

$$radius = 2\frac{GM}{c^2} \quad (2)$$

To summarize my calculations and algebraic manipulations, the Planck density that existed immediately after the Big Bang led me to the Planck mass, and this mass led to a black hole. As unbelievable as this seems, *the universe started as a black hole!* As you know, a black hole sucks energy: it doesn't explode into an expanding universe. Yet, this is what happened. There are a number of reasons why the black hole didn't remain:

1. Any quantum event lasts no more than one Planck time. So the black hole might have "disappeared" as soon as it came to be. Why would it have disappeared? See reason (4) below for a hint.

2. Since this was the very beginning of the universe, there was no energy outside of the black hole to be sucked into it. So, the black hole "died."

3. The idea that I had for searching for the Planck mass was to estimate the Planck energy density. The instant after the Big Bang didn't generate a quantum particle, but energy. A black hole, no matter how tiny, has to contain matter. So, since

there was no real matter at the beginning, there was probably no real black hole even though the initial universe did satisfy the formula (2) for the radius of a black hole.

4. The Singularity theorem (presented in the section below) states that an immense amount of energy came out of that black hole. This energy that came apparently from nowhere "burst" the black hole open.

Points (3) and (4) lead us into the section below.

The Singularity Theorem

Most scientists believe that the universe started at a single point called the *singularity*—that is, the entire energy of the universe was concentrated in that single point. Scientists theorize that this point was an infinitely dense black hole that contained the entire matter of the universe! This is what the famous physicist Stephen Hawking claimed to have proven in the 1970s. He demonstrated that every solution to the equations of general relativity theory guarantees the existence of a singular boundary for space and time in the past. This is now known as the *singularity theorem*.

According to Hawking's proof, the singularity was a black hole of infinite density. But as soon as I read that, I became suspicious for the following reasons. By definition, energy density is energy versus volume. So either the black hole had *no volume* or had *an infinite amount of energy*. Both cause difficulties:

1. *No volume.* At first grance, the concept of an absence of a volume doesn't appear to belong to the realm of physics. Or does it? There are two cases:

   a. The singularity point was a black hole, that is, it contained *matter*. Trouble is that matter occupies a volume. So black holes always have a volume, however small. Also if there was matter in that singularity point, then matter existed already at the very *instant* of the Big Bang.

But this contradicts head-on the Big Bang theory. Therefore the singularity *cannot* have been a black hole. Even if the Singularity theorem applied to the very instant after the Big Bang, the black hole would have been the one calculated at the end of the previous section, and that black hole lasted a mere instant.

b. The singularity point didn't contain matter—it contained *energy* only. Trouble is that energy has to occupy some space too because of its definition: *Energy* $= F \times d$, where F is the force, and d the distance. So, there is a distance. Let's be careful with this definition though. Because of Newton's principle of action-reaction, the force is a reaction to some force action. Consequently this definition is *not* a definition of energy, but rather the definition of energy *transfer.* So, the definition of energy has no distance or space associated with it. So it's quite conceivable that the singularity point contained energy, and possibly an infinite amount.

Also let's not confuse energy and temperature. The theorem doesn't state that the singularity point possessed a temperature. Temperature is a measure of heat, and heat is the result of a transfer of energy. A transfer requires a volume, however small. But the singularity point didn't have a volume. Therefore it had no temperature. Does that mean that the temperature was 0 degree Kelvin (or -273 celcius)? Does it mean that the temperature at the very instant of the Big Bang was that low, and then 10^{-43} seconds later, it jumped to a gigantic sizzle of 10^{32} celcius? Talk about a heat wave! What quantum event could account for such an enormous energy transfer?

A paradox arises from the discussion above. The very instant after the Big Bang (10^{-43} seconds), the universe had a volume. This caused a transfer of energy from the singularity point into the universe. So there was a source point, and a destination point for this

Big Bang energy. It's clear that the source was the singularity point. The destination point appears to have been whatever the edge that the universe had reached after that 10^{-43} second. Trouble is that the fabric of space had to be created *before* this Big Bang energy could reach that point. As I explain in the sections coming up, the fabric of space is itself constituted of energy. So the energy *of* space had to be created in the first place before energy could be put *into* that space. But then how was the destination point created if there was no space and energy there to go to? You can't reach a destination that doesn't exist. It's analogous to attempting to build a bridge without any kind of support to the ground. So, not only experts are unable to answer *why* the universe started, they can't even answer *how* it started. But few physicists will admit that limitation on their part.

2. *An infinite amount of energy.* There are three arguments:

    a. If that's the case, then energy would still be coming out of the singularity right now and will do so *forever*. If this was true, then the Big Bang is still going on at this moment. Yet scientists will all agree that the Big Bang is over (although in section "The Expansion of the Universe and Its Fate," I'll suggest that the Big Bang might not be over).

    b. Nature can't create an infinite amount of energy. If it did, it would violate the First Law of Thermodynamics that states that no closed system can create new energy. The closed system is the whole universe. But then, the singularity point didn't belong to that closed system. So, for this reason, the singularity point can certainly violate that law. For another reason, the singularity point doesn't possess any physical laws. This is the argument (c) below.

    c. Hawking's singularity theorem states that in a singularity point, all laws of physics "break down," indicating the definition of density has presumably broken down too. Can we refer to the concept of density at all then? I don't think so. So then the concept of infinite density—as stated by the singularity theorem—has no meaning. However because the laws of physics broke down, infinite energy is quite plausible. Hawking might be correct on this account because I stated a couple of times in this book that space is continuous. This is possible only if there is an infinite amount of energy in space.

There's no mystery to that breakdown of physics in a singularity point. The singularity theorem is supposed to describe what happened the moment *before* the Big Bang. So it's no wonder that the laws of physics broke down because these laws weren't even created yet! Physics started at the moment of the Big Bang.

Note that the singularity point contained the entire "universe" (I put the term in quotation marks because there was no universe yet) and not the other way around. That is, the singularity wasn't located in space. Space didn't exist outside the singularity, and didn't even exist inside the singularity point simply because the point has no volume. Then the Big Bang unraveled all of what makes the universe: its four dimensions, the various types of forces, matter, energy, etc. Because those things came from the same point, it suggests that they're related. You might recall that I already implied that earlier in this chapter.

I recall listening to a newscast in the 1980s in which Hawking was being interviewed. The physicist claimed to have proven that God doesn't exist because his singularity theorem supposedly showed that the Big Bang had a natural explanation, not a divine cause.

However brilliant Hawking may be he is mistaken. The actual creation of the Big Bang may be explainable by a singularity point. How did that singularity point come to existence? Can physics explain that? I don't think so because the singularity point constitutes the beginning of physics—energy. Anything before that is nonphysical and therefore belongs to the world of Consciousness that I talk about in Chapter 9 and 10.

So, some supernatural phenomenon was involved in the creation of that singularity point. Furthermore, at the end of Chapter 7 I provide an extremely simple philosophical argument that leads to Hawking's result: a singularity point started the universe. Yes, indeed. Moreover, I suggested a cause to the instant of creation of that singularity point. Note that I did say that I provided a *philosophical* argument, not a physical explanation. There's no physical explanation to the appearance of that singularity point. There's however a physical explanation to the *instant* of the creation of the Big Bang that followed. It's that antigravity that I proposed in the first section of this chapter. Of course the first moment after that instant is entirely explainable by physics.

In the section "Where Do Souls Reside" of Chapter 10, I'll propose an inspiring connection between the results of the singularity theorem and the afterlife.

The Expansion of the Universe and Its Fate

Physicists say that all observations to this day converge on one conclusion: the universe is expanding. This makes sense for at least one reason: the Second Law of Thermodynamics. If energy came out of a singularity point, according to this law, the energy *eventually* dissipates. We see that phenomenon with weather patterns all the time, with vapor coming out of a hot pot of soup, and so on. We'll delve into more details of that Second Law in the section "The Second Law of Thermodynamics" of Chapter 9.

A popular view is that the expansion happened at the speed of light because it's light that came out of the singularity point and was shot straight outward. This suggests that the expansion happened then and still happens at a constant speed. However, this isn't the case as we'll now see.

What's most surprising is that physicists have discovered that the expansion of the universe is occurring at an *accelerated* rate. Apparently the reason is a mystery. Physicists propose that some antigravity somewhere in the universe pushes galaxies away, thus accounting for this acceleration. However they admit to not knowing if antigravity exists in the first place! I argued in favor for the existence of antigravity in the first section of this chapter as well as in Chapter 6.

Using my own theory of gravitation as described in Chapter 6, I'll now propose an explanation for this acceleration (and in the process, in Appendices 5 and 6, I'll show that my theory of gravitation appears to work). My explanation requires mathematical work involving calculus. (If you're interested in studying that work, it's in Appendix 6.) The result of the mathematical derivations is illustrated with the following graph that depicts the acceleration over time:

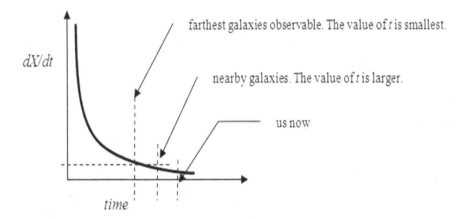

The vertical axis is for *dX/dt*, the mathematical symbol that I use for the acceleration of the universe (*X* stands for the speed of expansion). Appendix 6 contains a demonstration that the equation for this graph is

$$\frac{dX}{dt} = \frac{\Sigma}{t^2 \xi} \quad (1)$$

where Σ and ξ are important constants of *energy density* related to space. We'll come back to these constants in the next section.

A physicist looking at my equation would surely mock it because it's nowhere near the correct equation for expansion. But I'm not deterred. Despite the inaccuracies, in Appendix 6 I decided to keep this equation because it led me to values for the constants Σ and ξ

that turn out to be *very good* estimates of actual energy densities related to space. Moreover, my equation led me to Newton's famous gravitational equation

$$F = G \, \frac{m_1 m_2}{r^2}$$

I figured that if my theory of the acceleration of the expansion of the universe leads me to Newton's gravitational equation, there must be some truth behind my theory.

Note that because some of my mathematical derivations in the Appendix 6 are only approximations, inevitably the equation (1) is also an *approximation* of the actual acceleration of the universe over time. The next section coming up will use my formula to calculate an estimate to the acceleration for nearby galaxies (those with a large value of the the time t) and find that it's very accurate. As for galaxies farther away, I don't have any data to verify my equation against. Nonetheless observations obtained by astrophysicists tell that the accelation *increases* the *farther* away the galaxies are. My equation does reflect that trend! The reason is that those galaxies are seen by us as they *were* billions of years ago. Thus, their acceleration detected here on earth right now is from very far into the past, and hence the variable t is small for those galaxies and larger for galaxies that are closer to us. My graph does reflect this inversely proportional relationship between acceleration and time.

This doesn't imply that my equation provides accurate values for the accelaration of those far away galaxies for which the variable t is small. It's abundantly obvious that the "spike" near $t=0$ isn't at all accurate because it is associated with accelations that would have generated speeds well beyond the speed of light—this is physically impossible. Clearly if there was a spike into the past, it was in reality not that pronounced when it happened. My equation obviously provides very poor estimates of the acceleration for small values of the variable t.

The mathematical work in Appendix 6 is based on my own theory of gravitation. Recall from the section "Fighting Off the 'Void'" of Chapter 6 that my theory suggests that gravitation is a manifestation of nature's attempt to *neutralize* space density differentials. This effect takes place when an area of space becomes too thin of space density so nature sends energy and matter there.

I theorize then that the acceleration of the expansion started when the space density of the edge of the universe reached a critical minimal value. Indeed, as the universe expands, the surface of its edge increases, but the entire amount of energy available in the entire universe doesn't increase because the Big Bang ended a long time ago (note that this is assumption made by phycisists). My theory of expansion is itself based on my own theory of gravitation that suggests that gravitation is a manifestation of nature's attempt to *neutralize* space density differentials. Keeping that in mind, here is a picture of the expanding universe:

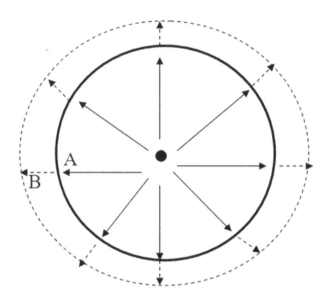

The inner sphere *A* is where the edge of the universe is now, and the outer sphere *B* is where the edge will be some short time later. If we suppose that the edge A has a minimal space density, then the edge B *cannot* have a space density any thinner. Therefore nature would send energy rushing toward the edge of the outer sphere B. This would create acceleration. I've used this model to develop a series of simple differential equations that led to the graph presented above.

You may recall that in Chapter 6, I noted that because the PED (potential energy density) around and inside a celectial body is thinner than farther out, the space density where gravitation is stronger is thinner. This causes the fabric of space to *stretch* where there's a gravitational field. Because my theory of the expansion of the universe is based on my theory of gravitation, it follows that the fabric of space at and near the edge of the universe stretches. The effect is similar to when you blow air into a balloon.

I've found via the Internet that the speed of galaxies extending away from each other occurs at a mere speed of 77 kms/sec per 3.26 million light-years. A simple calculation reveals an acceleration of a constant 7.5×10^{-13} kms/sec^2 or 7.5×10^{-10} m/sec^2. This acceleration is only tiny (a snail accelerates much faster than that). There's another way to believe that the acceleration of the expansion of the universe is so small. Our own galaxy isn't immune to this acceleration, yet we humans don't feel any acceleration at all as our galaxy appears to move along at a constant speed.

Note that the estimation of the previous paragraph of the acceleration given by physicists is for the *observable* universe: its edge is obviously not observable. If the constant acceleration of 7.5×10^{-10} m/sec^2 as estimated by physicists is correct, then the observable galaxies lie near the end of the curve of the graph where the curve has a nearly constant slope, reflecting the constant acceleration observed:

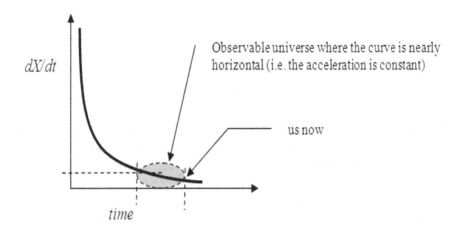

As the graph shows, the farthest galaxies are seen by us as having a high acceleration while nearby galaxies appear to have a lower acceleration. Note that near time $t=0$, the acceleration was extremely high, most likely then the speed of acceleration was higher than the speed of light.

Could it be that the speed of light was faster many billions of years ago than the constant c of today? All physicists agree that the speed of light is constant everywhere in the universe. Accordingly, either (a) this assumption is wrong, or (b) just before the time $t=0$ when this acceleration started, matter in the universe was expanding *very* slowly, possibly not expanding at all for some periods of time.

The second option regarding a very slow expansion is much more plausible. In fact, as will be argued in a few paragraphs, there was a deceleration before the universe started to expand with acceleration. In any case, the galaxies that happened to be near the edge of the universe that fateful day when the acceleration shot up as shown in the graph must have shattered under the strain of the acceleration. If there were any intelligent living beings then, their world must have collapsed as in the most horrific doomsday scenario.

Has the universe always expanded with acceleration? I doubt it for two reasons. The first was given two paragraphs ago: the speed of light that can't be exceeded. The second reason goes like this: let's assume that the entire energy that originated from the Big Bang had all come out within, say a few million years. In other words, the Big Bang ended billions of years ago. (No physicists claim otherwise.) When the Big Bang ended, the radius of the universe was obviously much smaller than it is now, yet it contained all of the energy the universe currently has.

We can easily deduce then that the space density of the then-smaller universe was definitely much higher than what it is now. Consequently, the edge of the universe had a very high space density as shown in the picture below:

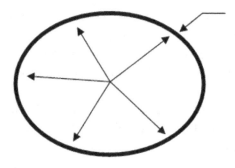

Space density of the edge is higher than the minimal possible

Because the space density of the edge was much higher than the minimal density, the edge didn't have to react the way it does now by quickly streaming more energy toward the edge in an attempt to keep the density to no less than some minimal value. Consequently, the graph shown earlier with the spike couldn't occur yet. Because the entire energy of the universe was confined to a much smaller sphere than it is now, it must have been many billions of years before this acceleration happened. I don't suggest that no energy was sent to the edge before this moment of acceleration as of course energy reached the edge all the time. It's just that the universe was in no hurry to send massive amounts of energy to the edge.

So the complete graph of the acceleration dX/dt versus the time looks like this:

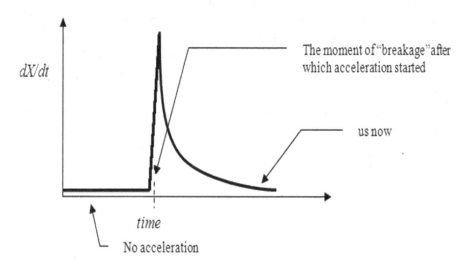

The nearly vertical line indicates the moment of "breakage" where mysteriously the acceleration started. At that point of "breakage," it probably was doomsday for many parts of the universe as I described earlier.

What caused the universe to suddenly expand with acceleration? What was the rush for? I can't help but notice that this graph looks very much like when a lightbulb bursts. At the moment of the burst, the luminosity of the bulb shoots up much higher than normal, and then quickly the light dies down. The graph also looks like one in which a balloon is inflated at a constant speed, then at a critical point, it bursts, and with time the air escapes with a sudden sharp acceleration that then slows down quickly. Should we conclude that perhaps just at the time that the universe reached that "breakage," the space density of the edge of the universe had gone below the critical minimal space density? Or that then the universe inevitably overreacted very quickly and consequently sent massive amounts of energy to the edge, thereby accounting for the spike in the acceleration dX/dt?

What could have caused the space density of the edge of the universe to go below the critical minimal space density? Space density is measured by the amount of pure energy (that excludes matter) within a volume. As I mentioned before, the energy density must have been very high for a while after the Big Bang. I propose the following series of events that perhaps led to the spike in acceleration:

1. As the radius of the universe increased, space density decreased. At that time, the universe was made of subatomic particles. Then atoms started to form. This led to the creation of matter.

2. As matter is concentrated energy, that energy had to have come from the surrounding environment (space), which was made of free-flowing energy (light). In a sense, energy was "sucked" out of the surrounding environment as matter was being created. Because mass is given by the equation $m = E/c^2$, an *immense* amount of energy had to be sucked out of

the space in order to create matter. Consequently, the density of free-flowing energy *decreased* dramatically throughout the universe.

3. In turn, this led to less outward pressure to expand the universe, and most importantly, the energy density might have decreased below the critical level. As a result, I suspect that the radius of the universe most likely *decreased* drastically for a period of time in an attempt to keep the energy density from becoming too low. Thus, the expansion of the universe *decelerated* rapidly. This would be analogous to putting a balloon full of air in the fridge. Energy is sucked out of the balloon, thereby making the balloon shrink. Of course this is only an analogy. In our situation, obviously no energy was sucked out of the universe.

4. Once the space density was back to normal, as more and more matter was being created, this matter eventually was pushed outward so that the radius started to increase once again. This expansion increased the volume of the edge of the universe, thereby decreasing once again the energy density of the universe and on its edge.

5. This happened as matter was being created, which further decreased the free-flowing energy density.

6. These two factors (4 and 5) together probably contributed to an *exponential* decrease of the energy density of the edge of the universe. Eventually the edge was too thin, and so the universe reacted by sending a *massive* amount of energy toward the edge. This would explain the spike in the graph above.

Based these deductions, I propose that the acceleration of the expansion of the universe went something like this over time:

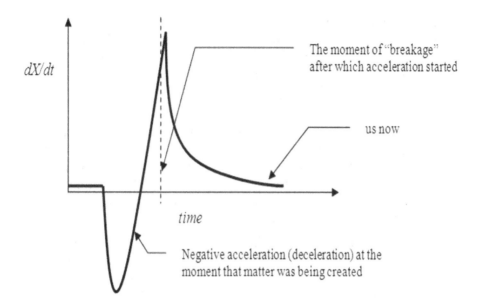

After the universe had passed that "breakage' point, there was no return, and from then on, the universe has essentially been very slowly dying, all the while struggling to stay alive by accelerating energy toward its edge to maintain a minimal space density. Is this a sign that the universe is starting to run out of energy? I think so.

OK—the purpose of this section is to investigate what might be the fate of the universe. I just suggested that the universe might be running out of energy. How bad is that? According to the Web site www.space.com, the universe is 156 billion light-years wide, and according to Wikipedia, it's 13.7 billion years old. There are 365 days in a year, 24 hours a day, and 3600 seconds in an hour. This calculates to there having been 4.3×10^{17} seconds since the Big Bang. As for the size of the universe, the calculations come to 7.4×10^{26} meters radius. So the calculation of the speed of expansion of the universe is only a matter of dividing 7.4×10^{26} meters by 4.3×10^{17} seconds, which gives 17.2×10^{8} meters per second.

But this speed is faster than the speed of light, and suggests that the speed of light was higher far away in the past. No way! The reason for this erroneous result might be that the age of the universe is calculated based on the *observable* universe. That number is obtained by multiplying how old the physicists *think* the universe is by the speed of light. The reasoning there is quite straightforward: we can only see out to that distance from which light can have reached us since the universe began. This explains why I arrived at a speed of expansion in the range of the speed of light.

As there will be no end (apparently) to the acceleration of the expansion of the universe, eventually the edge will reach the speed of light. Oops! What will happen then? As the speed of light is the limit, the acceleration will stop dead all of a sudden (I'm assuming here that the speed of light can't change and thus can't increase). What will be the consequences?

One consequence is that the edge will very quickly run out of energy, and this in turn will force the inner part of the universe to send more energy toward the edge (this is based on my own theory of gravitation that was found earlier to be equivalent to the principle of dissipation of energy). Essentially, the universe will eat itself to death. It's as if someone is asked to run nonstop for a whole week without eating or drinking. The body will have to find energy anywhere within the body to try to survive—and the body will literally eat itself.

This is what might happen with the universe: it will fade away very much like clouds do if there's not enough humidity present in the atmosphere. More precisely, space will no longer contain any energy other than the minimal energy possible as determined by the energy density constants $\xi = 2.2 \times 10^{-7}$ J•L^{-1} and $\Sigma = 2.31 \times 10^{22}$ J•L^{-1} (more on those two numbers in the next section). If you want to know how I calculated these constants, consult Appendix 6.

A view somewhat similar to that expressed in the previous paragraph is adopted by numerous physicists. It's named the *Big Freeze*, and is the most popular view. It's believed to begin 10^{14} years after the Big Bang. The universe is now about 10^{10} years old. So, humanity doesn't need to worry.

But hold it! Perhaps the universe won't fade away after all. As the universe will be expanding to no end, surely energy densities will reach numbers below these two constants. However, according to physicists, these numbers *cannot* be any lower no matter what. Thus, either:

1. The radius of the universe will continue to increase anyway and the energy density of dark energy will continue to increase... somehow (perhaps from the result of the singularity theorem of Chapter 8 that states that the universe has an infinite amount of energy). This view is called the *Big Rip* and is estimated to begin after the universe is 200 billion years old.

2. The universe will begin to collapse into itself due to gravitational pull among galaxies. It's estimated that this would happen after the universe reaches 100 billion of years. We are now at 14 billion years. This view is named the *Big Crunch*.

In my view, the Big Rip theory is flawed because it requires that the universe has an unlimited amount of energy which isn't the case unless the following view applies:

3. There's still energy coming out of the Big Bang to this day and it will continue coming out forever. I remind you that in the section "The Singularity Theorem," 'I stated that the singularity point was an *infinitely* dense black hole that contained the entire matter of the universe. If that is correct, then energy is still coming out of that black hole, in which case the universe will never run out of energy and so will expand forever!

However, you won't find a single physicist who believes that the Big Bang is still going on to this day. This leads to my own view:

4. If all energy created by the Big Bang has been released already, then instead of continually expanding, the edge of the universe will eventually stop expanding and will begin to shrink (decelerate) to maintain minimal energy densities. Then when the density of the edge is strong again, the expansion might resume again. The radius of the universe will start to decrease in size (despite my equation that states the con-

trary) in order to increase the space density of the edge. I theorize that the radius of the universe will shrink and expand, back and forth, like this forever. Don't mock my idea because I'll provide some simple mathematical work at the end of this section that suggests this possibility of an oscillating radius of the universe.

For the past sixty years, the popular theory regarding the end of the universe is that it will collapse into itself back into the singularity point due to gravitational pulls from galaxies. But physicists have never been able to prove this possibility. Moreover, don't these physicists realize that it's *physically* impossible for the universe to shrink back into a singularity point? Indeed, to be reduced back to a singularity point again, the volume of the universe would have to become zero, meaning that the space dimensions would have to vanish! This is physically impossible.

Some physicists suggest that the whole universe will eventually be sucked into a single black hole—this is the Big Crunch view. Keep in mind that a singularity point isn't the same as a black hole. The latter has a volume; the former doesn't. In a black hole, in principle the universe would still exist, only in a very small volume. A singularity point means the absence of a physical universe. So, the Big Crunch view is plausible.

In Chapters 7 and 10, I show that a system of thoughts created the physical universe. For the universe to shrink back into a singularity point would be equivalent to undoing that system of thoughts. Needless to say, it's impossible to undo a thought, and some things in nature also can't be undone. For instance, it's impossible to undo the birth of baby. That same principle applies to the universe. I see nothing that can undo the Big Bang. Consequently, the universe will exist forever—that is, at any given moment in the future, the universe will still be there! The existence of the universe will *never* end, more precisely, its end is indefinite.

You might recall that earlier in this chapter, I claimed that the universe isn't infinite and never will be. Am I not contradicting myself by suggesting that the universe will never end? No—*never end* means that at any given x moment in the future, the universe will still be there. As the universe had a beginning, at that moment x, the universe will have been in existence *y* number of seconds. Thus, the number *y* will never be infinite. Infinity isn't a number; consequently the universe will never reach infinity even if it exists forever!

The universe will never come to an end! This seems so illogical because all things eventually come to an end, right? Plants, animals, insects, and humans eventually die. Humankind will eventually disappear. The earth, our solar system, and our galaxy will eventually disappear. All galaxies eventually come to an end. The key word here is: *disappear*. If something has disappeared, it has obviously died. But the universe will never disappear even if it should become a black hole because the universe consists of the space dimensions, the time dimension, and energy, and these will never disappear. It's simply physically impossible to undo God's initial Creation. If the universe does come to an end, then it has to be of nonphysical causes. Perhaps only God can put an end to the universe.

There's another argument to suggest that the universe will never end: time. As I argued in the section "What Created the Universe?" of Chapter 7, God's thought of time triggered the creation of the universe. In the section "Our Virtual Universe" of Chapter 10, I show that once a thought is generated, it never vanishes. Consequently the thought of time will never disappear. Because that thought triggered the universe, the universe's existence can never come to an end.

Of course these opinions are open for debate because they are philosophical in nature ...

OK—what about my idea proposed earlier in this section in view (4) of an oscillating radius of the universe? Is it a crackpot idea? Let me develop some mathematical work that suggests that the idea is plausible. Recall the formula (2) in section "What the Big Bang is Not"

$$radius = 2\frac{GM}{c^2}$$

that gives the event horizon radius of a black hole. Just out curiosity, I'm going to apply it to the whole universe. Since the exact mass of the universe is not known, there's no point in using into the formula the precise values of the speed of light and of the gravitational constants. So, I'll use approximative values of 10^{-11} m^3•kg^{-1}•s^{-2} for the gravitational constant, of 10^{53} kilograms for the mass of the universe and of 10^8 m•s^{-1} for the speed of light. This gives

$$radius \approx 2 \frac{(10^{-11}\text{m}^3 \bullet \text{kg}^{-1} \bullet \text{s}^{-2}) \times (10^{53} \, kgs)}{(10^8 \, m \bullet \text{s}^{-1})^2}$$

or

$$radius \approx 2 \frac{10^{42}\text{m}^3 \bullet \text{s}^{-2}}{10^{16} \, \text{m}^2 \bullet \text{s}^{-2}}$$

or

$$radius \approx 2 \frac{10^{42}\text{m}^3 \bullet \text{s}^{-2}}{10^{16} \, \text{m}^2 \bullet \text{s}^{-2}}$$

or

$$radius \approx 2\text{x}10^{26} \, meters$$

This is an approximation of the radius of the event horizon should the universe be a black hole. Depending on where you take your sources, the universe is estimated as having a radius of between 10^{26} and 10^{27} meters. We have an almost perfect match here. Therefore it appears that the whole universe might reside inside a... black hole. What a shocker!

In section "What the Big Bang is Not", we found that the universe started as a tiny black hole, and now it seems that it's still a black hole today! At first glance, this appears to be *total* non-sense. So, what's to think of this? Two possible explanations:

1. A black hole has to obey the event horizon radius formula, but this requirement is perhaps not sufficient for something to be declared a black hole, or

2. The event horizon radius formula is a sufficient requirement. This implies that a black hole is not necessarily something with an extremely dense body.

I favor the first explanation as do most physicists. But what if the second explanation was correct? Perhaps the only requirement for something to be a black hole is that it doesn't let light espace. Don't you think that the universe as a whole meets that requirement? Indeed, imagine that you were outside the universe and looking at it. Would you see the universe? No—the universe would be totally black. Why? It's because, as I explain briefly in this chapter and in more detail in Appendix 6, at the edge of the universe, it's the fabric of space that is created, and the fabric of space doesn't radiate any light. Essentially the universe is wrapped into itself. Light radiates *inside* the universe, but not onto its edge.

With this view in mind, I suggest that the universe has perhaps always been a black hole. What a daring thought! Furthermore imagine this. If the universe is a black hole, then it might be a black hole within another universe. Then that universe might itself be a black hole within another universe, and so on. There's no way to know if this scenario of embedded universes is reality. It feels too crazy to be true, don't you think?

Regardless whether the universe is a black hole or not, physicists know that the universe initially did obey the event horizon radius formula used a few paragraphs ago. Then I performed a calculatation which shows that the universe obeys that formula today. So, perhaps the universe has always obeyed that formula. Let's assume so, and from this assumption, I wish now to combine this view and the concept of energy density of the universe (presented in the section below and in Appendix 6) to propose that the universe might perhaps expand and shrink indefinitely. Physicists know that the universe did at one point shrink then resume expansion again. The mathematical development below demonstrates that this could happen again in some distant future.

Let's start with the event horizon radius formula

$$R = 2\frac{GM}{c^2} \quad (2)$$

where R stands for the radius of the universe.

The energy density of the universe is simply its mass (matter and energy) divided by the volume of the universe. The volume is

$$\frac{4\pi R^3}{3} = volume$$

So, the mass M is

$$M = \left(\frac{4\pi R^3}{3}\right) \times D$$

where D is the energy density of the universe. Plugging that into the event horizon formula (2) gives

$$R = 2\frac{G\left(\frac{4\pi R^3}{3}\right) \times D}{c^2}$$

or

$$1 = \frac{8\pi G R^2 D}{3c^2}$$

Isolating the energy density D gives

$$D = \frac{3c^2}{8\pi G R^2} \quad (3)$$

Obviously, the variables R and D depend on time, and the other terms are constants. Taking the first derivate over time of this equation gives

$$D'(t) = \frac{-3c^2}{4\pi G} \bullet \frac{R'(t)}{R(t)^3}$$

or another way to put it,

$$D'(t) = -2D(t) \bullet \frac{R'(t)}{R(t)} \quad (4)$$

Because of the minus sign, the fluctuation of the energy density of the universe and the fluctuation of its radius are in reverse relation with one another. Let's study the four cases that arise from the equation. The first two cases assume that the entire energy content of the universe is constant (i.e. the Big Bang is over):

1. When the radius of the universe increases, $R'(t) > 0$, the energy density of the universe decreases, $D'(t) < 0$. This makes sense because if the amount of energy in the universe is constant, the energy density is bound to decrease as the radius increases. I use this observation to develop my theory of the expansion of the universe detailed in Appendix 6.

2. If the radius of the universe was to decrease, $R'(t) < 0$, the energy density of the universe would increase, $D'(t) > 0$. This makes sense too. Physicists believe that the universe did shrink at one point in the past.

Should the entire universe still be inside a black hole then perhaps the singularity point is still there. Consequently the entire energy content of the universe perhaps increases still to this day (recall that this is also what Hawking's Singularity theorem claims to have proven). Just out of curiosity, let's assume that to be true. There are two cases:

3. If too much energy bursts out of the singularity point then perhaps the energy density of the universe might increase, that is, $D'(t) > 0$. This implies that $R'(t) < 0$, that is, the radius of the universe *decreases*. This is counter-intuitive as you would expect the reverse effect, right? Physicists know that at some point in its past, the universe did indeed shrink.

4. If too little energy comes out of the singularity point then the energy density of the universe might decrease, that is, $D'(t) < 0$. This implies that $R'(t) > 0$, that is, the radius of the universe *increases*. This is also somewhat counter-intuitive. Yet this is precisely what physicists are observed.

You see that the equation tells that it's plausible that the radius of the universe might oscillate back and forth. This matches my own view (4) for the fate of the universe presented earlier in this section.

Note that the equation (4) tells that $D'(t) = 0$ if and only if $R'(t) = 0$. This case would be possible only if the Big Bang is over and that the radius of the universe remains fixed which is not at all realistic.

Out of curiosity, let's use the equation (3) to find an estimate of the energy density of the universe. This gives

$$D = \frac{3(3x10^8 \text{m} \cdot \text{s}^{-1})^2}{8\pi(6.7x10^{-11}\text{m}^3 \cdot \text{kg}^{-1} \cdot \text{s}^{-2})(10^{27}\text{m})^2}$$

or

$$D \approx \frac{3(9x10^{16}\text{m}^2 \cdot \text{s}^{-2})}{8\pi(6.7x10^{43}\text{m}^5 \cdot \text{kg}^{-1} \cdot \text{s}^{-2})}$$

or

$$D \approx \frac{2.7x10^{17}\text{m}^2 \cdot \text{s}^{-2}}{1.7x10^{45}\text{m}^5 \cdot \text{kg}^{-1} \cdot \text{s}^{-2}}$$

or $D \approx 1.6x10^{-28}$ kg/m^3. Let's convert this density to the units J•L^{-1}. Using the equation $E = mc^2$, the number of Joules is $(1.6x10^{-28}$ kg$) \times (3x10^8$ m/s$)^2 = 1.4x10^{-11}$ Joule. Using the conversion 1 m$^3 = 10^3$ liter calculated in section "What the Big Bang is Not", the energy density $D \approx 1.4x10^{-14}$ J•L^{-1}. This density is very close to the zero-point energy (the minimal density possible) of $2.2x10^{-13}$ J•L^{-1} that I use in the section below. You would expect the energy density D of the universe to be higher.

Why is the energy density of the universe so low? There are two possible reasons: (a) it might be that by now the universe has expanded so much that it is almost empty. It sure doesn't feel that way because in the night sky, you see countless stars. But those stars are very far and few between. In its early days, the universe was much smaller therefore much denser; (b) the estimate for the mass of the universe is based on an estimation of the number of stars in the universe. According to Wikipedia, cosmologists estimate that the matter from stars make up only 0.4% of the entire energy of the universe. So, in my energy density estimate, a correction factor of $100/0.4\% = 25000$ applies. This gives en energy density $D \approx$ $(1.4 \times 10^{-14} \text{ J} \cdot \text{L}^{-1}) \times (25000) = 3.5 \times 10^{-10} \text{ J} \cdot \text{L}^{-1}$. This number is about one thousand times the zero-point energy density. According to what I read on Wikipedia, this number sounds about right.

An Estimation of the Minimal Space Density

I mentioned in the previous section that the acceleration of the expansion of the universe occurred when the edge of the universe had reached a critical minimal space density. The fate of the universe depends on the ability of space to retain at the edge of the universe minimal space energy. Throughout the mathematical development of Appendix 6, the symbol Σ is used to symbolize the energy density of the edge. I'll now give the value of that minimal energy density. (If you're interested in the mathematical work required to arrive at that number, consult Appendix 6.)

The minimal value of Σ turns out to be $8.1 \times 10^{16} \text{ kg} \cdot \text{s}^{-2}$. Because one $\text{kg} \cdot \text{m}^2 \cdot \text{s}^{-2} = 1 \text{ Joule}$ then the value of Σ comes to $8.1 \times 10^{16} \text{ J} \cdot \text{m}^{-2}$, where J is one joule. The joule is a measure of energy. This is the density of energy at the edge of the universe, and seems extremely high for a *minimal* space density! When expressed in energy per liter, the number becomes a staggering $2.27 \times 10^{22} \text{ J} \cdot \text{L}^{-1}$. This conversion is based on the conversion factor $1 \text{ m}^3 = 10^3$ liter calculated in section "What the Big Bang is Not". The square root of $\Sigma = 8.1 \times 10^{16}$

$J•m^{-2}$ is $2.85x10^8$ $J•m^{-1}$. That number cubed gives $2.31x10^{25}$ $J•m^{-3}$. With the equivalence $10^{-3}m^3 = 1$ liter, we obtain the space density of $2.31x10^{22}$ $J•L^{-1}$.

This number would be the energy density of *empty* space! Can empty space contain so much energy? This is an extremely large space density considering that the energy density of the nucleus of helium-4 is $8.57x10^{30}$ $J•L^{-1}$ (this is the *binding* energy number obtained from Wikipedia). Something doesn't add up here ...

... Or perhaps it does! Recall that in the section "Can Anything Go Faster than Light" in Chapter 5, I speculated the existence of dark energy in space. That energy is inherent to the fabric of space and is the mechanism that allows a quantum particle to go from one float to another. Essentially dark energy allows motion to take place. In the section "Motion in Space–Time: The Plot Thickens!" of the same chapter, I suggested that quantum particles skip in space by being carried by energy in space very much like when a person is carried above a crowd of people. The person temporarily is slowed down, and then reaccelerated, giving an impression of constant, smooth speed. The section "Motion Finally Explained!" later in this chapter will provide the physical phenomenon that accounts for that quantum skipping. We'll see that the particle doesn't truly skip around.

Because of the principle of action–reaction, the force of reaction from space has to equal the force from the quantum particles. Based on these thoughts, I have a hunch that the number $\Sigma = 2.31x10^{22}$ $J•L^{-1}$ representing the energy density of empty space means that empty space can support or carry any particle with an energy density of $2.31x10^{22}$ $J•L^{-1}$ or less. What particles could those be? This will be answered in the next section.

Now let's ponder about what the energy density constants Σ and ζ really mean physically. Why did I feel the need to develop equations involving these two constants in Appendix 6? It's important to notice that the constant Σ reveals the energy density at the *edge* of the universe. As the universe expands, what do you think is first created at its edge? Understand that the edge can't be made of the three-dimensional space that we experience here on earth. This is because before the three dimensions of space can be created, the *fabric* of space has to be created. Only after the fabric of space is created can all three dimensions be created. Thus, the constant Σ gives the energy density *of* space—essentially the energy density of that dark energy mentioned in the previous paragraph. Once the fabric of space is created, then the three dimensions of space are created.

Once the three dimensions are created, energy can go into that brand new space, which is what the energy density constant ξ is all about. If you study Appendix 6, you'll learn that this constant represents the *volume* expansion at the edge of the universe. As it is brand new space near the edge of the universe, inevitably initially it's the *minimal* amount of energy possible that moves into that new space. This is what the constant ξ is about: it gives the minimal energy density *in* space, not *of* space.

Mathematician John Baez at his Web site http://math.ucr.edu/home/baez/vacuum.html used observations from a famous experiment, namely the Wilkinson Microwave Anisotropy Probe, to calculate a minimal energy density in space of 9×10^{-10} J•m^{-3} (material used with permission). In Appendix 6, my estimate turns out to be $\xi = 2.2 \times 10^{-10}$ J•m^{-3}, which is very close to that.

Note that when expressed in units of J•L^{-1}, the constant $\xi = 2.2 \times 10^{-13}$ J•L^{-1}. Physicists call this number the *zero-point* energy of space. This is the energy that cannot be removed from space, even from vaccum space. For instance, suppose a sealed box containing air. Vaccum is created inside the box by removing all the air, that is, by removing all molecules and atoms, meaning that no matter is left at all. Yet there will still be energy left—this is the zero-point energy.

Note the dark energy density, $\Sigma = 2.31 \times 10^{22}$ J•L^{-1}, is billions of times denser that the minimal energy in space. This suggests that the universe is made up almost entirely of dark energy.

But not quite: the term ξ specifies the *minimal* energy density in space. As we move away from the edge of the universe, this density inevitably increases. Physicists estimate that the universe is made of 73 percent dark energy. So 73 percent of the energy from the Big Bang is used to create the fabric of space, otherwise known as the constitution of space.

Before finishing off this section, let us see how well my equation

$$\frac{dX}{dt} = \frac{\Sigma}{t^2 \xi}$$

estimates the expansion of the universe. After all, this equation was derived based on (a) my own theory of gravitation, and (b) my own theory of the expansion of the universe. So it better work! We need an estimate for the value of the variable t, the time that the universe started to expand with acceleration. The universe is estimated to be some 13.7×10^9 years old, but that isn't the number sought. So I searched the Internet and found that physicists estimate that the universe started to accelerate some $t = 7.5 \times 10^9$ years $= 2.4 \times 10^{17}$ seconds ago.

Next we need to change the units of $\Sigma = 2.31 \times 10^{22}$ J•L^{-1} to J•m^{-2}. That yields 8.1×10^{16} J•m^{-2} as was calculated at the beginning of this section. Let's plug the numbers into the formula:

$$\frac{dX}{dt} = \frac{\Sigma}{t^2 \xi} = \frac{8.1 \times 10^{16}\, J \cdot m^{-2}}{(2.4 \times 10^{17}\, s)^2 \cdot (2.2 \times 10^{-10}\, J \cdot m^{-3})}$$
$$= 6.4 \times 10^{-9}\, m/s^2$$

Given that the estimated acceleration as observed by physicists is 7.5×10^{-10} m/s^2, my equation gives an acceleration that's not that far off the mark.

Here are a few known facts that my theory of expansion has checked out:

1. Newton's gravitation formula could be derived from my theory (check it out in Appendix 6).

2. The minimal space density $\xi = 2.2 \times 10^{-10}$ J•m^{-3} derived from my theory does correspond closely with the known estimate.

3. The minimal space density $\Sigma = 8.1 \times 10^{16}$ J•m^{-2} derived from my theory does correspond closely to the energy density of neutrinos. (See the section "The Energy Dimension" of Chapter 8 for the details about this.)

4. My theory provides an estimate of the acceleration of the expansion of the universe that is close to the observed value for nearby galaxies.

I hope you'll agree that my modest theory of gravitation and of the acceleration of the universe should deserve some attention.

The Energy Dimension

Based on the many things that I've discovered so far about space and space energy density, I suggest that the fabric of space is somewhat like a sea. The quantum particles that reside in space "float" at the surface of that sea. The term *float* that I first suggested in Chapter 2 to explain motion of particles was a very good hunch because it's very fitting now.

To further explain quantum motion, I'll continue to use the analogy of the surface of a sea. This will help you have a visual reference for what happens. Then in the next section, I'll slowly bring you toward the real physical explanation for motion.

For a particle to travel in space, space has to be able to sustain the energy of that particle. With the analogy of the sea, this is like a ship that is carried by the water. The surface of the sea is able to carry the ship because it applies a force upon it equal to the weight (i.e., energy) of the ship. In the case of the nucleus of an atom such as helium, the reaction from the fabric of space has to equal the energy of the nucleus. My number calculated above, $\Sigma = 2.27 \times 10^{22}$ J•L^{-1} of space density, is nowhere near the energy density of the nucleus of helium 8.57×10^{30} J•L^{-1}. The nucleus would "sink."

Is my number Σ wrong? Not at all—remember that this number is the *minimal* space density. It's just that the nucleus of helium simply can't exist near the edge of the universe where energy density is indeed minimal. Sure enough, we won't find helium at the edge of the universe as the space energy density is just too thin there.

What particle then can exist near the edge of the universe? Let's try with other particles.

By using the formula $E = mc^2$, we can estimate the amount of energy in a proton. Its mass is estimated at 1.67×10^{-27} kg. So the proton contains $(1.67 \times 10^{-27}$ kg) x $(3 \times 10^8$ m/s$)^2 = 1.5 \times 10^{-10}$ kg•m^2•s^{-2} or 1.5×10^{-10} Joule. The volume of the proton is given by the formula $4\pi R^3/3$. This is expressed in m^3 units. The proton has a radius of 8.25×10^{-16} meter. Thus, its volume is $4\pi/3$ x $(8.25 \times 10^{-16}$ meters$)^3 = 2.36 \times 10^{-45}$ m^3. Let's use the conversion factor 1 m$^3 = 10^3$ liter calculated in section "What the Big Bang is Not". So the equation for the volume expressed in liters is $4000\pi R^3/3$. So the proton has a volume of $(2.36 \times 10^{-45}$ m$^3)/(10^{-3}$m^3/liter$) = 2.36 \times 10^{-42}$ liters. Finally, the energy density is given by the formula $mc^2/(4000\pi R^3/3) = 3mc^2/(4000\pi R^3)$. Let us use the symbol δ to represent the energy density of a particle. So then

$$\delta = {(3mc^2)}\big/{(4000\pi R^3)} \quad (1)$$

Thus, the proton has an energy density δ of

$$3 \cdot (1.67 \times 10^{-27}\ kg) \cdot (3 \times 10^8\ m/s^2)^2 \big/ (4000\pi \cdot (8.25 \times 10^{-16}\ m)^3)$$

or

$$3 \cdot (1.5 \times 10^{-10}\ kg \cdot m^2 \cdot s^{-4}) \big/ (7.1 \times 10^{-42}\ m^3)$$

or 6.3×10^{31} J•L^{-1} which is much higher than the minimal energy density 2.27×10^{22} J•L^{-1} calculated by my equation. So the proton would "sink" near the edge of the universe.

What about the electron? According to Wikipedia, it has a mass of 9×10^{-31} kg and a radius of about 10^{-15} meters. But I've found Web sites that give it a radius of 10^{-18} meters. If we average the two exponents, we obtain about 5×10^{-17} meters. So its energy density δ is

$$3 \cdot (9 \times 10^{-31}\ kg) \cdot (3 \times 10^8\ m/s^2)^2 \big/ (4000\pi \cdot (5 \times 10^{-17}\ m)^3)$$

or

$$3 \cdot (8.1 \times 10^{-14} \ kg \cdot m^2 \cdot s^{-4}) \Big/ (1.6 \times 10^{-45} \ m^3)$$

This ratio is equal to 1.5×10^{32} J•L^{-1} which is again much higher than the minimal energy density 2.27×10^{22} J•L^{-1} calculated by my equation. So the electron would also "sink" near the edge of the universe.

We won't waste time trying every single quantum particle like this until we find one that fits the minimal energy density of 2.27×10^{22} J•L^{-1}. Instead, let's work at it the other way around by isolating the mass in the formula (1):

$$\frac{4000 \pi \delta R^3}{3c^2} = m$$

We're searching for a particle that has an energy density of about 2.27×10^{22} J•L^{-1}, or one billion times lower than that of the electron. This is easily achieved if the particle sought is $(10^9)^{1/3} = 10^3$ times larger than the electron while having the same mass. But then being larger, the particle will likely have a larger mass. The proton is such an example, and its energy density is higher than what we are searching for. The likely particle must have a slightly smaller size than the electron and a much smaller mass. Let's assume $R = 10^{-18}$ meter. The mass obtained is

$$\frac{4000 \pi \cdot (2.27 \times 10^{22}) \cdot (10^{-18})^3}{3 \times (3 \times 10^8)^2} = \frac{(2.9 \times 10^{26}) \times (10^{-18})^3}{2.7 \times 10^{17}} =$$

$$\frac{2.9 \times 10^{-25}}{2.7 \times 10^{17}} \approx 10^{-42} \ kg$$

This mass is indeed much smaller than that of the electron as it's close to that of a neutrino! The lightest neutrino has a mass of about 10^{-39} kg, but its radius isn't known. I assume here a radius of 10^{-18}

meters because that is indeed slightly smaller than all subatomic particles such as the electron, the neutron, and the proton.

As I said in the section "The Quantum World" in Chapter 5, physicists don't believe that quantum particles exist that are smaller than the neutrino. Thus, the calculations above can only point to the neutrino. Therefore, the neutrino has the lowest energy density of all quantum particles, with its energy density probably being either equal to or slightly lower than the minimal energy density of space, which my formula calculated to be 2.27×10^{22} J•L^{-1}.

Physicists have been speculating about a connection between dark energy (the fabric of space) and the neutrino. This is precisely the connection that my theory of the expansion of the universe has just found. Exciting! Therefore, my constant $\Sigma = 2.27 \times 10^{22}$ J•L^{-1} calculated in Appendix 6 has an air of truth about it, don't you think? As dark energy is the energy of space, the calculations made above suggest that dark energy density is about equal to the energy density of neutrinos.

Physicists found it very difficult to detect the neutrino, and it took over forty years to do so experimentally. One explanation for this difficulty is that perhaps the neutrino is so tiny as to be nearly impossible to detect. Yet, I suggest that the neutrino isn't much smaller than the electron which was discovered over 100 years ago. I believe the reason the neutrino was so difficult to detect wasn't due to its tiny size, but rather to its lack of an electric charge. Consequently, it doesn't react with anything and thus isn't deflected by anything.

My calculations with the neutrino suggest that the minimal energy density of space at the edge of the universe has to be just dense enough to support neutrinos. Actually, this makes sense because when new space is created, it only needs to be dense enough to support the smallest particles. This means that the edge of the universe would contain just light and then neutrinos:

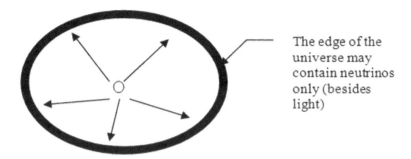

The edge of the universe may contain neutrinos only (besides light)

Then as the universe expands further, more energy is sent to where the neutrinos are, and so the space density increases and can allow for the support of other energy-heavier particles.

For curiosity, I calculated the energy density of our earth. Its radius is 6.4×10^6 meters, and its mass is 6×10^{24} kilograms. This gives an energy density of 5.1×10^{17} J•L^{-1}. Since the earth of made of atoms, you would think that the earth's energy density should be at least that of protons, that is, 6.3×10^{31} J•L^{-1}, as calculated earlier. What's the earth's energy density so low then?

The earth may feel dense to us, but if you were to look down at the subatomic level, you would see mostly empty space, and electromagnetic energy (light) all around. These have a low energy density. Even the core of the earth, with a temperature of about 7000 celcius, has a low energy density. Using the conversion of 1.9×10^3 Joule per celcius, the energy density at the core is only $(1.9 \times 10^3$ J/°C$) \times (7000$ °C$) = 1.3 \times 10^7$ J or an energy density of 1.3×10^7 J •L^{-1}.

I suggest that dark energy—the fabric of space—is the fourth true, real, and physical dimension of the universe because:

1. It's independent of the other three space dimensions.

2. It appears to be equivalent to the time dimension. Indeed, recall the graph in the section "The Time Dimension is Nonphysical" of Chapter 4:

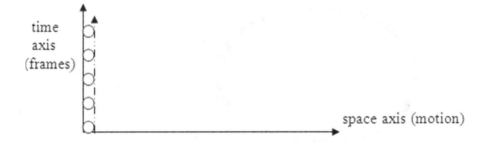

where I concluded that anything sitting strictly on the time dimension is no longer in space (i.e., it's no longer in the three dimensions of space). I suggested that if it isn't in space then it can't have energy. What I didn't realize at that time is that the fabric of space itself is made of energy, but it is energy that's never emitted. So it appears as having no energy in the three-dimensional space.

This is why I suggest that dark energy appears to be a good candidate for the entity that drives the time dimension. But can anything sit on the time dimension? No, simply because time isn't a physical dimension. As the fabric of space is physical, it can't be sitting on the time dimension.

The only way out of this is to conclude that the fabric of space is itself a dimension, the *real* physical fourth dimension. Do you recall that in Chapter 3, I suggested that the time dimension is needed to account for motion? But again, as time isn't a physical dimension, how could it account for motion? It's clear now that the fabric of space (dark energy) is a better candidate for that fourth dimension that accounts for motion. The section below explains how dark energy makes motion possible.

Motion Finally Explained!

This section explains the mechanism of quantum motion. Going back to the analogy used at the beginning of the previous section (i.e., that the fabric of space is somewhat like a sea on which quantum particles "float"), I propose that space is made of four dimensions: the three XYZ dimensions, and then the energy density dimension. To illustrate, here are the XYZ dimensions:

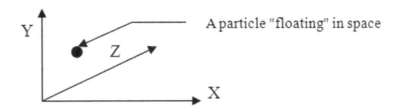

Take a particle in space, say the one shown above. Recall that in the section "Space–Time Fuzziness" of Chapter 2, we discovered the fuzziness formula $\Delta s \cdot \Delta v$, and it was determined that this fuzziness occurs in two-dimensional space. So as to a quantum particle, it "floats" on a flat surface. Let's look at it as it "floats" in that flat two-dimensional space:

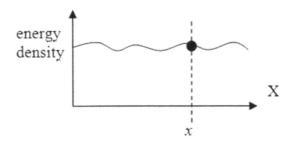

The wavy shape in the graph is to depict fluctuation in the energy density of the fabric of space onto that flat two-dimensional surface. For simplicity, only one dimension (the X axis) is shown. The fabric of space is the energy density on every quantum spot on that two-dimensional space. As the particle floats in space, the flat two-dimensional space changes direction, thereby accounting for motion is three-dimensional space. But the quantum particle isn't aware of that: it always views itself in a two-dimensional space.

By the way, humans can't see the fabric of space as it's at a level we can't perceive. But given the description above, it appears that a quantum particle "sees" that fabric. It "floats" onto it like floating onto a magical carpet. It's also like a passenger on an airplane at night. The passenger can't see anything outside and thus has no point of external reference. When the airplane turns, the passenger can't see that his body is turning in three-dimensional space as inside the plane, nothing has apparently moved. The passenger in his seat feels as though he is traveling in a two-dimensional space. I hope that this analogy helps explain what might be happening at the quantum level.

The energy density of that flat two-dimensional space has to be such that the particle doesn't "sink," say at point x. Should the particle sink, then it would disappear from view in the three-dimensional space! It would still be in fabric of space, but would no longer be visible. Would it reappear should the energy density at the point x increase sufficiently? It's probable.

This new view of space leads me to my final attempt to explain motion. This final attempt will constitute the correct physical explanation. How exciting! We are minutes away from discovering how matter moves in space. But before proceeding, I wish to remind you of our four previous attempts:

1. In the section "A Fourth Dimension is Necessary" of Chapter 2, I suggested that quantum particles skip in space, thereby allowing their motion. I explained that the skipping was due to the particle moving up in the time dimension although I didn't know for sure if that dimension existed at all.

2. In the section "Space–Time Fuzziness" of the same chapter, I suggested that nature periodically loses track of where the particle is, and by the time it finds it again, the particle has skipped to a new location.

3. In the section "Motion in Space–Time: The Plot Thickens!" of Chapter 5, I suggested that space pushes the particle from one quantum step to another. When the particle is being pushed, it' not visible; otherwise it *is* visible.

4. In the section "Quantum Steps Taken by Particles" of Chapter 5, I suggested that motion is made possible by the idea that nature sometimes sees the particle as waves, then as a particle, and back and forth. The particle skips when it's seen as a wave, but stays put when it's seen as a particle.

The third attempt was the one to come closest to explaining motion because it directly implicated space. I had a hunch that space was implicated. Well, that makes perfect sense because motion takes place ... in space, of course! I hope to have persuaded you that the time dimension is the wrong candidate to explain motion.

The first attempt, although exotic, relies on the assumption that motion in the time dimension is possible. But the section "Motion in Space–Time: The Plot Thickens!" of Chapter 5 showed that motion in time happens in the human mind only.

The second and fourth attempts are very closely related because they are both based on the "fuzziness" of motion at the quantum level. I have a hunch that both attempts have a tie-in with the upcoming fifth and final attempt at explaining motion using space density, but I'm unable to prove that mathematically.

With my latest discovery in this section that space density is the fabric of space—the real fourth dimension—it becomes clearer what goes on. Motion is possible by the particle "swimming" on the surface of space. Keep in mind that space doesn't have a surface in the sense used here as this is just an analogy to help us visualize it. For reasons that will become clearer a bit later, it has to be that the particle surfaces in space periodically, a period equal to the Planck time, P_T. When it surfaces, it's visible in space. When it's under the surface of space, it's invisible. This would account for the quantum steps described so many times in this book.

We finally found the physical reason why a quantum particle skips in space. But be patient: the explanation isn't quite complete yet because I'm still using the analogy of the surface of a sea. Later, I'll provide the real physical explanation for this phenomenon of particles "swimming" in and out of the fabric of space.

You may want to stop me here and request that I verify if my new view of space explains why a particle's Planck length P_S and

Planck time P_T vary depending on the speed of the particle. Recall that we looked at this subject in Chapter 4. But let me answer this question again by using the analogy of the sea to provide a short explanation. Imagine a small boat advancing on the water. Suppose that the water is unable to support the boat when it's at rest perhaps due to a leak. But if the boat picks up enough speed, it will be able to float on the water. If the boat advances *slowly*, it will tend to sink for a *long* period of time (and long distance), then resurface. Here's a picture (for simplicity, the dots represent the boat as it advances):

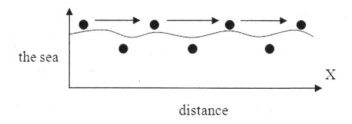

distance

The effect of sinking is exaggerated in this picture in order to make it clear that the boat has surfaced only four times. You can see that the Planck length is *long*, and so the "tick" of time of the boat is *long*, implying that time goes *faster*. Indeed, as we saw in Chapter 4, time speeds up when the particle slows down.

Now compare when the same boat goes *faster*:

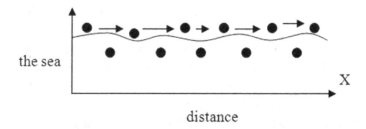

distance

The boat has surfaced six times. It sinks for a *shorter* period of time. You can see that the Planck length is *short*, and so the tick of time of the boat is *short*, implying that time goes *slower*. Again, as we saw in Chapter 4, time slows down when the particle speeds up.

Now imagine if the boat goes so fast that it never sinks:

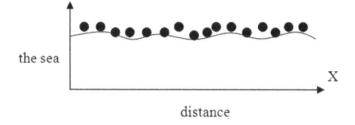

distance

Guess what? If the boat were a particle, the particle would be a photon—that is, light! Indeed we saw in Chapter 4 that photons have a Planck length of value 0. That is, photons have no space between their "floats," or in other words, they never "sink" into space. The reason has to be because light has the same cosmic pressure as dark energy. Energy is energy, after all. So then if it is photons (free-flowing energy) that are on the surface, this implies that in the picture given earlier:

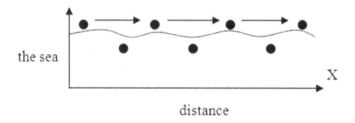

distance

the dots above the "surface" are *not* the particle, but rather of *emitted* energy from the particle. This goes well with the intuition that I had in the section "Emitted Energy Versus Internal Energy" of Chapter 4 that brought me to suggest that a "float" is a place where *emitted energy* lands, and I quote what I suggested there: "It's the energy emitted itself that creates the 'float.'" To remind you, I copied at the top of the next page the picture from Chapter 4 that illustrates the explanation. Consult that picture before reading on. The energy symbols in black are the ones that are on the surface of the fabric of space (the sea surface in my analogy). The particle itself is under the surface of the "sea."

We're getting very close to a complete physical explanation for quantum motion.

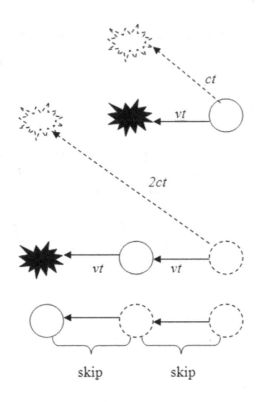

skip   skip

Why does the particle sink under the surface? To answer this, we have to recall what was said in the section "Emitted Energy Versus Internal Energy" of Chapter 4. There I asked the question: how can nature tell the difference between the emitted energy and the internal energy? The answer given was that emitted energy radiates outward whereas internal energy stays within the particle. Energy within the particle doesn't radiate. Most interesting!

It turns out that the "sea" is the surface of the fabric of space, and the fabric of space is made of energy that *does not* radiate. Conclusion: the particle's internal energy *merges* with the fabric of space. It's as if we put a black circle onto a black sheet. The circle can't be seen.

Now you might ask: if the particle is under the surface of space, how are we able to see matter? Answer: we don't actually "see" matter—we see its emitted energy.

But what's that emitted energy anyway? Where does it come from? It's important to understand that this emitted energy is *not* energy that a charged particle, such as an electron, emits when it's excited by a passing wave of light. We're talking about energy that can only come from *within* the particle itself.

But then what's the difference between the internal energy and the emitted energy? Here's the shocker: there's no difference! It's the *same* energy, but in different forms. This is a consequence of the result that I found in the section "Derivation of the Famous Equation $E = mc^2$" of Chapter 1. You may recall that I derived there the equation that gives the energy within a quantum particle: it's the famous equation $E = mc^2$. In that section, we found that this is the amount of energy emitted when a quantum particle "bursts" into a wave. This burst is simply the emitted energy of the particle. Internal energy— the tangled-up waves—simply becomes emitted energy.

How does this relate to the motion of quantum particles? Answer: recall in the section "Motion in Space–Time: The Plot Thickens!" of Chapter 5 in which I found that the amount of energy taking place during a Planck space is $\Delta E = 2mc^2$. This is the amount of energy when a particle skips from one "float" to the next one.

Drum roll, please … the following completes the physical explanation for quantum motion.

Using both equations $E = mc^2$ and $\Delta E = 2mc^2$, we now have the following physical explanation for what happens when a particle performs a quantum step. This isn't an analogy; this is the *real* physical phenomenon that explains motion.

As stated a few paragraphs ago, a particle surfaces out of the fabric of space when it emits energy. The particle inside the fabric of space bursts with emitted energy equal to $E = mc^2$. Let's call that energy E_w. At that point, the particle has become a wave *in* space. That wave hits the "surface" of the fabric of space. This "surface" has a cosmic pressure that causes the emitted wave to curl back into itself, thereby reconstructing the particle. The energy $E = mc^2$ is essentially put back into the particle, and the particle sinks back into the fabric of space. Let's call that energy E_p. So during a single quantum step, the amount of energy involved is $E_w + E_p = mc^2 + mc^2 = 2mc^2$, which is the same as $\Delta E = 2mc^2$. Here's the picture:

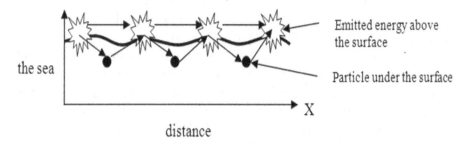

the sea

Emitted energy above the surface

Particle under the surface

X

distance

An analogy to this motion is when a fish has a burst of energy and leaps out of the water, but is pushed back into the water by gravity. The fabric of space is made of energy that billions of times a second "sinks" the quantum particle back under the fabric of space whenever the particle bursts into emitted energy.

Motion takes place only when the particle is a wave (i.e., free-flowing energy). The particle doesn't move at all while it's a tangled-up ball of waves. It doesn't get any more bizarre and yet fascinating: at the quantum level, a particle bursts and breaks loose into a wave, then the wave curls back into a particle back again, a particle identical to the previous one. This scenario of "breaking" and "reconstructing" the particle takes place many billions of times per second. Physicists have a name for this reconstruction of the particle event: the *wave function collapse*. Every quantum particle is associated with a wave function that describes the shape of the wave that dictates the behavior and motion of the particle. Put simply, when the function collapses, the wave becomes a particle.

Every piece of matter in the universe jumps in and out of the fabric of space the way explained in the previous paragraph. As our human body is made of matter, our body also moves in and out of the fabric of space! What we see is the stuff out of the fabric of space. You may recall that in the section "Quantum Steps Taken by Particles" of Chapter 5, I already suggested that quantum motion took place the way presented here. My hunch is now proven correct.

Note that because photons don't sink into the fabric of space, they are in a continuous state of energy bursts. This explains why photons show up as light (light is simply pure energy).

Continuing with the analogy of the sea, the particle swims in and out just like a fish. A fish doesn't swim in a straight like: it zigzags through water. Just the same, from the figure above, if we look at it from the top (i.e. above the "sea"), the particle will move along in a zigzag fashion:

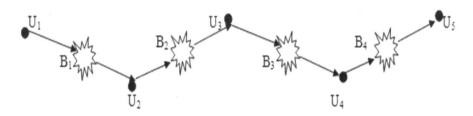

The motion takes place in three dimensions, but only two dimensions are shown here. The energy bursts labeled B_i are the places where the particle jumps out of the fabric of space. The spots U_i are where the particle is stopped by the fabric of space which forces the particle to collapse (curl up) into itself into the fabric of space. Then the particle bounces back (hence the zigzag motion) and bursts into energy again and comes out of the fabric of space, and so on. Note that the particle is matter only where the spots U_i are. Everywhere else, it bursts into pure energy. The particle moves only when it's in form of pure energy. Because energy travels at the speed of light, then the particle moves at the speed of light then stops at the spots U_i for a while. Therefore overall, the speed of the particle is much less than the speed of light.

You may want to stop me here and request that I verify that this new view of motion explains why a particle's Planck length P_S and Planck time P_T vary depending on the speed of the particle. This is easy to do:

1. If the particle goes *faster*, it will have more energy, so it will burst into a wave *sooner* than if it were going slower. So there would be *more* waves per second (in other words, there would be *more* quantum steps per second). Recall that in the section "An Absolute Frame of Reference" of Chapter 5, I proved that a quantum step corresponds to a wave. So the more waves there are, the more quantum steps there are. Likewise the fewer waves there are, the fewer quantum steps there are.

2. If the particle goes *slower*, it will have less energy so it will burst into a wave *later* than if it were going faster; thus, there would be *fewer* waves per second (in other words, there would be fewer quantum steps per second).

These two scenarios match what we found in Chapter 3 and 4. The mystery of motion seems to be solved ... at last!

I recall learning in my physics class about an experiment made in the early twentieth century in which physicists shot a beam of electrons through a small hole. They were stunned to find that the effect was the same as when light was shot through that hole! Yet electrons aren't a form of light. This result eventually led to the duality principle proposed by De Broglie a dozen years later whereby particles sometimes behave like waves, other times as particles.

With my explanation of motion given above, the explanation becomes clear. The electrons shot through the hole behave like waves because they *do* move in space as waves and *are* waves in motion while giving off their energy in quantum lumps. You may recall that in Chapter 5 (mostly in the section "An Absolute Frame of Reference!"), I proved that waves give off their energy in lumps.

Essentially, electrons are waves. This explains yet another mystery about atoms: why do electrons, which have a negative charge, don't collapse into the nucleus, which has a positive charge? In the case of planets, it's the centrifugal force that holds a planet away from the sun. But this can't be the correct explanation for electrons orbiting the nucleus because gravitational forces at the atomic level are millions of times weaker than electromagnetic forces that attract the electrons toward the nucleus of an atom. So the electrons should definitively be strongly attracted to the nucleus and collapse into it.

Why don't they collapse? The answer is that since electrons move as waves, the length of their orbits around the nucleus *has to* be a multiple of the wavelength associated with the electron. For instance, the electron of the hydrogen atom has an orbit's length equal to—perhaps—one wavelength associated with the electron. Despite the fact that the nucleus attracts the electron with great force toward the center of the atom, the electron simply *cannot* follow an orbit length less than a complete wavelength.

With this view, it's possible to estimate the closest distance that the electron can be from the nucleus. In section "The Quantum World" of Chapter 5, I calculated that the wave length of the electron is 2.2×10^{-12} meter. This is also the smallest circumference possible of the electron's orbit. Using the formula *circumference* $= 2\pi \times radius$, the distance from the nucleus is no less than $(2.2 \times 10^{-12}$ meter$)/2\pi = 3.5 \times 10^{-13}$ meter. We can infer that the size of the nucleus of hydrogen is smaller than that. Indeed physicists estimate the radius of the nucleus of hydrogen to be 1.6×10^{-15} meter. So, the electron is at least 200 times farther away from the nucleus compared to its size. You realize now that the electron is always very far from the center of the atom. Bigger atoms with numerous electrons will see some of their electrons follow orbits much farther away from the nucleus, but the orbits' length will still follow a multiple wavelengths.

Another discovery can be made from the picture above of the zigzag motion. It was said that the energy bursts take place at the speed of light. Let's mention the speed of light with a "c" on the picture:

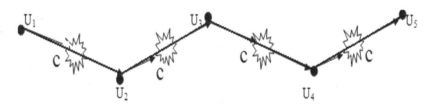

Consider the first segment U_1U_2. Don't you find that the imaginary right-angle triangle that can formed:

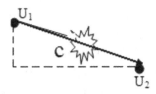

looks like the figure below

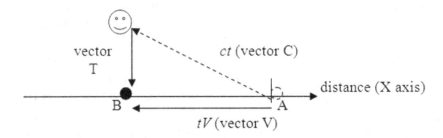

taken from section "The Clock of the Universe" of Chapter 4? They both are the *same* right-angle triangle. This proves that the time vector T is actually a distance. I have to conclude that at the quantum level, *time is a distance*—you may recall that I had already expressed that hunch in Chapter 4. As hard as this is to believe, at the quantum level, time does *not* exist! Consequently Einstein's space-time theory has no physical reality. You may also recall that I already expressed in Chapter 6 my doubts about the existence of space-time as a physical model.

Now, let's return to the discussion about the fabric of space. It's made of energy called dark energy. Because dark energy enables motion in space, it enables motion of matter and energy as well. Therefore, dark energy is at the source of the reason light travels at 3×10^8 meters per second. But I lack the knowledge to prove it.

Also does space density of dark energy vary in space? It all depends on the nature of light. If light behaves the same way everywhere in the universe, then because the picture earlier shows that light and dark energy have the same cosmic pressure, the implication is that dark energy has the same density of 2.31×10^{22} J•L^{-1} and cosmic pressure everywhere in the universe. This is what physicists believe to be true at the moment. However, if physicists ever discover this to be false, then the implication could be that the speed of light hasn't always been what it's now!

The idea of an energy dimension is fascinating, but it leads to a puzzle: if space is made of energy, then why does that energy not radiate? Why can't we see that energy in any way at all? On the one hand, if space energy did radiate, it would become particles *in* space and not *of* space. It would be like a particle sunk into the sea that moves to the surface where it can be seen. On the other hand, the answer might be that space energy can't move to the surface due to some space pressure preventing it. That pressure would be caused by the energy density of the dark energy itself. So part of the energy of dark energy would be used such that dark energy would always remain "under." An analogy would be if we built a bridge of very tightly packed rocks:

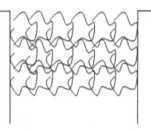

The pressure serves two purposes: (a) to make the bridge strong enough in order to support weight on it, such as a bicycle or a small car, and (b) to keep any single rock from falling.

Also, as the fabric of space is made of energy, that energy is either in the form of matter or pure energy. It obviously can't be of matter! So space is made of pure energy. I mentioned in the section "A Formula for Discrete Energy" of Chapter 5 that pure energy is carried by photons. Thus, dark energy is carried by photons. But because these photons don't emit energy (as explained in the previous paragraph), I'll call them *virtual photons*.

It's now clear why outerspace is black: no energy is emitted by space itself. This realization that the fabric of space is made of energy that doesn't emit light goes well with the section "The Time Dimension is Nonphysical" of Chapter 4 where I suggested that the time dimension contains "dormant" energy. But now I realize that it's the fabric of space that's made of that "dormant" light, not time. This conclusion will lead me a startling idea in the section "Where Do Souls Reside" of Chapter 10.

Again you may see that there's one little problem with this picture of the pressure of space: if it prevents virtual photons from surfacing into space, how would it then be possible for particles in space (such as in the pictures above) that are "sunk" under the fabric of space to resurface in space? Well, this was already answered a few paragraphs ago: the particle "bursts" into pure energy described by the equation $E = mc^2$. Essentially a quantum particle *explodes*, thereby reappearing in space. This happens billions of times per second.

In addition, as dark energy is absolutely everywhere, it's also inside our bodies. This implies that my body weight also includes the weight of the dark energy within me! How much additional weight could that be? Suppose a person weighs 70 kilos. The average body density of a human body is roughly 0.001 kilo/cm^3 or 1 kilo/liter, so a body occupies about 70 liters. The dark energy *minimal* density is $\Sigma = 2.27 \times 10^{22}$ J•L⁻1, meaning that the amount of dark energy in the body is more than $(2.27 \times 10^{22}$ J/L$) \bullet 70$ liters $= 1.6 \times 10^{24}$ Joule. Using the equation $E = mc^2$, and isolating $m = E/c^2$, the mass of dark energy in the human is

$$m = 1.6 \times 10^{24} \text{ J} / (3 \times 10^8 \text{ m} \bullet \text{s}^{-1})^2 = 1.6 \times 10^{24} \text{ J} / (9 \times 10^{16} \text{ m}^2 \bullet \text{s}^{-2}) = 1.8 \times 10^7 \text{ J} \bullet \text{m}^{-2} \bullet \text{s}^2.$$

The conversion factor for a joule into mass is one Joule = one kilo•m^2•s^{-2}, so one kilo = one Joule•m^{-2}•s^2. Consequently 1.8×10^7 Joule•m^{-2}•s^2 = 1.8×10^7 kilograms, or 18 million kilograms inside a human body. Now, this is what I call a weight gain program! The obvious question is: why can't I feel that huge weight within my body? There are two reasons:

1. Dark energy has *no* weight! Gravitation is what gives a weight to something. But dark energy isn't affected by gravitation simply because dark energy isn't *in* space, but rather *of* space. Only the things that reside in space are affected by gravitation. In fact, dark energy is part of what creates gravitation, so it certainly can't be affected by what it creates!

2. Dark energy doesn't emit its energy into space. If it can't emit energy, its energy can't radiate, so there's no physical way to feel the energy. Note that this isn't a limitation of our bodies in feeling dark energy but a limitation of nature itself.

An analogy to this would be the little creatures that live two miles deep down in the ocean where the water pressure is enough to crush a huge ship. Why don't these creatures feel that huge pressure? It's because the inside of their bodies is soaked into the same pressure, therefore creating pressure equilibrium. So the creatures just slide along the heavy water as if it were nothing. It's the same idea with dark energy: we slide in it as if there were nothing there. Keep in mind also that it's dark energy that enables motion. It doesn't exert any friction against our bodies, but we are literally "soaked" in dark energy.

Gravitons Revisited

Gravitons were introduced very briefly in the section "The Graviton" of Chapter 6. Why do I bring this up again? It's because I believe they're related to the fabric of space, which was the subject of the previous section. Note that a graviton can't have mass because mass generates gravitation, so if a graviton did have a mass, it would generate some gravitation of its own! In turn, this implies that the graviton generates gravitons, which, in turn generate gravitons and so on, to no end. In other words, gravitons would be self-referential, which clearly is a physical impossibility. Most physicists believe in the existence of gravitons—but don't they see the impasse this creates? The only way beyond this impasse is that gravitons are energyless (mass by definition includes matter and energy, and gravitons can't have either one). Is there such a thing in physics as an *energyless* entity? Two possible answers:

1. *No*. But then the graviton could only exist in some *nonphysical* dimension. Could there be a gravitational dimension? I doubt it. The only nonphysical dimension we've encountered so far is the time dimension, which I argued against at great length in Chapter 6.

2. *Yes*. But then the graviton couldn't *be* energyless—it would only *appear* so. But wait a minute! Didn't we encounter a similar situation already? Yes, with the fabric of space, which is full of energy, but none of it emits, thereby giving the impression of being energyless.

The second answer is the only one that makes sense. Therefore, gravitons are virtual photons inside the fabric of space! I searched on the Internet, but I couldn't find a single Web site where this view is proposed. But I remain convinced by my logic, and you'll see below that my view is consistent with experiments performed by physicists.

That fabric constitutes space, and it contains a lot of dark energy. I also call it "God's Light" because it has to be the very first type of light that came out of the Big Bang, and it's also the kind of light that expands the universe. But, although it's light, none of it can emit light. Now how do we explain that? In an attempt to answer this question, I used in the previous section the analogy of a bridge of tightly compacted rocks. Let's now consider a more realistic explanation.

As a graviton is related to the fabric of space, let's focus on where in the universe space is being created at this very moment: *onto the edge of the universe*. What's the first thing we expect to find at that edge? It has to be the light that creates the fabric of space, the "God's Light," then some more energy follows behind to fill the newly created space.

Now imagine that "God's Light" as it "pushes" the boundaries of the universe. You may recall from the section "Light with a Spin" from Chapter 5 that a free-flowing photon travels as follows:

Energy-wise, light also travels this way. Note that I stated this is the path for a *free*-flowing photon. What do you think the wave pattern looks like at the edge of the universe? As follows, perhaps?

Edge of the universe

At the edge of the universe, could such a half of a wave be produced? No—there's no such thing as a half wave, so this situation is impossible. So then, what instead does happen to the end of the wave as it "hits" the edge of the universe? It does the same thing that a wave does when it hits a wall: it bounces back. It reflects backward just slightly, perhaps the length of a few wavelengths. It's difficult to depict this reflection, but it would be like a spring that's looped back into itself—something like this:

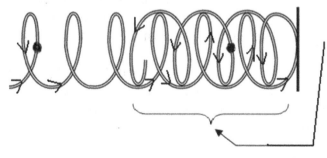

In this region, the wave loops back into itself

The arrow points to three waves that "hit" the edge of the universe and they were reflected back (looped back) into themselves. As a result, there are now *twice* as many waves near the edge. The resulting effect is that the photons there now have twice the spin. You can clearly see that there are twice as many waves at the edge.

In the section "Light with a Spin" in Chapter 5, I explained that quantum particles have a spin, and that light (the photon) has a spin of one (1) because the photon spins a full circle. The picture above shows that photons at the edge of the universe have twice that spin: a spin of two (2) simply because there are twice as many loops there.

Note that the "bouncing back" probably spans just a few wavelengths as shown above because new waves come toward the edge constantly, thereby breaking the bouncing back. Also the photons can't lose that double spin because they become entangled as more photons come from behind. It's this effect that creates the fabric of space.

Think of it this way: it's very much like waves washing onto the beach. The waves reflect back, but only just a little because oncoming waves make them all pile up instead. In a sense, the edge of the universe *absorbs* some of the bouncing back of photons, much like a beach does with water. Just as water piles up at the beach, photons pile up at the edge of the universe, making them "spin" more than photons elsewhere in the universe. The consequence is that pure *energy* piles up at the edge of the universe. Thus, the edge has a lot of energy stored.

Now what do you think is first created at the edge of the universe? It's the fabric of space. Consequently the fabric of space contains *a lot more* energy than the energy *in* space. To no surprise, this is precisely what the constants of minimal space density $\xi = 2.2 \times 10^{-10}$ J•m^{-3} *in* space and of the minimal space density $\Sigma = 8.1 \times 10^{16}$ J•m^{-2} *of* space as derived by my own theory of gravitation show!

Recall that in the previous section, I concluded that the fabric of space is made of virtual photons. It's become clear that gravitons are themselves virtual photons. In fact, the following two results observed by physicists prove me right:

1. Gravitons are impossible to detect in particle accelerators. Physicists have tried for many years to no avail as when they force the collision of two protons; the energy released is the

sum of the energy of both photons, except for some tiny amount of energy missing. But they can't see any quantum particles to account for that residual amount. They offer as an explanation that gravitons might exist in some new gravitational dimension that can't be seen, but I disagree. My theory is that the residual energy is stored in gravitons that are firmly "glued" to the fabric of space, which can't be detected. Therefore, gravitons can't be seen. It's that simple.

2. A quick search on Wikipedia revealed to me that gravitons have a spin of 2. Ah—just what I deduced! It shouldn't come as a surprise that gravitons have a spin higher than 1 because their task is to carry matter through space. Since matter is heavier than light, it tends to "sink" into the fabric of space. This effect was explained as few times before. So, it takes a lot of "spin" to dislodge matter form that fabric.

Based on my theory presented above, I propose that gravitons don't cause gravitation. They are simply the mechanism that ensures the motion of quantum particles in space, and that's what the fabric of space is for, isn't it? I stick to my theory presented in Chapter 6 that gravitation is caused by a differential of energy densities in space. That differential of densities causes motion of quantum particles—motion ensured by gravitons.

In Appendix 6, it will be shown that the edge of the universe doesn't expand evenly, but rather more like a wave:

The crest of the wave comes from the energy from behind that caught up with the edge and hence is now moving faster. Given what's been deduced in this section, waves at the edge actually curl up inward causing photons to curl up as well and giving them a 2-spin rather than the normal 1-spin that light has. It's that change into 2-spin that makes the light at the edge turn into dark energy making up the fabric of space. Again, the effect is somewhat similar to waves washing onto a beach.

Now, how to explain that gravitons don't emit energy? With a spin of 2, they are looped inward in such a way that they can't move forward unlike photons with a spin of one. Their energy is "stuck." You might see here a parallel with matter that is made of tangled-up looped waves. So why doesn't a graviton produce matter? I've no answer other than to note that quantum particles of matter such as electrons and protons have a spin of ½. Thus, a quantum particle of spin 2 produces a virtual photon in the fabric of space; a particle of spin 1 produces a photon of free flowing light; a particle of spin ½ produces matter. I leave the rest for physicists to clarify.

In section "Space Density" of Chapter 6 I proposed that a concentration (a cloud) of pure energy causes antigravity. On the one hand, I suspect that graviton is also the mechanism that executes antigravity. This seems to make no sense. How can graviton serve both gravitation and antigravitation? I proposed in this section that the graviton is the quantum particle that enables motion. On the other hand, because I argue in the first section of this chapter that antigravity came out of the Big Bang a mere instant before gravitation, the fabric of space was not created yet, and so gravitons didn't exist yet.

So, perhaps antigravity isn't carried out by gravitons but by some other quantum particle. Because antigravity caused the Big Bang, and that only pure energy (which is just light) could have been present, I suspect that the antigravity quantum particle has a spin of one—the same as light. I remind you that these quantum particles aren't the cause for those forces. So, there aren't any quantum particles that are at the source of gravitation or antigravitation. As I explained more than once in this Chapter 6, gravitation and antigravitation are caused by space density differentials. Gravitons and the antigravity quantum particle only carry out the effect of those differentials which translates into motion.

Note that because physicists have not yet been able to prove the existence of antigravitation, much of what I said in the past two paragraphs is debatable.

Chapter 9: What Created Life?

A proper attempt at answering this question would require an entire book, or perhaps an entire library! Many books and research papers have been published over the last few decades. By now reasearchers know a lot about the evolution of life, but are no closer to answering the big question: what created life?

I suggest that they might be looking in the wrong places. You'll find my answer to this question in the section "Molecular Structures and Life." But for a better appreciation of what led me to the answer, I strongly recommend you read this chapter from its beginning.

In fact, to acquire a good understanding, this chapter should be read in parallel to Chapter 10! Of course this isn't feasible. Why should these two chapters be read at the same time? It's because this chapter sometimes uses material from Chapter 10. Both chapters are closely related. At first glance, this seems ridiculous. How can the questions "What Created Life?" and "Is there an afterlife?" be related? Aren't the concepts of life and afterlife the opposite of one another? Actually, no! Read on, and you'll eventually see how they're related.

The subject of this book is about the quantum world. So you might be wondering: what does quantum physics have to do with the question of life? It's not obvious, yet this chapter claims that there's a connection.

The subject of this chapter is very touchy. I prudently avoid religious arguments which, for the most part, are in favour of Creationism. Having said this, this chapter does *not* dismiss this theory outright and entirely. I tried my best to neither take the side of Creationists nor the side of Evolutionists. Both theories address good points, and in the last section of this chapter, I even suggest that the theories complement each other!

Scientists have discovered that the conditions favorable for the creation of life—even at its simplest forms—are very complex. The answers to this question largely depend on who you ask. Some scientists and amateurs have turned to the calculations of the probability of life occurring by chance. The calculations (which are taken with permission from Kurt Johmann, a computer scientist, from his Web site www.johmann.net) are in the order of 1 chance out of 10^{2906} that the *simplest* bacterium was a result of chance. As the universe has been in existence for "only" 10^{10} years or so, we could conclude that there must have been divine intervention, but this is jumping to a conclusion too hastily.

Probabilistic Argument Is Weak

I've examined some calculations and probabilistic models from authors such as the one above who have concluded that the existence of life has to have involved the hand of God, and other authors who claim that life happened through evolution as per Darwin's Theory of Evolution.

Any way one looks at it, the probabilistic argument is weak for five reasons:

1. How reliable are the probabilistic models? (i.e., how closely do they really reflect the mechanism necessary for life in the universe?). Do the models take under consideration such factors as molecular forces or biological processes that increase the odds of life happening by chance? I discuss this point further later in the section "Darwin's Theory of Evolution and DNA mutations" of this chapter. Some people attempting probalistic models consider in their calculation all the different combinations of arrangement of molecules in the DNA. However, nature is such that many arrangements are impossible for both molecular reasons (such as binding forces) and biological reasons. So the probability of formation of the DNA by chance is much higher than calculated, although still tiny.

2. Some attempts assume that life can only occur on earth, neglecting to take into account that there are over 10^{22} stars in

the universe, thereby greatly increasing the probability of life *somewhere*.

In a series of my own probabilistic calculations in the section "The Probalistic Model" later in this chapter, I'll take these two factors under consideration, but find that the odds aren't improved by much.

3. However small the probability, it's no proof that God intervened. To explain my point here, consider the following simple analogy. Suppose that our world leaders decide to play a simple game with the world's population. Each citizen is given one die (that is 6×10^9 people), and then everyone throws their die at the same time every second with the goal being that all 6×10^9 people get the same number. That probability is 1 in $10^{4668907502}$! Suppose that the successful outcome happens just a week later (i.e., over 600,000 tries later). Is this proof of a divine miracle? We have three reasons to argue that it's not:

   a. No matter how small the probability is, as long as it's not a definite 0, we can't conclude that God had a part in it, simply because the outcome is *possible*— although very highly improbable. Divine intervention means that the outcome has to be a *miracle* in the sense that the outcome is *totally* physically impossible but happens anyway. It has to be something beyond not only human knowledge and understanding, but beyond the *laws of nature*.

   Some would refute my argument by saying that the outcome of this game of dice is so astronomically remote that it would most likely take longer than the entire duration of the earth to accomplish it. That's true, but the outcome has the possibility of happening at *any* time with always the same tiny probability for each attempt.

   b. Why would God care to intervene in this dice game anyway? Likewise, why would God have intervened in the creation of life? People of religious faith claim that God did create life. Fine—but probabilistic calculations can't be used to prove that. In fact, later in this chapter in the section "Codes and Algorithms for Evolution" I'll develop a simple probabilistic model that suggests life might possibly have come about on its own, but this still doesn't imply that God wasn't somehow implicated. If we wish to investigate whether God was implicated in the creation of life, we have to ask questions beyond the natural including: What's the purpose of life? Why is there life in the universe? How necessary is it? I'll investigate the answers to these important questions of purpose in various sections later in this chapter.

c. The likelihood of the outcome of that game of dice is extremely remote, but so are *all* other combinations of the dice rolls. Why don't we think of those too? Answer: *because we don't care about them.* We humans concentrate on the likelihood of life because we care about its significance. There may be a lot of other events that are equally unlikely to occur that remain in the shadow of our collective consciousness. Undoubtedly countless events take place that are even less likely than that of the appearance of life, but we're unaware of them.

As a simple example, suppose that I'm at the checkout at the grocery store. I discover that by pure chance, I've purchased the same items as the stranger at the next checkout, and moreover that we both wear the same shirt, and that we and the checkout clerks all share the same birth date. The probability of such a sequence of events is extremely remote. Neither I nor that stranger at the next checkout will ever know that such an unlikely event took place. However, should I be made aware of it, then I might attempt to attach a meaning to the coincidence. There are numerous unlikely events that take place every day of which we are unaware.

d. The question is asked after the fact: There *is* life, right here on earth! No matter how unlikely its appearance might be, life did occur. It's normal to wonder how this took place. It's just like any other phenomenon in nature in which we wonder: why? Why has life evolved here? Why not on another faraway planet? Did God intend to put life on this planet *specifically*? I doubt it. It just happens that life occurred here. In a few billion years, it might come up somewhere else. But we don't—and probably never will—have the benefit of knowing that.

4. Most probabilistic calculations—including my own presented in a series of sections coming up—give the likelihood of life occurring by chance to be incredibly tiny. Some people say that if life exists only on the earth in the whole universe, this is proof of a divine miracle. But the *correct* reasoning is the other way around! My calculations in the section "Origin of Life and DNA Structure" show the likelihood for life is incredibly tiny. This is why:

Probability theory says the less likely an event is to occur, then the less often it will occur. Based on this principle, the sample size of planets with life should be very tiny—perhaps comprised only of this earth. Actually should life (especially intelligent life) exist in a *great many* planets all over the universe, this would be a strong indication of divine intervention because such a large sample would indeed be very unlikely to happen in any way at all.

Our dilemma is that we don't know if there's life elsewhere, so we have no scientific argument to conclusively say that God created life. Scientists strive to find evidence of life on Mars in the hope this would prove that life is just a natural process and that God has nothing to do with it. Again, the implication should be the other way around: should scientists ever find present or past life on Mars, then perhaps divine intervention might deserve consideration.

5. Many people believe that Earth is the only planet to contain life. This is possible, but we'll likely never know. If indeed ours is the only planet with life, then it has got to be a divine miracle, right? Not really. We can't conclude that God intervened for two reasons given earlier that are worth repeating:

    a. Because the likelihood of life appearing by chance is so tiny, probability theory says that the sample size should be tiny, with perhaps Earth being the only planet with life.

    b. Why would God have chosen this planet as the sole place where life exists? Thinking that God placed the earth at the center of the universe when it comes to life is an argument no better than that of those who believed 500 years ago that the earth was at the physical center of the universe.

To indicate divine intervention, the process of creation of the event has to have happened beyond natural causes (i.e., beyond the laws of nature). For instance, suppose I broke my arm today, but by tomorrow, it had completely healed already. That would be a miracle and thus indicate divine intervention.

Our DNA (deoxyribonucleic acid) structure within each of our cells contains molecules found in the universe, with all of those molecules being bound according to the rules of chemistry well understood by scientists. So we can't immediately say that divine intervention is involved just on the basis that the DNA structure is too complex to have happened by chance. I'm not suggesting that life isn't a result of divine intervention but just that probabilistic models are the wrong approach to this issue. Scientists just don't know enough yet about the origin of life to draw a conclusion one way or the other whether life is the result of natural causes.

Origin of Life and DNA Structure

My understanding of Darwin's theory is that it's based on these two assumptions:

1. That life happened by chance on earth rather than by design.

2. That survival of a species is based on natural selection, a means by which good designs are preserved and bad designs are eliminated. The result is called evolution of the "fittest." Note that the designs evolve by means of random choices of changes in features of the species.

Let's assume that the first assumption is true (i.e., that development of DNA happened by chance). If so, we have to calculate the probability of its design to have occurred by blind trial and error. Some calculations I mentioned at the beginning of this chapter suggest the probability to be 1 chance in about 10^{2906} (give or take) for the simplest bacterium to have formed by chance. The calculations assumed that all molecular structures were already created (a big assumption). Accordingly, nature just had to juggle blindly with the arrangements until the right one came to be. How many arrangements could this be? Assuming that nature tried an arrangement every 10^{-43} seconds (this is the Planck time, the shortest time possible in nature), that comes to $3x10^{50}$ tries a year.

Now assume that nature tried that arrangement for every bacterium-sized surface on the surface of the earth. A bacterium is about 10^{-6} meters in radius. The radius of the earth is $5x10^{6}$ meters, so the surface of the earth is about $3x10^{14}$ meters squared. A bacterium occupies about $1.2x10^{-11}$ meters squared, so there could have been about $2.5x10^{25}$ bacterium-sized spots on the surface of the earth. For good measure, suppose that bacteria could exist 4000 meters underground. All together that accounts for 10^{29} spots on the earth where nature could try arrangements for a bacterium's DNA.

Why did I choose the volume of a bacterium as a factor? Why limit myself to bacterium-sized spot rather than a smaller size? The reason is that DNA develops within a cell, so it's reasonable to expect that at least that much space was needed to allow for DNA to develop. DNA didn't develop without some molecular environment that would have acted as a catalyst.

Back with our numbers, there are $(3 \times 10^{50}) \times (10^{29}) = 3 \times 10^{79}$ tries per year for a successful DNA strand. Now suppose that the earth existed since the beginning of the universe, some 1.5×10^{10} years. That makes for $(3 \times 10^{79}) \times (1.5 \times 10^{10}) = 4.5 \times 10^{89}$ tries in total (but note that the earth has actually been in existence for only 4×10^{9} years).

Next suppose that every single star in the universe had a planet orbiting it like our earth (another huge assumption). How many such planets are there? Astronomers using the Hubble telescope estimate that there are about 10^{11} stars per galaxy and about 5×10^{11} galaxies in the universe. This totals $(10^{11}) \times (5 \times 10^{11}) = 5 \times 10^{22}$ planets. So all together, nature had $(4.5 \times 10^{89}) \times (5 \times 10^{22}) = 2.8 \times 10^{102}$ tries in total for the entire universe since its beginning. Let's round that up to 10^{104} tries since the beginning of the universe for the entire universe.

Next we have to figure out the total number of arrangements of DNA strands that are possible *mathematically*. My calculations must take into account the influence that molecular bonding may have on the *improbability* of various strands of DNA (more on that later in this section). But first I had to gather the following facts about DNA from the Web sites Wikipedia.com and the www.ornl.gov one (used with permission) that hosts the Human Genome Project.

It turns out that DNA is composed of a very long helix-shaped string of molecules made of those very atoms. There are four types of molecules in the DNA: adenine (A), thymine (T), guanine (G) and cytosine (C) as seen below:

Guanine   H   Cytosine

Adenine   Thymine

(This picture was taken from the Wikipedia Web site.) Their individual structure is fairly simple. Note that hydrogen, carbon, nitrogen, and oxygen atoms contain respectively 1, 6, 7, and 8 electrons and protons. Thus, their atomic makeup is relatively simple compared to other atoms such as iron that has 26 electrons or uranium that has 92. The fewer electrons an atom contains, the simpler its bonding to other atoms can be usually. So it's conceivable that nature had plenty of time to create the molecules A, T, G, and C as the earth in its first billion years had plenty of those atoms in the atmosphere anyway. Indeed, it's known that the earth contained in its atmosphere a lot of methane, ammonia, and later water. The molecular structure of those gases contained carbon, nitrogen, and oxygen respectively, and all contained hydrogen, which was very abundant in the beginning. Here are the structures of the methane, ammonia, and water gases:

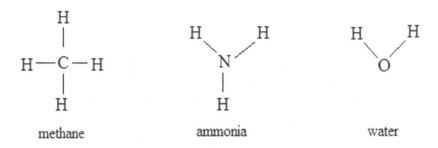

methane ammonia water

Inside the adenine (A), thymine (T), guanine (G) and cytosine (C), we can still see parts of the ammoniac molecular structure, but not the other two. Also all four types of atoms are present in the A, T, G, and C molecule bases.

What could have contributed to the creation of these more complex molecules? There must have been some violent encounter between the methane, ammonia, and water that caused their bonds to rupture and new bonds to be created that involved a mixture of all four types of atoms. It's known that the early earth had lots of powerful lightning storms and that lightning has the power to ionize gazes around it. The result is ruptured molecular bonds. These "loose" molecules are then free to bond with other "loose" molecules. It's estimated that the earth had up to one billion years to "experiment" with various bonds until it got these four molecules above (we will study in more details the events surrounding the early earth in section "Exponential "Learning"").

It's conceivable that these four molecules were formed by nature without any divine intervention. This is because many molecular structures are chemically *impossible*. In fact, the methane, ammonia, and water molecules can be formed in only *one way*: the ones in the above diagrams! Now when these molecules are mixed with other molecules, still there are only a few ways that a carbon atom binds with hydrogen atoms, that a nitrogen atom binds with hydrogen atoms, and that an oxygen atom binds with hydrogen atoms.

So what makes DNA such a complex molecular structure as to invoke divine intervention? Answer: it's not the structure of the adenine (A), thymine (T), guanine (G), and cytosine (C) molecules, but rather the very long *sequence* of such molecules that make up the helix structure of DNA. That sequence is composed of many *mil-*

lions of different arrangements of those molecules A, T, G, and C such as these:

(This picture was also taken from the Wikipedia Web site.) This one short sequence represents the code ACTG. An adenine (A) molecule can only hook up with a thymine (T) molecule, and a guanine (G) can only hook up with a cytosine (C) molecule. Note that the hookup is assured by a bond that involves a sort of modified H_2O bond— water!

In 1953, a famous experiment called the Miller-Urey experiment, named after the two scientists who conducted it, tried to recreate the conditions that are believed to have existed at the time that life first appeared on earth. The experiment started with a mix of water, methane, ammonia and hydrogen. The researchers then heated the water to simulate the Sun and ran electrical sparks through the mix to simulate lightning. They wound up with many organic compounds such as amino acids—basic building blocks of life. However the experiment fell very short of creating DNA and life. Fifty years have passed since then and scientists still don't know how life came about. In section "Molecular Structures and Life," we'll discuss further the origin of life.

The Probabilistic Model

It's well-known that DNA contains the blueprint of life. The order in which those molecules are linked determines the information contained in DNA. The sequence of those adenine (A), thymine (T), guanine (G), and cytosine (C) molecules in DNA for an organism is referred to as the organism's *genome*.

Genes are DNA chains made up of hundreds or thousands of simple molecules such as these above. It turns out that only a small fraction of the genes are there to give us our unique visible characteristics such as the color of our hair, shape of our heads, unique tone of our voices, etc. Most genes contain instructions to make another type of crucial molecule, a *protein*. Proteins include everything from hormones such as insulin (which regulates blood-sugar levels) to *enzymes* that help digest the food we eat. Some proteins turn other genes on and off, which then affect still other genes, creating complicated feedback loops. I theorize later in this chapter in the section "Codes and Algorithms for Evolution" using some simple mathematical manipulations that *feedback* systems driven by some sort of cosmic algorithms are possibly at the source of the existence of DNA.

Individual proteins are tiny cogs in incredibly complex biological systems. Consider the immune system, in which thousands of genes and proteins work together to field an army of cells and antibodies (another type of protein) against intruders. The DNA in each of the body's cells contains all the genetic information to produce a person. But in any given cell, only some of the genes are switched on, with the rest being dormant. That's what makes a liver cell different from a skin cell—different sets of genes are turned on in each.

Scientists found that the human has over 25,000 genes that flip on and off in the amazing molecular dance that leads to a human being. Moreover, the longest of all human genes is 2.4 millions links long of A, T, G, and C base molecules! This is just one gene.

The entire human genome is some 3164.7 millions links long (i.e., over 3 billion links). So, the odds of the creation of DNA of a human (or most animals) to have happened by pure chance is literally out of the universe, as will be calculated shortly. Note that the genes comprise only about 2 percent of the human genome. The remainder consists of noncoding regions whose functions may include providing chromosomal structural integrity and regulating where, when, and in what quantity proteins are made.

We now have all the information needed to calculate the odds of the human DNA to have happened by chance. Hold on to your chair: this is going to be a mind-boggling number!

Each base of the DNA strands has four possible molecules: A, T, G, or C as explained earlier. There are 3164.7 millions such codes in length in our DNA. Let's round it off to 3.2×10^9 links. That makes for a staggering 4 to the power 3.2×10^9 possible arrangements or

$$4^{3.2 \times 10^9} \ arrangements \ = \ 10^{1926591972} \ arrangements!$$

Finally, let's put all of this together: there are that many arrangements possible, but we calculated earlier that nature had time for only 10^{104} random tries since the very *beginning* of the universe for the *entire* universe. Think of it this way: it's like tossing 3.2×10^9 dies all at once the number 10^{104} times and hoping to create a good working and complete strand of DNA! Good luck won't be enough as it would take nature roughly $10^{1926591972} \div 10^{104} = 10^{1926591868}$ more "universe-time" to create the human DNA by chance only. Keep in mind again that this number takes into account *all* stars of the universe, and *all* moments since the beginning of the universe. This is about $10^{1926591868}$ times longer than the age of the universe. Another way to express this is that the odds are 1 against $10^{1926591868}$ that human life appeared by random chance!

Note that the actual *odds against* are certainly much higher still because I assumed two things in my calculations:

1. That all the planets of the universe already have all necessary molecules in place for the creation of DNA.

2. That each molecule's binding time is 10^{-43} seconds, the Planck time—the shortest time possible in nature. The actual time is certainly many times higher than that.

In fact the actual time for molecule binding is about 10^{-10} seconds per atom. So this is bad news because the odds against now go up to 1 against $10^{1926591901}$ that life appeared by random chance! These odds assume also that DNA molecules bind immediately without any time between trials, which is *not* at all realistic.

You might suggest that perhaps the universe is a lot older and a lot bigger (i.e., contains a lot more stars) than what scientists think. OK—suppose that the universe is one billion times older and one billion times bigger than is thought. This is a factor of $10^9 \times (10^9)^3 = 10^{36}$. So now the odds against are reduced just a bit to 1 against $10^{1926591865}$. Did you even notice the tiny change?

It's *abundantly* obvious that the universe didn't have the time to create by *chance* alone our DNA—very far from it. Yet most scientists believe that life appeared on earth from random events alone without any sort of source of consciousness. But my calculations make that seemingly impossible.

We must realize that something other than pure chance—perhaps some sort of "sorting-out" algorithms—played a role in the creation of our DNA by greatly diminishing the number of DNA bases that had to be tried until nature got it right. This view will be explored in more details in later sections. Moreover, it's estimated that the first bacteria appeared on earth about one billion years after its creation. This appears to be a very short time considering the odds calculated earlier of 1 against 10^{361236} for a bacteria's DNA to form by chance. This calculation is based on the fact that the simplest bacteria's genome has 600 thousands links or $4^{600000} = 10^{361236}$ arrangements.

But hold it! If you're mathematically inclined, you might spot a possible flaw with my calculations. Referring back to the figure above of the A, T, G, and C base molecules, note that in the helix chain, the adenine (A) molecule can only bind with a thymine (T) molecule and a guanine (G) can only bind with a cytosine (C) molecule. Does this greatly increase the probability of DNA forming via a random process?

No, because the molecules adenine (A), thymine (T), guanine (G), and cytosine (C) have the *exact same* kind of binding to the phosphate backbone of the DNA strands (as can be seen in the picture shown earlier). So all four codes A, T, G, and C have an *equal* chance to appear anywhere in the DNA strands. This implies that all combinations of the molecules adenine (A), thymine (T), guanine (G), and cytosine (C) are possible *chemically*, not just mathematically, and my mathematical calculation reflects that reality. It's like saying that any one number has an equal chance to appear on any one die of a toss of say, 10^{361236} dice. In other words, the probability of the binding of a base molecule A, T, G, or C is totally independent of the molecules around it.

Of all combinations of these molecules A, T, G, and C, my genes that make up my entire DNA are only *one* set of those combinations! You might respond that there are billions of people in the world and many more billions people who lived before us. As no two persons have the same DNA (except twins), millions of other DNA are possible. This is quite true. However 99.9 percent of the links are exactly the *same in all* people who ever existed and will ever exist. So, the many billions of people who ever lived don't increase much the probability of random DNA creation. We're still stuck with insurmountable odds.

My DNA replicates itself every time a new cell is created. Let me give you an idea of how accurate my DNA replication has to be. Genes can go horribly wrong, and a "misspelling" in just *one* letter of A, T, G, or C (i.e., an improper or missing link in the DNA chain) is a *mutation*. A protein called DNA polymerase copies each strand of DNA to create two double-stranded DNA molecules. A change in a single link of the thousands in a gene can produce disease, thus making me sick. Every day new cells are created in my body, yet I don't get sick because when a "misspelling" of the DNA code occurs, every single one of my cells has a feedback mechanism that can repair the DNA! The "repairman" is a complex system of proteins that come by and fix the error.

Mutations result when the DNA polymerase makes a mistake, which happens about once every 100,000,000 bases or only 0.000001% of the time. So, mutations are very rare. Actually, the number of mistakes that remain incorporated into the DNA is much lower than this because of the repair mechanism. Nonetheless, because the DNA of humans is huge, 3164.7 millions bases, this comes to about $3164.7 \times 10^6 \div 10^6 \approx 3165$ mutations per cell. Most of these are repaired as explained above. Also most of the mutations that can't be repaired take place in what's called "junk" DNA. This consists of DNA strands that are believed to serve no fonctions, and so are neutral.

Some 95 percent of the DNA is believed to be junk. Therefore only $3165 \times 0.05 \approx 158$ mutations affect functional areas of the DNA. This includes the genes. But there's nothing to worry about. Given the huge size of our DNA, the mutations that take place are negleable over the short term. Over a life time, they may cause tumors or cancer. Over a very long term—meaning a few million of years—, the mutations may lead to physiological changes in the human species. The same phenomenom applies to animals in general as well.

Now how can a protein *know* what the proper code sequence of my DNA should be, a DNA that is a few billions of links long? Answer: the protein doesn't know the whole sequence—it only reads and fixes small segments of four bases or so at the time. So not only did nature have to find time to create the DNA, it had to figure out a system to repair it. Surely you'll agree that this was an absolutely amazing feat accomplished in so little time in the evolution of life!

This mechanism resembles the feedback mechanism of our immune system. When our bodies are attacked by a virus, the brain senses that, and sends out white blood cells to fight off the invader. This mechanism requires communication to and from the brain, which is done with signals made of molecules. But in the case of the occasional DNA "misspelling," what's the signal made of? It can't be made of molecules as we're already at the level of molecules. *Solution*: the "repairman" system of proteins doesn't wait to be called to repair a strand: it checks *every* single pair of four bases of A, T, G, or C when a DNA strand is replicated. That's a lot of work—well over one billion checks for *every* single cell replicated. Absolutely mind-boggling! Now you know why protein is such an important part of a healthy diet.

As mentioned earlier, scientists know that changing the arrangement of even a very small portion of the DNA strand can result in a mutation that results in a species that dies. So it's clear that *most* of the DNA strands that can be imagined mathematically or even chemically *cannot* result in a life-form. So the astronomical calculation of a few paragraphs ago is very far off the mark of the actual number of DNA strands that are *biologically* possible.

How many combinations is that? In the case of animals, it's as many animals that can be! For instance, there are over six billion people on earth, so there are that many strands of DNA right now. If we consider also past people, that is many more billions of strands of DNA. However, as mentioned before, 99.9 percent of the links are exactly the same in *all* people that ever existed and will ever exist. In other words only 0.1 percent of the $10^{1926591901}$ different arrangements are biologically possible. But that still results in a huge number: $10^{1926591898}$ different DNAs. So humanity will not run out of DNAs any time soon. Keep this huge number in mind because it will be needed later in this chapter in the section "Codes and Algorithms for Evolution."

According to Wikipedia and other web sites, about 95 percent of the human genome has once been designated as "junk", as mentioned earlier. While much of this sequence may be an evolutionary artifact that serves no present-day purpose, some junk DNA may function in ways that are not currently understood. Moreover, the conservation of some junk DNA over many millions of years of evolution may imply an essential function. Some consider the "junk" label as something of a misnomer, but others consider it appropriate as junk is stored away for possible new uses. In any case, if that 95 percent of the human genome is ignored in my calculations, the number of different DNAs remains very high: $10^{1926591898 \times 0.05} = 10^{96329595}$. So, taking 95 percent of our DNA out of my probabilistic calculations still leads to the conclusion that the universe didn't have the time to create by chance our DNA.

Darwin's Theory of Evolution and DNA Mutations

Imagine this. The odds of life appearing by chance alone are one against $10^{1926591898}$. This indicates how incredibly complex the DNA molecule is, and how incredibly more complex is the atomic and subatomic world. People look up at the sky and are in awe of the vastness of the universe. I recall when I was a six-year-old looking instead down at the small world of insects and wondering how they viewed the world, then trying to imagine what world is hidden in the infinitely small world inside their bodies. Little did I know that the molecular structures hidden inside a single insect are billions and billions of times more complex than that of the entire universe! Indeed, my calculations in the section "The Quantum World" of Chapter 5 hints at that complexity.

We can assume by now that life didn't occur by chance, at least, not entirely. But is that proof of divine intervention? *No.* What about nature's ability to think up a way to create the DNA? Perhaps nature developed intelligence to help create the DNA structure. I'll develop this idea some more in others sections later on.

The structure of DNA is made of hydrogen, oxygen, carbon, and nitrogen atoms. Because electromagnetic forces bind these molecules in specific ways, it's quite plausible that a great many combinations of molecular structures *cannot* happen in the first place for *biological* reasons. Consequently the probabilities of DNA's molecular bases A, T, G, and C happening might be greater than we might suppose, but still be mind-boggling.

As stated two sections ago, Darwin's theory is based on these two assumptions:

1. That life happened by chance on earth rather than by design, and

2. That survival of a species is based on natural selection, a means by which good designs are preserved and bad designs are eliminated.

The first assumption has been refuted in the previous paragraphs. Let's now move onto Darwin's second assumption. Darwin's theory states that the ability to evolve is the determining factor in the survival of a species: the "survival of the fittest" as it's popularized. Most people don't question that wisdom—but I do! Darwin's assumption is flawed as it's not just the fittest that survive.

To quickly give you an idea of how flawed Darwin's principle is, consider an international tennis competition in which there's only one winner of many contestants: the "fittest" player. As Darwin's theory would have it, the other contestants wouldn't survive. But do they die? Of course not—they simply move on with their lives. Some might try the next year to win the competition. Others might change their careers and adopt other life choices; in other words they *learn* to do other things than playing tennis.

In the same way, in most cases there's no doomsday scenario with evolution or "law of the fittest." There's however a "law of the weakest" as indeed species that can't figure out how to adapt to their environment are likely to go extinct. Consider these thoughts:

1. Any species that can *learn* to adapt to a changing environment can survive, and this requirement fits *most* species. There are relatively few species that go extinct over time (except in extraordinary catastrophic events). For instance, millions of years ago the horse was no larger than a dog. It evolved, but the fact that it's large today is no indication that it's "fitter" than the small modern bird that's the descendant of the huge dinosaurs. There's also no king of the jungle: the lion doesn't evolve any better than any other species.

What Darwin seems to have failed to abstract is that a species develops a level of *consciousness*, and this consciousness is elevated out of the world. It's a nonphysical process. I'll come back to this nonphysical idea in the section "On Consciousness" and other sections. When we look at the theory of evolution, it appears that nature is able to think—and thinks up new tricks all the time.

Now I am fully aware that most scientists say that there's *no* evidence whatsoever that evolution is driven by some sort of that consciousness or learning processes. Instead evolution is driven by random events, such as random genes mutations. Nevetheless I'll propose later in this chapter (mainly sections "Codes and Algorithms for Evolution" and "Molecular Structures and Life") some arguments suggesting that evolution is partially driven by awareness on the part of the living species.

Nature had to think about how to create a living cell. Of course, this must have been a very long process with most certainly countless trials and errors along the way as nature most certainly didn't get it right the first time. Nevertheless, randomness alone simply can't cut it when it comes to evolving toward living cells. In the section "Codes and Algorithms for Evolution" I'll develop some calculations that suggest such a thinking ability might possibly be implicated. Nature is constantly trying to come up with species that have a better chance of survival. And when the idea isn't a good one, the species doesn't last long. Scientists insist that there's *no* evidence that some sort of consciousness is implicated in the process of the evolution of life. But how do they know? Consciousness is a nonphysical thing elevated out of the domain of science.

2. The fittest species are *not* the ones that win a competition *against* all other species. In fact, the reality is the opposite: the fittest species are the ones that developed *cooperation*

with other species. Here are a few examples that I collected from the television channel Animal Planet:

a. On the plains of Africa, baboons and impalas often work together. The impalas' good sense of smell complements the baboons' keen eyesight, making it hard for predators to approach undetected. A similar partnership exists between ostriches (who have keen eyesight) and zebras (who have acute hearing).

b. Soil is a complex medium for growth, bursting with organisms. Just two pounds may contain well over 500 billion bacteria, one billion fungi, and up to 500 million multicellular creatures from insects to worms. Many of these organisms work together breaking down organic matter—such as leaf litter and animal waste—while extracting nitrogen, which they convert into forms that plants can absorb. They also change the carbon into carbon dioxide and other compounds that plants need for photosynthesis.

c. When a bee alights on a flower, it enters into a symbiotic partnership with its host. The bee receives nectar and pollen while the flower gets a dusting of pollen from other blossoms of the same kind. This alliance enables flowering plants to reproduce. After being pollinated, flowers cease producing food. How do insects know that the "diner" is now closed? Flowers "tell" them in various ways: they might lose their scent, drop their petals, or change their orientation or color, perhaps becoming duller. This may disappoint us, but it's an act of great "courtesy" to hardworking bees that can now focus their efforts on plants that are still open for business.

d. Some tree ants are ferocious and will kill any insect that climbs their tree. Yet they let the aphid (a tiny insect) live on their tree. When gently stroked by the ant's antennae, aphids secrete sweet honeydew that the ants collect as food. In exchange, ants protect the aphids from predators. Just as a dairy farmer might put his cows in the barn overnight, ants often carry aphids to the safety of the ants' nest in the evening and return them to the "pasture" in the morning, usually to younger, more nourishing leaves. And we are not talking about just a few aphids, as ants may have "herds" that number in the thousands in a single nest!

e. Coyotes and badgers are usually enemies, but they decide to work together to find food. They are both fond of squirrels. The badger can easily dig squirrels out of the ground, which coyotes can't do. The trouble is that the badger is too slow to catch up with a running squirrel, but the coyote can easily outrun one. So they devised a scheme whereby the badger digs the squirrel out of the ground, then the coyote runs after the squirrel for the kill.

f. A tarantula can kill animals bigger than itself such as small birds and frogs. Yet the tarantula lets a type of frog live in the tarantula's nest. The frog need not worry because it's most useful to the tarantula. Indeed, the frog keeps the tarantula's nest free of insects that might otherwise eat the tarantula's eggs. In exchange, the tarantula protects the frog from predators. (Note that the frog's skin contains a toxic substance that would kill the tarantula if it tried to kill the frog.)

g. Our own bodies rely on cooperation. Quietly at work in our digestive tracts, there's an army of friendly bacteria helping us to stay healthy by destroying harmful invaders and by aiding in digestion and in the production of essential vitamins. In return, we as hosts provide the bacteria with food and a supportive environment, as well in the outer world the carbon dioxide that plants need for photosynthesis.

Mutual support can be seen at every level of life, from microbes to man and between similar and dissimilar species. This cooperation suggests a sort of *global* consciousness among species—a network of interconnected and interdependent organisms! If we search the Internet we can easily find numerous other examples of cooperation in nature. From these many examples, it's clear that evolution involved a required ability of the species to learn about their *interaction* with other species or organisms, big and small, and improve on those interactions. There's is little randomness about that learning process.

Looking at the enormous odds against life, it's abundantly clear that Darwin's Theory of Evolution is incomplete. Or is it? Either Darwin's theory has been refuted by my calculations or it's my calculations that are flawed. I stated earlier that Darwin's theory assumes random events, but does it really? To verify my understanding (or misunderstanding) of Darwin's theory, I decided to look up on the Internet in various sites such as Wikipedia to find out what his theory really says. I discovered that Darwin's Theory of Evolution is based on five key observations and the inferences drawn from them:

1. Species have great fertility—they make more offspring than can grow to adulthood.

2. Populations remain roughly the same size, with modest fluctuations.

3. Food resources are limited, but are relatively constant most of the time.

From these three observations, it may be inferred that in such an environment there will be a struggle for survival among individuals. Here are two additional observations:

4. In sexually reproducing species, generally no two individuals are identical. Variation is rampant.

5. Much of this variation is heritable.

Notice that randomness isn't even mentioned in any way. Clearly random events have contributed greatly to the process of evolution of the species, but it seems clear from my research that natural selection (which evolution is based on) doesn't rely on random events *alone*. Therefore, there are two fallacies with using probability theory to refute Darwin's theory of evolution:

1. That the process of evolution is random.

2. That life happened randomly (i.e., DNA happened randomly, and so life came about from "tossing DNA bases" until the right ones came along).

My own probabilistic model that my calculations are based on therefore does *not* reflect in any way the events and mechanisms that favored the appearance of life. It should be clear to you by now that the development of DNA didn't occur by chance alone like the game of tossing dice. I hope to have made it clear that probability theory is the wrong tool to employ for shedding light on the question of whether divine intervention was involved in the appearance of life or in its evolution.

The next few paragraphs on random DNA mutations seem to prove me wrong though! Not so, in my opinion. Random DNA mutations do suggest that the process of evolution is random. However, it says nothing about the life itself. DNA mutations explain how life evolved, but it says nothing about how the DNA came about in the first place, that is, how life came about. In any case, below is a simplified description of a famous experiment conducted in 1952 that placed at the forefront the theory of random mutations as an explanation for evolution.

On that year, Esther and Joshua Lederberg performed an experiment that helped show that many mutations are random, not directed. Directed mutations say that mutations that are useful under particular circumstances are more likely to happen if the organism is actually in those circumstances. In other words, mutation is directed by what the organism needs—some sort of consciousness causes the mutation. For instance suppose that the climate gets colder, and the species need to find a way to keep warm. Their fur would gradually grow thicker over the years and generations—this could be directed mutation. There's little evidence to support directed mutations. I maintain nonetheless that some mutations are directed, and I'll make a case for it in section "Codes and Algorithms for evolution."

In this Lederberg experiment, they capitalized on the ease with which bacteria can be grown and maintained. Bacteria grow into isolated colonies on plates. These colonies can be reproduced from an original plate to new plates by "stamping" the original plate with a cloth and then stamping empty plates with the same cloth. Bacteria from each colony are picked up on the cloth and then deposited on the new plates by the cloth. Esther and Joshua hypothesized that antibiotic resistant strains of bacteria surviving an application of antibiotics had the resistance *before* their exposure to the antibiotics, not as a result of the exposure.

Despite my probabilistic calculations of the previous section, experts in the theory of evolution persist and claim that laboratory experiments have proven that evolution is carried out almost exclusively by *random* mutations of the DNA. I believe too that random mutations played an important role in the evolution. Moreover I'll present arguments later on that make this role clear and obvious. My point nonetheless is that there's more to the story than that as the next few sections will argue.

Wikipedia defines mutation as "changes to the nucleotide sequence of the genetic material of an organism." The nucleotide consists of molecules that comprise the structural units of the DNA. So then it goes from its definition that the DNA of species was changed over millions of years by *mutations*. The question is did these mutations occur almost entirely at random, or is there another mechanism involved here that isn't random? I'll gradually develop my arguments in some of the upcoming sections. For now, I wish to suggest that the word "random" mutations may be a misnomer. I'll clarify my view in section "Codes and Algorithms for Evolution."

Note that mutations can be beneficial, neutral, or harmful for the organism, but scientists claim that mutations don't "try" to supply what the organism "needs." In the section "The Probabilistic Model," I calculated that there are some 158 mutations per cell replication in the human body. Because we don't get sick, it's clear that nearly all those mutations are neutral and totally harmless. Even the mutations that are either harmful or useful may in some cases not manisfest any effect until some other random external event—such as a change in climate—triggers the effect. An example will be given in section "Codes and Algorithms for evolution."

I propose that some sort of intelligence or consciousness is involved in the theory of evolution and the DNA. OK—I know that researchers have discovered a lot about how life evolved, and they *never* needed to introduce this mystical notion of consciousness to explain the processes involved. Nonetheless you'll agree that human beings do possess some level of consciousness and intelligence. Even some animals possess such capabilities to a lesser degree. According to the theory of evolution, these capacities have evolved over the many thousands of years. When was the last species that didn't possess consciousness? Where do we draw the line? How can we know that a species doesn't possess awareness and intelligence? These questions should also belong to the theory of evolution.

So, in the next few sections, I wish to shift your attention from the "hardware" of life, that is, the DNA structure to the "software" of life, that is, its purpose and the consciousness that it possesses. The intent is to make you aware that there's more to this universe than its physical content i.e atoms and molecules. The universe is more than the some of its physical parts.

A Purpose Led to Life

Even though the probabilistic argument can't be used to prove or disprove that the appearance of life in the cosmos is of divine intervention, it certainly puts serious doubt into the theory that life appeared in a *random* fashion in spite of that famous Lederberg experiment described in the previous section.

As the evolution toward life is so astronomically unlikely, if only scientists could find complex life forms on another planet, then the case of creation by God could be strong because the existence of complex life forms on two planets would be so improbable that it would strongly suggest divine intervention. Having said that, note that single-cell life forms might not be as unlikely as first thought as my other calculations in the section "Codes and Algorithms for Evolution" of this chapter will suggest.

Perhaps rather than calculate probabilities—or rather improbabilities—we should ask: why would God create life? If you don't believe that God is implicated then ask instead: why would nature create life? We need to search for a *purpose* for life as there must have been a purpose for this creation. As stated in the section "What is God?" of Chapter 7, God is logical and so wouldn't create something for no reason. So what could that reason be? One element of an answer stems from the realization that life couldn't be possible without a reduction of entropy in a localized region such as the earth (and perhaps in other solar systems as well).

What's entropy? This question leads into a complex discussion that we'll deal with in more detail in the section "Entropy Versus Organisms" later on in the chapter, but for a start consider the following: in order to create life, a planet on which life could flourish had to be created and made favorable. A sun had to evolve so that this planet could revolve around it in a favorable orbit. A galaxy had to evolve so that this solar system could exist inside it.

Entropy is a measure expressed by the Second Law of Thermodynamics that roughly says *"Energy spontaneously disperses from being localized to becoming spread out if it is not hindered from doing so"* (taken with permission from the Web site http://www.entropysite.com/students_approach.html). But this law isn't favorable to the possibility of life because in a global sense of the universe, it implies that galaxies, solar systems, and all will be destroyed eventually *if* not hindered from doing so. However, via the creation of celestial bodies such as described above, an *orderly* system could be created some time after the Big Bang.

I suspect that the Second Law of Thermodynamics prevailed in the first few thousands of years after this event—it most likely had to in order to disperse energy from the Big Bang—but then electromagnetic forces hidden inside light came into play. These forces eventually created subatomic matter, then atoms, then molecules, then lumps of matter, then matter large enough to generate gravitation, which led to the evolution of solar systems, etc. and finally life.

These forces would have been rather useless if they appeared in such a way as to cancel each other out, so difference in forces had to be ensured. This must have been the beginning of *diversity* in nature. God's thought must have been that diversity in nature (not just in life but in the physical laws also) was necessary to ensure a reduction in entropy. For instance:

1. As you learned in Chapter 8, in the first second after the Big Bang, the four forces of nature, gravitation, electromagnetism, the weak and strong nuclear forces were just one. Then they split into separate forces, thereby assuring "force diversity" in nature.

2. We see on earth a wide variety of species, and this is what ensures the endurance of life. If only a couple of species of plants existed, should they be attacked by a species of insects, their survival as a species would be in jeopardy rather quickly.

I might add this regarding the diversity by way of the creation of physical forces. There was a purpose in God's thought to create light with *electromagnetic fields*. Without electromagnetic fields, the universe would never have shaped to anything. Most physicists will agree to that. Moreover, in the section "Our Virtual Universe" of Chapter 10, I theorize that electromagnetism might be the manifestation of a thought from the Creator of the universe. I know that this seems very absurd, but there are arguments behind this idea. Read on.

On Consciousness

Why do I bring up the notion of consciousness? One reason is that the many examples of cooperation between species presented in the section "Origin of Life and DNA Structure" suggest a global consciousness (i.e., that species have an understanding of their environment). Another reason is that life and consciousness are closely related. This will be explained mainly in section "Molecular Structures and Life" coming up. A last reason has already been given at the end of the section "Darwin's Theory of Evolution and DNA Mutatio": consciousness is implicated in of the theory of evolution.

What's the requirement for having a consciousness? Before I attempt an answer, recall that in the section "Reality Versus Abstraction" of Chapter 2, I suggested that size, velocity, and time are abstract. The notions of mass and matter are also abstract. Now don't tell that to a construction worker pushing a wheelbarrow of gravel! Again it all has to do with the fact that matter is made of waves and that nature is sensitive only to them. Essentially, matter isn't a "thing" but a *state of organization*. This is much like an ice cube. At the molecular level, it's water all the time, but in a state of organization different from liquid water when it becomes ice. At the molecular level, nature doesn't realize that ice and water are different.

Really, matter is just a *manifestation* of the mind (i.e., an invention of the mind very much like when watching a three-dimensional movie). As stated in Chapter 1 and other places, physically matter is just a bunch of tangled-up waves. So to nature, matter is at the same level of reality as wireless (invisible) communication with our cellular phones. What makes the difference is the state of organization of the waves of either case. This requires an ability of interpretation on the part of nature of what it "feels." It might be that nature has realized that need.

Think about it: matter is made of electromagnetic forces. Suppose we hold a magnet over a nail. The nail lifts up, but we can't see or feel anything as electromagnetic force is invisible. So if nature never created matter as "balloons" of tangled-up electromagnetic waves, the universe would be visibly empty. I believe that this abstraction of matter, in a philosophical sense, is the source of the Heisenberg uncertainty principle that we discovered in the section "Another Formula for Fuzziness" in Chapter 5.

Velocity, size, length, and mass are all things that humans experience, but nature's view of these boils down to waves "interacting" one way or another. As an analogy, consider the software development division of IBM and all its computer programmers. Some of the software products that IBM produces are so large that not a single engineer or programmer has a complete understanding of the end product. Programmers might only be aware of the coding they produce every day without having a concept of the big picture. That is the way it is with nature: it feels the tiniest waves, but has no concept for the bigger picture. Actually, this isn't entirely true: nature has a concept of the bigger things, but this occurs at a higher level of consciousness. There are different levels of nature: nature at the quantum level, nature at the molecular level, nature at the biological level, nature at our level, etc.

As another analogy to explain abstractions, suppose a two-dimensional world of random dots:

The dots are the "things" for a being that exists at that level. Suppose a superior being were to arrange them as follows:

The being in the two-dimensional world would have a hard time deciphering the pattern. But suppose that they recognize the pattern *N*, and attach a meaning to it. The letter is then no longer just a set of things, but rather a state of organization expressing a *concept*. However, a concept isn't a "thing." It's the same with matter, velocity, and so on—they are concepts elevated out of the physical world.

There's also a unit of measure to determine the distance between the dots. This unit may seem rather useless and become clear only after the pattern is regarded completely *out* of the system. This is the case with motion in quantum steps. The significance of the quantum unit, the Planck length, becomes evident once we realize that it's necessary for motion to be possible. At the quantum level, nature is barely aware that motion takes place, hence the fuzziness equation $\Delta s \Delta v$ that I derived in the section "Space–Time Fuzziness" of Chapter 2.

So concepts, representations, or ideas are all more than the sum of their parts, just as the letter N is more than a set of dots. But what makes them this way? Answer: their *purpose*—and a purpose is something elevated *out* of the physical world. The purpose of the letter N is to compose a word. A word is more than the set of dots that make it. Once a purpose is detected, it's evidence of a level of consciousness of the system. Here are two other examples of purpose:

1. I can't help but draw a parallel with DNA, which is a sequence of bases represented by the letters A, T, G, and C. The meaning of these codes is also *out* of the physical world: they are at the biological level—biology is not physics even though it depends on that. Just the same. The existence of DNA has nothing to do with physics, although its complex molecular structures are entirely based on the laws of physics. The evolution toward DNA is clearly evidence that nature elevated itself out of the physical world. So then there's equally a purpose to the creation of DNA.

2. In just the same way, there's a purpose to matter. If nature didn't interpret those "balloons" of tangled waves as matter (recall Chapter 1 and the section "The Quantum World" of Chapter 5), the universe would be nothing more than intangi-

ble and shapeless energy. Clearly, nature had to realize the need for that purpose.

Here's another analogy regarding state of organization. Suppose I traveled back 500 years ago and took an airplane with me. People looking at it will see two wings and a few other parts associated with a plane, but they'll have no clue what it is until I tell them what it's used for—that is, its purpose. Once they know that, they might deduce that the wings resemble those of a bird and therefore that it's an object used for flying. Thus, the object is more than the sum of its parts. Once people know that the plane is for flying, they most likely would also deduce without me having to tell them what the wings and the propellers are for. How would they be able to do that? By drawing a parallel to the shape of birds, as making connections with already-known ideas allows people to elevate themselves to a higher level of consciousness.

The examples above bring me to propose that the same process of elevation of consciousness occurred with nature as well. But clearly such a process requires a minimal level of intelligence. Indeed—and I'll give arguments later to support my theory that nature possesses intelligence! This sounds crazy, but read on and you'll see that it makes sense. Also, nature isn't physics. Consequently, nature isn't the universe—nature is a part of the universe. (You might recall that I deduced that already in the section "The Beginning of Nature" of Chapter 8.)

The physical environment is itself abstract, or at least relative. For instance, to a bird the air is solid because it's sufficiently solid to enable it to move around. Water in a lake is solid as far as a fish is concerned, but to us, none of these two environments would allow us to move or live in them. So if you ask a bird, a fish, or a human what solidity means, you'll get different answers. Finally, as stated in Chapter 1, at the quantum level, solidity has no meaning at all. Solidity is the result of an interpretation; therefore, solidity is an abstraction.

So it seems that nearly everything is abstract as far as both nature and humans are concerned. But surely there has to be something that's real! Energy is real, isn't it? Yes, it's real. But wait a minute:

as it depends on the speed of the object and its mass—both of which are relative and abstract—shouldn't energy be abstract too? No, because energy is at the *lowest* state of organization, very much like the dots making up the letter *N* are at a lower state. The only thing that nature truly feels as real is energy *transfer*, and this happens when energy passes from one location to another at the quantum level (but never above that level). This realization has profound consequences such as regarding the nature of gravitation. Quantum energy is the only thing in nature that isn't the result of some interpretation.

Note that, because to nature the size or matter of an object is abstract, nature isn't terribly interested in *objects*. Nature is more interested in *events* that occur as a result of a transfer of energy. When it comes to objects (matter), nature is all butterfingers! Nature doesn't grasp matter very well and thus it tends to "drop" matter once in a while (or "sinks" into the fabric of space). This is a philosophical explanation as to why quantum particles skip in quantum steps, and the consequence is the Heisenberg uncertainty principle (fuzziness) that I talked about many times in Chapters 2 to 4 mostly.

I just conveyed the opinion that reality is relative, not absolute. As we imagine our reality based on our senses and interpretations from them, there are as many realities as there are people in the universe! This doesn't imply that people's realities don't have things in common as they do. For instance, other than those with vision problems, every single person in the world sees the sun and the moon the same way. So the sun and moon are real as far as people and animals are concerned, not imaginary. However nature's interpretation of the sun and moon differs from ours. Our world of actuality is made up by our *mind*. Is there a "real" reality? Consider all people in the world, each having a mind. There's clearly an overlapping of the realities as perceived by those minds. The intersection of those is what we might call "reality."

As for humans, when it comes to social interpretations, reality isn't clear-cut at all. How does Joe Smith know that his interpretation of social reality is correct? What's accepted as reality is nothing more than what's considered the social norm. Statistics theory says that the norm is determined by the 95 percent of the population that belong to a specific group. The other 5 percent of the population are out of the norm. Without realizing it, every day people use statistics to determine what moral values or behaviors are considered normal. It would be interesting to conduct a study to determine if the social norm as *perceived* by humans corresponds to the statistical norm of 95 percent. Our perception of our environment shapes our consciousness.

Animals and insects too interpret what they sense. Their reality is surely not the same as ours. As there's no real reality, we have to imagine reality by *interpretation*. The act of interpretation is evidence of a level of consciousness because the subjects have to realize that a *proper* and *logical* interpretation is necessary for making sense of their environments in order to even *survive*. A person unable to make sense of anything would become so stressed as to die quickly. So consciousness is very necessary for survival, and any organism that shows evidence of not wanting to die demonstrates a level of consciousness. If a unicellular organism tries to fight off an invader, it shows some consciousness. Survival and evolution is *not* entirely dictated by random DNA mutations.

Evolution of a species requires a consciousness not only at the level of the individual of the environment in which the creature lives, but also at the level of the group. Evolution is the cultivation of consciousness and intelligence. A creature that understands more of its environment increases its ability to adapt and evolve. This observation greatly explains why life evolved so quickly which is what the calculations made in the section "Codes and Algorithms for Evolution" will suggest. More will be said about reality versus illusion in Chapter 10.

Of course since a species has no control over its environment, awareness on the part of the species for the need to gain an ability to adapt isn't enough. Certain amounts of luck (such as already possessing favorable random DNA mutations that agree with the environment) play a very important role.

Ever since I learned in my college biology course school that a living being is made up of proteins, enzymes, and DNA and that all of them are simply constituted of complex molecular structures, I wondered: *what is then the definition of life?* My body is a factory of chemical reactions. Then what could be the difference between those chemical reactions and the ones between inorganic molecular structures in a laboratory? Is it the complexity of the molecular structures of the proteins, enzymes, and DNA and their interaction that defines life? I don't think so. What distinguishes the living from the inorganic is that the living has a level of consciousness. I believe that *all* living things have a level of consciousness. (I'll discuss this idea some more in the section "Molecular Structures and Life" coming up a few sections from now.)

A living thing has consciousness if it meets one or more of these criteria:

1. It shows evidence of making sense of its environment and interacting with it. This means that the living has to interpret input which then becomes information.

2. It demonstrates a will to live, such as wishing to consume food.

3. It demonstrates a will to replicate (or reproduce).

4. It evolves.

5. It's able to produce thoughts.

6. It demonstrates some level of intelligence.

A species doesn't necessarily have to have all six characteristics, but the more it possesses the more aware it is and the better its chances for evolution or survival. Note that criterium (4), the capacity to evolve, is actually a consequence drawn from the other criteria. Based on this list, it's safe to say that a rock on the beach isn't a living thing. The water washing on the shore of that beach isn't a living thing either even though it's the basic for the creation of life. Finally, the complex DNA molecule isn't a living thing even though it's essential to life!

Earlier, I proposed that nature has intelligence and a consciousness. Should that mean that nature is a living organism? The jury is still out on that one, but I dare to suggest that indeed nature might be an organism, depending on the level. Most people will consider my suggestion as totally absurd. Keep in mind that we are part of that organism, so it's very difficult to be objective about it. It's indeed a problem to be objective about a system when we are deeply embedded inside it.

Remember though that there are different levels of nature, and my suggestion here of living doesn't apply to all levels. It's obvious that nature at the quantum level isn't a living organism. However, it may manifest some very low level of thought processing. How dare I suggest such a silly idea? Consider this:

1. I already gave an example in Chapter 4 of consciousness at the quantum level: the "ticking" of time. In Chapter 7, I proposed that time is a thought from the Creator. In Chapter 8, I suggested that the time dimension was created by nature at the time of the Big Bang. This is evidence that nature, at the quantum level, captured that thought of time from the Creator. As to capture a thought requires an understanding of it, the quantum world possesses some level of consciousness, although only an extremely primitive one. Einstein's space-time theory states that the time dimension truly exists. If so, then there *has* to exist some consciousness of the quantum level to interpret that time.

2. In the section "The Thoughts–Energy Dimension" in this chapter, I'll provide a philosophical proof that thoughts manifest themselves first at the quantum level. Based on the list of six criteria above, nature at the quantum level does meet one of the criterias. So quantum events follow some level of consciousness. Note that the list of six criteria doesn't imply that a conscious entity is a living thing. It's the other way around: a living thing possesses consciousness. So if it lives, it has some level of consciousness, but not necessarily the other way around.

For instance, a set of thoughts doesn't constitute a living entity of course, but it can demonstrate a level of consciousness, which is something *elevated* out of the physical world where the living reside. As in Chapter 10, a set of thoughts will be proven to be nonmaterial (i.e., not part of the physical world), it may then manifest a consciousness even though it doesn't have life in a physical sense.

Scientists have discovered that unicellular organisms can learn about their environment. This requires a certain level of consciousness. Is consciousness possible without a brain? Yes, apparently so. Therefore, not only are the brain and consciousness not the same thing, consciousness doesn't require a brain! What crackpot idea, right? Not so.

Keep reading, and eventually you'll see that, yes, thoughts may be generated without a brain. This will be made more evident in Chapter 10. For a start, consider this: scientists have discovered that bacteria have a sex. "Sex is crucial because it scrambles genes much more quickly than random DNA mutations and helps new species to evolve and emerge as the environment changes. Even bacteria have a primitive form of sex, where they exchange genes to add variety to their entire genetic complement, or genome" (taken with permission from http://www.telegraph.co.uk/earth/).

There are plenty of examples in the fields of immunobiology, genetics, and neurobiology that cellular organisms (not only neurons) do have a form of memory and an ability to learn hence some primitive "thinking" capability. I'll take one example only: the immune system.

Certain types of immune cells called memory B and T cells store information on certain foreign structures (usually proteins called peptides) that they've encountered. If they're challenged with this very specific piece of information (that is, from bacteria or viruses) later on they jump-start a rapid and adaptive (hence an ability to "learn") immune response. But they can't do it all by themselves because other cells have to provide information by presenting antigen through specific receptors on their surfaces, or they have to contribute specific activating factors such as soluble interleukins or various kinds of surface molecules.

So, cellular organisms do have a level of consciousness, but much more so as a *group* that individually. This proves that *cooperation* takes place at the cellular level.

What this system has in common with the brain is its ability to learn which also requires "memory" to compare a current input to a previous one. This memory is stored in one specific cell (in the pattern of active genes in either the immune cell or the developing organism) but has to be activated (immune system) or established by the interaction with other cells.

This is, to a degree, similar to what happens in the brain although neurons are part of a network that is much more prominent.

There's another thing that has no brain: nature. Yet, nature has to interpret what it senses. In particular:

1. It had to create time and motion of matter (not to be confused with transfer of energy, which happens without requiring any consciousness). So nature has a consciousness for the necessity for these things without which the universe wouldn't go very far.

2. Particles are nothing more than nature's interpretation of energy in matter. So can we infer from this that nature has a consciousness? Does nature realize that if it didn't properly interpret energy that matter couldn't exist and that nothing tangible could move other than energy?

   Reflecting back to the section "Skipping through Space" of Chapter 2, we found that nature had to interpret energy in such a way as to allow it to skip in space; otherwise motion would never be possible. Where does this ability of nature to interpret come from? It surely can't be from the physical world. Could God be the source of nature's consciousness? Are the laws of quantum physics the blueprint of nature's consciousness? As we'll see later, it appears that quantum physics might be a repository of thoughts and consciousness.

It's clear that there are different levels of consciousness. It seems that the more complex biologically an organism is, the higher its consciousness. But it's not that simple as the level of intelligence definitely has a part to play. Learning is impossible without consciousness, and the latter is impossible without the former: they go hand in hand. Consciousness is like a tutor that supervises the living being.

However, our consciousness can become limited in two ways:

1. When we decide that what can't be measured or proven with experiments has no validity (in other words, the sum is no more than the sum of its parts), as this limits our possibilities for interpretation. For instance, as some of my ideas in this book can't be proven or have never been proven scientifically, most scientists would surely immediately discard them as crazy. This reaction may get in the way of a higher consciousness.

2. When the learned knowledge becomes so automatic that it enters the unconscious. Here are two examples:

    a. The way that we are raised leads to core beliefs that become hidden in our unconscious minds. This can limit our awareness. Another example is that humans are now so used to live with electricity as a power source that this dependence has entered our unconsciousness. Should electric power disappear (from some catastrophic event), humans would be quite lost to the detriment of our survival.

    b. Animals including humans have no conscious control over the structure of their DNA. However, scientists have lots of evidence that evolution gradually changed the DNA of animals. It must have been that in a very distant past when only unicellular organisms existed, they had some conscious control over their DNA. At that time,

DNA was most likely a relatively simple molecular structure—the DNA we know didn't create itself in one instant! Its molecular structure was itself a result of evolution. Consequently, some level of consciousness existed at the molecular level. Now one billion years later, this consciousness is lost on animals and humans. Our cells replicate themselves "mechanically" without thinking, without consciousness. It remains though that this replication process *had to be learned* in a very far distance in the past. I know that this idea seems totally crazy to scientists, but I have arguments in support of this later in this chapter, mainly in sections "Codes and Algorithms for Evolution" and "Molecular Structures and Life".

As an analogy, consider the invention of the automobile. While it was in the process of being invented, the inventor had to think deeply about how to construct it. The creation of the first automobile was the result of a high level of consciousness. However, the replication process of automobiles in a factory requires a much lower level of consciousness. The factory workers just have to learn how to put the pieces together. The worker doesn't need to know why the automobile is replicated in a specific way. The worker doesn't even need to know what a car is for! Nowadays, automobiles are replicated by robots. Therefore, there's no consciousness at all involved in the construction of automobiles and the process is now confined in the domain of the unconscious.

I hope to have given you some reasons to believe that some level of consciousness exists at all levels of nature. More will be said about the topic of consciousness later in this chapter and the next one.

The Second Law of Thermodynamics

In the section above, I introduced the idea that nature has consciousness. One purpose of consciousness is to fight off extinction. But how can nature fight it off? Answer: by reducing entropy of a localized system where conditions for life are favorable ... and this idea takes us back to the topic of the section "A Purpose Led to Life." Entropy is closely related to the Second Law of Thermodynamics, so before proceeding any further, we need to spend some time learning in more detail the concept of entropy and the Second Law of Thermodynamics.

That law isn't well understood by many people, even by some scientists! The worst description I've seen stated "A way of looking at the second law for non-scientists is to look at entropy as a measure of chaos or disorder" (from the web site http://en.wikipedia.org/wiki/Laws_of_thermodynamics). The Second Law doesn't mean disorder, nor does it necessarily imply it; the word *disorder* is misused. However, the reference to entropy by Wikipedia somewhere else on that same Web page is correct: "The entropy of an isolated system not in equilibrium will tend to increase over time, approaching a maximum value at equilibrium." Wikipedia defines *entropy* as "a measure of the *un*availability of a system's energy to do *work.*" Essentially, entropy quantifies what the Second Law describes. Entropy measures the dispersal of energy: (a) how *much* energy is spread out in a particular process, or (b) how *widely* spread out it becomes.

It may be easier to understand the Second Law if we bring in the First and Third Laws as well, which state:

1. *First Law*: Energy can be changed from one form to another, but it can't be created or destroyed. The total amount of energy and matter in a system remains constant, merely changing from one form to another.

2. *Second Law*: In all energy transfers, if no energy enters or leaves the system (assuming an isolated system), the *potential* energy of the state will always be less than that of the ini-

tial state. In the process of energy transfer, some energy will dissipate as heat. The potential energy is the one that is available to do work. An implication is that as entropy increases, the potential energy decreases.

3. *Third Law*: As temperature approaches absolute zero Kelvin, the entropy of a system approaches a constant minimum greater than 0. In other words, the system can never reach zero Kelvin.

These three laws work together. Essentially:

1. The First Law is a statement of the conservation of energy.

2. The Second Law is a statement about the direction of that conservation.

3. The Third Law is a statement about never reaching absolute zero (0 degree Kelvin).

These three laws should now feel a bit more intuitive to you. In fact, the Second Law of Thermodynamics is so much a part of our everyday experience that we never notice its involvement! Here are a few manifestations of that law:

1. A hot pan cools down. This is a case of an immediate dispersal of energy because the pan tends to disperse the energy of its fast-moving particles (commonly called heat) in its metal to anything it contacts such as the cooler room air (slower moving molecules that then increase their speed somewhat).

2. Gasoline explodes. This is a case of a hindered dispersal of energy from a chemical reaction because it tends to react with the oxygen in air, but does so *only if* the mixture is ignited (i.e., given a little extra energy from a spark or frame). Then the gasoline (carbon-containing substance) and oxygen O_2 can spread out some of the energy in their chemical bonds (those that hold atoms together in molecules) in forming car-

bon dioxide (CO_2,) and water (H_2O) that have lesser energy in their bonds. The *difference in energy* is dispersed to *all* the molecules in the gaseous vapor (leftover oxygen, nitrogen, carbon dioxide, water, carbon monoxide, etc.).

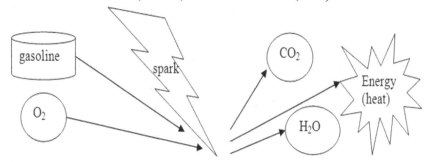

This makes them move extremely fast (due to heat) and the pressure in a confined space immediately increases. Such a high pressure in the small cylinder volume further spreads out the energy of the hot molecules (Second Law again) by pushing the car's pistons down forcefully so that the piston rods disperse their kinetic energy by turning the crankshaft, etc.

3. Rust develops on an old nail. The molecule of oxygen (O_2) reacts with the iron (Fe) in the nail. The result is that the oxygen bonds are broken, thereby releasing electrons. As a result a lot of energy is released. Some of that energy and those electrons are captured by the iron atoms to produce the "rust" molecules Fe_2O_3 or Fe_3O_4 depending on the availability of the oxygen. But there's still energy left over that is dissipated.

4. A tire blows out, or cream mixes in coffee. Those are cases of substances spreading out their energy in space without any change in that original energy. So, this seemingly disparate example involves no change in the temperature or change in the initial energy (no sparks or explosion). (Actually, in the case of the tire, real air does cool down on expansion.) Scien-

tifically, these two spontaneous events are in the same category:

a. Expansion of a gas such as air involves volume increase (a spreading out of the original energy in a larger space).

b. A mixture involves a separation of molecules of one type from its kind (a sort of dispersal of the initial energy because all the molecules of the same kind are no longer adjacent to one another).

The sense of "if not hindered" in some of these examples should not be lost on you. It's the omission of this idea of "hindrance" or "obstruction, momentarily or for a long time" that leads most people astray in their understanding of the Second Law. So much emphasis is usually placed on the immediacy and the inevitability of the dire effects of the Second Law that it can be made to seem threatening to almost every aspect of our lives.

The Second Law of Thermodynamics isn't an instantaneously obeyed law. It accurately predicts the *probability* of the dispersal of energy that is localized or "concentrated" in a group of molecules or atoms—and that can result in undesirable events ranging from serious accidents to disastrous forest fires or to our ultimate deaths. In this sense, the Second Law is one of disorder and decay.

However, the law:

1. Is completely silent about how long it may take for its predictions to be fulfilled.

2. Is silent about the *mechanism* that tries to dissipate energy, thereby increasing entropy. In other words, execution of the Second Law may happen in so many different ways and at different levels of nature.

3. Is silent about the origin of the energy that's involved in the mechanism. This implies that the Second Law applies to *all* types of energy (e.g., it doesn't apply only to vapor).

The Second Law isn't always about doomsday scenarios because we have to look at the open systems rather than just the localized, isolated system. This was the case with the example earlier with the gasoline tank explosion of the gases inside it. The tank is the localized system, but the open system includes the pistons that make the car move.

The Second Law can manifest itself in advantageous ways, in particular when we look at global scenarios on the scale of the whole earth. Consider these:

1. In the open system of earth and sun and outer space, we have the privilege of taking advantage of the Second Law for human benefit, as nature does too for maintenance of its high-energy content ecology on the earth. We do this by *diverting* part of the energy to our purposes as it's *dispersing* when a spontaneous process follows the second law. An example has been given in the list above (item 1: our use of gasoline in our machines). The payoff is our use of the Second Law for our human goals. It's not just diversion of some of that dissipating energy from the burning gasoline but diversion of the energy flow of the fossil fuel to make engines and machines that transformed our material world.

   Of course, we aren't able to divert more than a portion of the energy obtained from combustion for our use. Some of any energy dispersion continues immediately on its way to complete dissipation in the environment out of the tailpipe of the car following the initial explosion of the fuel. The Second Law may be delayed but it's never violated: eventually there's dispersion of energy. It's just that it doesn't include *all* the available energy.

2. The sun's radiant energy disperses as it strikes water molecules in the ocean and causes them to move more rapidly (i.e., the water becomes warmer and evaporates more readily). This is called the water cycle that we learned in science class. In this process of dissipating the sun's energy, many

tons of water is raised in the air, which creates clouds as some of the water molecules spread out part of their energy to the cooler upper atmosphere. When the sun's energy is dispersed in striking the earth's surface and heating it, some of it is shadowed by clouds.

The uneven warming of land and water causes variable columns of warm air rising and increases random air motion. The results are winds that further diffuse the original energy of the massive air movement. Water in the air that was in the form of clouds cools radically as it starts to flow over high mountains or encounters cold air and precipitates as rain, adding to lakes and creating stream sources at high elevations. Of course, this gives potential energy to such streams because they are far above sea level. Water flowing from heights dissipates its potential energy by flowing downward (for instance by cutting ravines and/or forming canyons).

Here are two ways we humans profit or are impacted by the sun's radiant energy dispersion:

1. We take advantage of water flow in rivers (dispersing their potential energy as they flow down toward sea level) to turn turbines connected to electrical generators that produce electrical power for us (further diffusing the potential energy of the flowing water). Winds dissipating their energy in turning windmills attached to generators also produce some electrical power. These are a few of the actions by which nature, in following the Second Law, provides us with fresh water, variable breezes, snow-capped mountains, and higher-than-sea-level water that drives our turbines and generates electricity.

2. Occasionally, movements of wind and warm moisture from a tropical ocean can cause a concentration of energy to form a hurricane. (More heat in the rising moisture from the warm ocean surface has been fed into the incipient circling vague wind pattern, and by chance the weak Coriolis force from the earth's rotation aids the circling a bit and more moisture is

sucked up. Thus, a huge vortex begins and sweeps more warm moisture into it. Of course, the observer of a destructive hurricane can't sense the basic contributions of solar energy or the complex energy dissipation from the warm ocean surface to cooler upper air that coincidentally formed it.)

The "death" of a hurricane is a more obvious example of the Second Law in action. Unless this kind of ocean-originated storm is continuously fed thermal energy from warm waters to maintain its high-energy existence, a hurricane spreads out its energy and dies down. The Second Law always is a valid tendency and—in dynamic cases like this—demonstrates that tendency in a relatively short time rather than years.

We use or take advantage of the Second Law by diverting energy from its "downward" (dispersing) flow to run engines that aid our transportation across the earth and into space. However, neither we nor nature ever are actually *defeating* the Second Law. Energy spontaneously disperses *if* it's not obstructed from doing so, such as when diverted by us or by nature. Note that nature makes use of the Second Law by means of developing *cyclic* processes. This was clearly shown with the water cycle picture above. In the next section, we'll see that cyclic processes are also used by living organisms to make good use of the Second Law. Have these organisms learned this trick, or has it all happened through random events?

Entropy Versus Organisms

The previous section covered a few phenomena in which the Second Law may be to our advantage. In this section I present phenomena where that law may be a constant threat to *life forms*, but how nature turned it into an advantage.

Many people including even scientists state that entropy is a measure of disorder. They note that living cells are not disordered and so have *low* entropy. So the Second Law of Thermodynamics is therefore *contrary* to the appearance of life. As the examples below will show, the flow of energy in living cells maintains order and life—it has to. The Second Law of Thermodynamics is hindered by biological mechanisms. Entropy wins when organisms cease to take in energy and die. There's some truth in associating order to low entropy when it comes to living forms. Nonetheless, the examples below demonstrate that there are cases where the Second Law isn't hindered but rather taken advantage of by organisms developing *cyclic* processes.

To achieve reduction of entropy, order must be created out of disorder characterized by high entropy. Order is somewhat the reverse of entropy. So, in the evolutionary process, I propose that a *conscious* decision must have been made by nature to create organisms that have low entropy. A unicellular organism (which is highly organized) attempts to lower entropy. How does an organism lower entropy? To answer, I sought inspiration from the Web site www.entropysimple.com/content.htm (used with permission).

Our bodies are made up of tens of thousands of chemical substances (compounds) that are essential to our functioning. However, the oxygen in our atmosphere that we need to live also tends to destroy almost every one of those essential biochemical compounds. Why? Oxygen plus any of our essential organic compounds have total energy content higher than the oxidized compounds of carbon dioxide and water (CO_2 and H_2O) that would be formed from them. Thus, if the Second Law were not somehow obstructed, almost all the substances in our bodies would *immediately* react with oxygen in the air so that some of the bond-energy in those essential substances would spread out as heat:

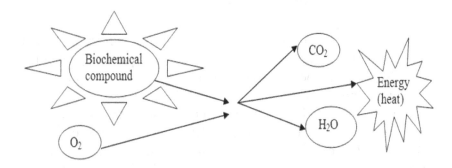

From concentrated or localized energy to diffused or dispersed energy—that's the pattern in nature that the second law sums up. This is exactly similar to the scenario in the previous section in which gasoline and oxygen had higher energy in their bonds than do their products, carbon dioxide and water. However, the reason we never spontaneously oxidize (combust) as rapidly as does gasoline in oxygen is the large amount of *water* H_2O throughout our cells that restrains such a process. Wood in the trunks of living trees burns (oxidizes) slowly and with difficulty because it's both solid and wet—in contrast to faster burning leaves and branches with less moisture (and more surface area). Nevertheless, even if our whole body didn't quickly oxidize, we might have a sufficient number of cells in us (say a hundredth of a percent of our total of critical cells to our life continuance) that could randomly oxidize and follow the second law in dispersing their concentrated energy. That could be enough to cause serious dysfunction and death.

Fortunately, there's a profound reason (more profound than the reason given above of the large amount of *water* H_2O throughout our cells) that our cells and their chemical constituents resist the threat of the Second Law (i.e., that they "must" react with oxygen because then they would follow the law by dispersing their energy). The reason is the existence of activation energies, an innate *obstacle* to the Second Law of Thermodynamics in chemical reactions. We've seen it present in our illustration of gasoline and oxygen: no reaction occurs until a spark or flame is first injected in the mixture to give a little energy "push" to start it. This is typical of almost all reactions of oxygen with the biochemicals within us. Thus, even though the Second Law is a fundamental threat to our lives in our atmosphere that contains oxygen, it's equally fundamentally *obstructed* by *activation energies* from causing our oxidation and death.

This oxidation process occurs in astoundingly complex ways and in many steps (so any energy that is spread out as heat is slowly and moderately released unlike the seemingly "one-step" instant and explosive dispersal of energy when gasoline reacts with oxygen). Furthermore, any heat that is dispersed in our bodies isn't wasted because it keeps our bodies warm to function optimally even in a cold environment. Some of the energy flowing downhill from food oxidation is captured by "coupled reactions" so that a medium-energy substance, adenosine diphosphate (ADP), is *raised in energy* to become a greater energy-containing substance such as adenosine triphosphate (ATP), a complex molecular structure. ATP is in every cell in our bodies to disperse energy for a multitude of different reactions while it becomes ADP, then it's regenerated by another coupled reaction, and the cycle goes on:

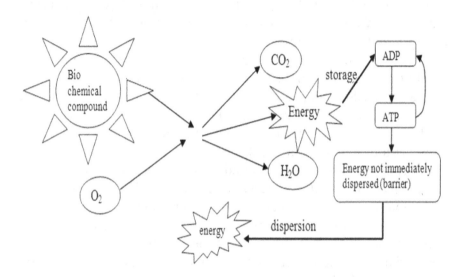

Of course, the storage of ATP is contrary to the predictions of the Second Law, but we know how it "beats" the Second Law: the energy within the bonds of those ATP molecules is kept from being dispersed by *activation energy barriers* until our cells need it for some action. ATP and similar energy-storage sources are what give us the instant conscious choice of using our arm muscles for work or our eye muscles for looking in a particular direction—or our brain for thought.

Many of our about 30,000 chemical substances and the complex cells from which they are made must continually be destroyed and the residues excreted as new ones are synthesized. There can't be minutes in which oxygen is *not* supplied to the energy-requiring cells of the heart or pumped to the energy-requiring brain: we die from a heart attack if adequate oxygen isn't given to its cells and our brain will be permanently damaged (or if too many minutes elapse without oxygen for energy dispersal and ATP synthesis, we'll die). The Second Law—or better, the energy flow predicted by the Second Law—is essential to all life.

So we now have an answer to the question of how an organism lowers entropy: it devises a pair of complex molecules ADP and ATP that store energy, preventing it from dispersing. Note that these molecules don't defeat the Second Law; they just delay its execution. You'll note that the ADP/ATP pair is involved in a cycle. It seems that wherever there's life, *cycles* are devised by nature to sustain that life. We saw that in action earlier with the planet's water cycle. Without that cycle, could life flourish? The next paragraph shows another example of a cyclic process.

In general, the photosynthetic process uses Second Law dispersal of the sun's energy similar to what we humans do with fossil fuels. Plants take some of the sun's dispersing energy (plus carbon dioxide CO_2 from the air or animals as they breathe out and water H_2O from the air) and make new chemical compounds in the plant that are more complex and more energy-containing than the original carbon dioxide CO_2 and water H_2O. Meanwhile oxygen O_2 is released:

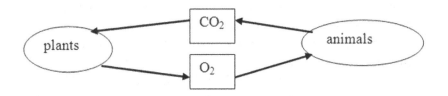

Subsequently those new active chemical substances in the plant, in complicated processes form carbohydrates, some amino acids, fatty acids, and thousands of other compounds by a myriad of other reactions—but also dissipate some energy in all of these secondary syntheses as heat. Overall in the plant, the "downhill" process of energy being dispersed from the sun is diverted and then coupled with an "uphill" process of concentrating energy in new plant substances but there's no violation of the Second Law: only about 30 percent of the downhill solar energy has been captured to synthesize new substances in the primary process of photosynthesis. The net overall dispersion ("loss") of energy (70 percent) is *still greater* than the concentration ("gain") of energy (30 percent).

The overall energy pattern is similar to our driving a car uphill. This may seem to be contrary to the second law for a moment because we've "created" great potential energy by ending with a heavy car at the top of a hill. However, calculations quickly show that far more energy has been dispersed from changing the chemical bonds in the gasoline and oxygen to carbon dioxide, water, and heat (to make the pistons, gears, and wheels move) than the potential energy that the car acquires by being at the top of the hill. In the huge number of processes more complex than driving a car uphill, photosynthesis uses or diverts only some of the downhill Second Law energy flow to create the "uphill" substances and supply the energy for the growing plant to continue to function.

As with many of the examples described so far, living organisms have developed mechanisms to *recycle* energy forms, thereby reducing entropy or at least keeping it in check. This isn't limited to living cells. An ecosystem takes advantage of recycling as well as was shown in the previous section with the water cycle. The recycling ability applies also in outer space. For instance, when a star dies, it often explodes. The dust thus produced is eventually recycled to make new stars and solar systems. This is how our own solar system was born. So when something dies, it's used as an energy source by another organism. This recycling ensures that entropy remains low and that both parties survive.

At the same time that we realize the Second Law of Thermodynamics to be a constant threat, we should not see it as our "greatest enemy": What if the direction of energy flow was *not* always from being concentrated to being dispersed? What if the process was often erratic or precisely 50–50 —with energy flowing in reverse, that is, from being dispersed to being concentrated half the time? No organism could live in such a universe. As organisms, we are basically energy-processing machines.

To maintain the multitude of different "automatic" biochemical processes in our body that require energy to force substances "up an energy hill" to have more concentrated energy within them, we must continually have energy supplied to us from *outside* ourselves in the form of oxygen and food. It's the always-dependable direction of spontaneous energy dispersion from that oxidation of food that makes possible the total range of our energy-demanding activities as well as our very lives themselves.

So, living forms have found ways to use the Second Law to their advantage whenever possible, as well as ways to hinder the Second Law when it's not to their advantage. This suggests that living forms developed a level of *consciousness*: they had to realize that dissipation of energy had to be *controlled* as otherwise the living organism dies. It seems that nature had to learn this trick. This evolutionary result can't be attributed to pure luck, not just because of the complexity involved, but because of the *meaningful* result obtained: life elevated out of the physical world. In other words, life isn't of this physical world! Section "Molecular Structures and Life" later on will develop an argument that should convince you of that.

Let me use an oversimplified analogy to explain entropy versus organisms. Suppose you're playing a game of bridge. You thoroughly shuffle a deck of cards, and then deal them randomly. This is equivalent to dispersing energy. Entropy is at its maximum if the shuffle is totally random. To reduce entropy, you would have to *consciously* sort out the cards using some "sorting-out" algorithm. The algorithm in this case consists of the rules making up the game of bridge. The rules of the game tend to favor cards to become somewhat sorted out as the game progresses.

By creating celestial organisms (e.g., galaxies) and biological organisms as per the many examples given, entropy at the celestial level as well as the molecular level is reduced. Then, conscious decisions lower entropy or at least keeps it in check. In the case of living cells, the "sorting-out" algorithm involves those "activation energy barriers" described earlier.

Sometimes it's impossible to reduce entropy. For instance, suppose that an artist creates ice sculptures in a park. The park has low entropy because its structures were created with a purpose in mind. When spring comes, those sculptures—however beautiful—will melt away. Entropy has increased to its maximum. In this case, this disorder isn't reversible. So order doesn't always win over entropy. For instance when we die, entropy wins.

In the case of the creation of the universe, at the beginning, the universe was terribly chaotic with extremely energetic quantum particles flying from all directions. It has to be that this high entropy must have been *reversed* at the correct time (and not too late) so as to ensure that the entropy effect *could* be reversed. At this point, you should begin to realize that there was a conscious entity involved here to fight off entropy. Do you remember the fundamental constants of nature listed in section "Should the Universe Even Exist" of Chapter 8? It is the *very precise* values of these constants that enabled the reversal of the high entropy of the early universe. These constants ensured the evolution toward an orderly universe.

I suggested earlier that the electromagnetic field hidden in light was the property that enabled nature to create order out of disorder. Light somehow bent and created closed-in loops to create electromagnetic particles: matter (recall the discussion on matter in Chapter 1). These loops created both positively and negatively charged particles. These oppositely charged quantum particles (protons and electrons for instance) attracted, thereby attempting to increase entropy. But by having electrons buzz around the nucleus of the atom rather than scatter all over, nature obtains controlled entropy: the Second Law is hindered. When electrons are excited, the result of which is either the release of visible light or the production of electricity, the Second Law wins but then looses back again when the electrons return to their original energetic state. Overall, the energy level of the atom is never lost and remains available for work. Atoms are nature's way to keep the entropy in check.

Electromagnatic polarity is such that it wants to attract positively and negatively charged particles. This ensures motion. But wait a minute—doesn't motion increase entropy? It depends on what the motion does with the energy in the system. The answer is in the affirmative if that motion causes reductions in the amount of energy that can still do work within the system (such as when sturring a cup of coffee). Otherwise the answer is negative: entropy is reduced (this is the case for a unicellular organism or any electronic gadget that runs on batteries). The entropy would be increased if the oppositely charged particles were to merge, nullifying the charges. Remember that the Second Law wants to reduce the energy available for work.

We find so many things in the universe that are based on polarity:

1. Electric charges (as mentioned above)

2. Magnetic poles

3. Nuclear forces

4. Sexes

All these forces control entropy *if* they are in states of organization. If there is disorder, then polarity looses the fight against the Second Law. Note two things:

1. There exist only two poles, positive and negative, and females and males. There's never an extra pole. Why? Because the addition of another pole would complicate the system, potentially adding to the entropy. Imagine how complicated society would be if there existed a third sex. But note that neutral entities exist. Some particles have no charge (e.g., neutrons), or have no mass (e.g., photons), or some animals have no sexual organs. These entities neither contribute to entropy nor take away from it. They can serve however as a catalyst to the system which may help it run better. For instance pouring lubricating oil into an engine helps it run more efficiently. But the oil is a neutral agent in that its purpose isn't to run the engine.

2. Gravity isn't in the list. As far as physicists know, no antigravity force exists, so gravity isn't really a polarized force (this doesn't surprise me because in Chapter 6, I say that gravity is a differential of space density). Should I conclude that gravity doesn't play a role in reducing entropy? No, because, as I explained in Chapter 6, gravity is the result of a difference in energy densities in space. Such a difference reduces the entropy (disorder) of a system.

If there were no difference in energy anywhere in the universe, nothing would ever happen; nothing would ever have been created. It would be like going to a conference where everyone has the exact same knowledge and expertise as everyone else so all talks would be useless. But if someone entered the conference with either a higher level of expertise (or the reverse, a lower level of expertise—a sort of differential of energy or polarization), dynamic discussions among members of the conference would slowly emerge with that new person. Polarization is a necessary *stimulant* for physics and for the evolution of life as well. Note that the evolution of life and variety of species increased greatly once sex was invented by nature. Sexual poparization served as a stimulant.

These sorts of polarizations are crucial in reducing entropy because it creates differences in energy levels that are necessary for constructing organisms. Polarization helps keep things rolling. Take for instance a battery. Its negative and positive poles ensure that the current travels through the car, keeping it in "shape." Energy would be useless without these polarizations. As mentioned earlier, entropy is a measure of the amount of energy no longer available to perform work. Thus, the purpose of polarization is to reverse the effect of entropy, and vice versa.

Electromagnetic forces are also needed to form molecular bonds. DNA is made of molecules—anything organic is made of molecules! So, without electromagnetic forces, there would be no life. Put another way, without light, there would be no life. Light is the source of life. Doesn't the Bible state that or is it in my old high school biology book? Electromagnetic forces make bonding of atoms easier. For instance hydrogen, carbon, nitrogen and oxygen atoms can't be bonded in very many ways. There are only a few specific arrangements of these atoms thatn can be put together. This sort of "sorting out" of bonding must have greatly reduced the number of bases in the DNA structures that nature had to try before it got it right. Don't think that nature created the four bases G, C, A, and T in an instant. There most likely were untold attempts with other bases, but they failed.

Therefore, DNA wasn't entirely created by pure chance: electromagnetic forces were perhaps one of the "sorting out" algorithms invented by nature to make life more possible. But there was also a biological "sorting-out" algorithm that helped even more. As will be calculated in the section below, without such *conscious* "sorting-out" algorithms, life would probably *never* have occurred. These algorithms are based on feedback and cooperation between species. Many examples were given earlier in this chapter in the "Darwin's Theory of Evolution and DNA mutations."

Reducing entropy is necessary for more than *survival*—it's necessary for *evolution* as well. In fact, survival is rather uncertain without evolution. The more varieties of species there exist, the better for evolution. Variety is the reverse of entropy, the sorting out of the melting pot.

But what mechanism could encourage variety? Answer: sex. I provided a hint a few paragraphs ago. For a few billion years, single-cell organisms split up into two cells to procreate. There were no sexes. Other than the odd mutation, this didn't allow for a variety in the genes. Evolution led to the invention of sex to assure a wide variety of species and at a lot faster rate than mutations alone could do. Nature gave plants sexes as well so that plants could grow in a varied number of environments. Note that sexed plants appeared on earth at the precise time that an increased growth of plants was most needed: during the period of the giant plant-eating dinosaurs. This suggests that there existed a feedback mechanism between the two:

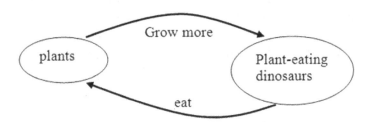

Even though nature had figured out a way to create organisms (celestial or biological ones), nature wasn't done yet! These organisms had to be in *equilibrium* and remain that way. This is clear because a system out of equilibrium would inevitably increase entropy and possibly collapse. Think of a house of cards: if one card out of equilibrium falls, everything collapses. To ensure equilibrium, a *feedback* system has to be developed (more on that later below).

This recycling is analogous to cooperation between organisms. In the next section I'll use the concepts of equilibrium and feedbacks to propose another feedback system to help use see how life and the DNA might have come about.

Yet another mechanism that reduces entropy is the ability for different organisms or species to *cooperate*. Numerous examples of this were given earlier in the section "Darwin's Theory of Evolution and DNA mutations."

In summary, the Second Law of Thermodynamics says that "Energy spontaneously disperses from being localized to becoming spread out if it isn't hindered from doing so." I propose a Law of Consciousness that says: "Nature's level of consciousness gradually increases, creating more complex structures that if not hindered from doing so, eventually lead to life and intelligent life."

It's clear that the Second Law and my Law of Consciousness are in a constant struggle against one another. But as this section showed, the Law of Consciousness can never eliminate the Second Law—it uses it to its advantage. Nevertheless, in the end, the Second Law wins: life on the earth will eventually vanish.

There appears to be a constant struggle between physics and life: physics wants life out of this physical universe while, with its ability of awareness, life keeps fighting back. In Chapter 10, I propose the existence of a hidden universe of Consciousness. That nonphysical universe and this physical universe are in constant struggles with one another. But the will of life to manifest itself into the physical universe is so strong that it eventually wins hence life here on earth.

Codes and Algorithms for Evolution

Being a college professor in information technology, I see a similarity between an organism or an environment and a computer. The hardware corresponds to the physical makeup of the organism. The coding in forms of 0s and 1s correspond to chemical messages sent to various components of the organism. These messages convey different meanings just like different sequences of 0s and 1s convey different meanings inside the computer. The software is what makes the hardware function. The software uses a language and a set of algorithms that tell various components of the organism how to co-operate for its well-being. The execution of the software depends on the information available in the computer.

This information corresponds to the DNA of the organism. Indeed, DNA is represented by a sequence of letters that dictates how the organism functions. DNA is coded information, and the software (algorithms) controls organic chemistry and structure. DNA isn't the software; it's more a structure used by the software to tell the organism how to function.

It's interesting to note that, while the DNA differs from one species to another, its backbone structure is the same for all and, it's theorized, since the very beginning of life on earth. DNA is based on a language unchanged for a few billion years! Could it be a subset of a language nearly as old as the universe itself, a sort of cosmic language? Or could DNA develop entirely differently on another planet?

Suppose as a computer programmer, I developed software that was capable of learning from the input I gave it. It would be software with artificial intelligence. It seems that DNA profited from such intelligence that came from nature's consciousness. There's plenty of evidence that species are capable of adapting to the environment, provided that environmental changes occur *very slowly*. Some parts of a species (toes or whatever) that are no longer needed eventually disappear and they no longer figure in the DNA or would be turned off (essentially dormant). These are genes that are no longer needed.

This means that DNA must have gradually (indirectly) received messages from the environment, and this translated into understanding that a certain gene was no longer needed. Such learning is also crucial in keeping entropy low. Indeed, suppose otherwise: that the species couldn't learn from its environment. It would become out of phase with it and so would have difficulty living in it. Everything would be phased out, and so entropy would increase. The consciousness of the species dictates that it should be able to learn so as to ensure survival. So the DNA continuously receives feedback from its environment. This whole system is pictured below:

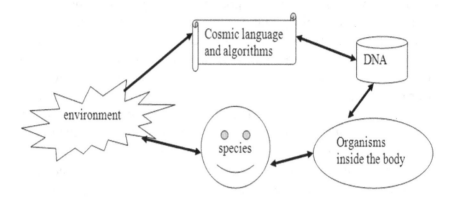

With such a feedback, the "cosmic algorithms" can ensure that species are in "tune" with their environment. Note that by the very nature of a feedback system, nature can't create species "out of the blue"—to the dismay of creationists, no such thing as instant creation of complex life-forms is possible. Feedback systems rely on continuous flow of energy (information) a *little bit* at the time. This feedback system allows the species to evolve, but *very slowly*. Should a huge catastrophe occur, such as a comet slamming into the earth, most species wouldn't have time to go through this feedback system fast enough to survive. The only species that would remain would be the lucky ones that happened to be equipped with genes that are tuned to the changed environment.

Now, experts in the theory of evolution will straightaway refute my model of conscious feedbacks on the basis that a species that adapts to an environment does so, not from consciously changing its DNA, but from favorable *random* DNA mutations that just happen to be present into their DNA. So, it's just by pure luck that the species' DNA was tuned into the changing environment. Species with a DNA that isn't tuned in to the environment just went extinct or had to move somewhere else where the environment was more agreeable with their DNA. This is called undirected DNA mutations, and there's a lot of evidence in support of this view. Essentially a species has no control over its DNA. I am in total agreement with that. I nonetheless continue to believe that some cosmic language is at play here, and that some kind of very subtle consciousness drives that language. I'll provide later examples to support my view.

The cosmic language is essentially mathematical and is used by nature in the form of thoughts to structure everything in the universe such as DNA codes. Everything is influenced by thoughts in some sense as will be demonstrated in Chapter 10 with my idea of what I'll call the "hidden universe" which is an undetectable universe of thoughts and consciousness.

The dynamics of the system above goes in three layers roughly as follows:

1. Consciousness of the macroscopic species level:

    a. The species observe the environment.

    b. The species learn about their environment.

    c. The species live in their environment, that is, interact with it.

2. Consciousness of the nature level:

    a. This physical interaction is fed into the organisms inside the body.

    b. The biology of the body realizes that further adaptation is needed (evolution).

    c. This leads to changes in the coded DNA. Note that the species are not conscious of these changes.

3. Unconsciousness:

    a. The new DNA is replicated in every cell.

    b. The new DNA is created every time a new offspring is born.

    c. Mutations (mostly random ones) change the DNA sequence.

Of course the whole process is a lot more complicated than this. I'm not a molecular biologist so I'm not able to provide further details.

Note that the only part where the species is conscious in this evolution process is in layer 1. The species is *surely* unconscious as to what happens at the DNA level. The downside of this unconsciousness is that if the environment changes, the DNA can't adapt rapidly and so the species is vulnerable. For this reason, it's reasonable that species best adapted are the ones with a DNA that just happen, by pure luck, to already be adapted. I said "by luck" because scientists possess a lot of evidence that such favorable DNA mutations occur randomly. In other words, there's no consciousness whatsoever in the construction and modification of the DNA. As you may have noticed by now, I don't agree entirely agree, and reasons will be provided later in this section and the next two sections.

The whole process of adaptation to the environment is *faster* at layer 1, the macroscopic level, and extremely *slow* at layer 3, the unconscious level. Given that the "modern human" evolved some 60,000 years ago, it probably takes at least many more thousands of years for the DNA to change even the slightest. However, not all changes are microbiological ones as some are based on learned behavior at the macroscopic level. For instance, when an animal hides for the winter or migrates, this is a learned behavior communicated from the previous generation to the next one. But then perhaps it's a DNA mutation that explains this behavior. We don't know. It's difficult to tell the difference between a learned *logical* behavior and one that is the result of a DNA mutation.

Note the interactions in the system depicted above:

1. *Chemical interactions*: these are the interactions between the environment and the species, and the organisms (cells) inside the body and the DNA inside the nucleus of the cells. I theo-

rize that at this level, directed DNA mutation might possibly occur. The awareness for DNA mutations would come from the cellular level because this is the environmental layer closest to the DNA molecular structure.

2. *Logical interactions*: These are the interactions with the cosmic algorithms (essentially the consciousness of the species and nature). This includes the interaction between DNA and the cosmic language and algorithms.

Now, why do I suggest that DNA is implicated with the logical aspects of the system? Because scientists have discovered that the DNA is a coded system of its own. DNA is a molecular structure assured by connectors labelled adenine (A), thymine (T), guanine (G), and cytosine (C). These are coupled in various combinations making the familiar helix structure. These codes ATGC are arranged inside the DNA in billions of different ways making up the genes. These arrangements not only can be viewed as codes, but *are* indeed codes stored using molecular structures. These are no different from the codes stored inside a computer. Another example of logical interaction (this one at the macroscopic level) is the cooperation between species that I talked about earlier in this chapter in the section "Darwin's Theory of Evolution and DNA mutations."

It stands to reason that these codes were created based on some level of consciousness of the system depicted above.

In the picture above, note that the arrow goes both ways between the DNA and the species: (a) random mutations in the DNA may affect the species such as their behavior or their physiology. These are called *undirected* DNA mutations; (b) changes to the physical environment affect the species, their behavior or physiology thereby eventually causing changes into their DNA. These are called *directed* DNA mutations. It was expressed in section "Darwin's Theory of Evolution and DNA Mutations" that researchers have never found any evidence whatsoever in support of directed mutations. Here are five examples of evolution of species, the first three clearly reflecting undirected mutations whereas the last two might be cases of directed mutations, in my opinion:

1. Where I live, there are wild rabbits with brown fur in the summer, then white fur in the winter. Here's one way to look at this. The rabbits were conscious for the need to adapt to the environment and this led to the necessary change into their DNA. This would be an example of directed mutation of their DNA.

    But there's one problem with this view: the necessary changes into their DNA could not possibly have occurred quickly enough. Imagine it. The rabbits were not adapted to their environment for perhaps a few generations until the DNA finally caught up with the change. How could the rabbits learn to change their DNA? The theory of evolution states as a rule that species that are better adapted to their environment increase their odds for survival. A species, that has to wait generations until its DNA mutates favorably, doesn't meet that rule, and so it will likely not survive. Species always try to learn to adapt to their changing environment such as changing their behaviour if need may be. But in the case of a physiological change, survival to the changing environment is possible only if that physiological change is, just from pure luck, *already* present into the DNA. This is a case of undirected DNA mutations—the rabbits didn't learn to turn their fur white for the winter. What happens is that the mutation caused the rabbits to run out of melanin, a substrance that gives color to their fur, when the temperature plummets.

    As for the other species of rabbits without this mutation, they remained easily visible on the snow, and easy prey. So they either went extinct or moved where the climate was warm. There's a very easy way to realize that species don't learn to change their DNA. Many species with fur, such as cats, don't have this ability to turn white in the winter. It's not that they've never thought of that strategy, it's just that their DNA doesn't have the necessary mutation.

2. Another example is with the giraffe. Its neck got longer not because it wanted to reach the top of trees to eat leaves but because a mutation in its genes caused its neck to grow over many generations. Then the giraffe slowly realized that a longer neck allows it to reach the leaves. Physiological changes led to changes in the animal's behavior and its understanding of the surrounding. It's easy to realize that this is a case of undirected DNA mutations. Suppose the opposite: directed DNA mutations. This would imply that the giraffe realized that, to reach the leaves on top of the trees, it needed a longer neck. Then this consciousness eventually led to a favorable change to its DNA. This makes no sense. How can the giraffe have conscious control of its DNA? There's another way to see that directed mutations didn't occur: many other species of animals around would likely also have realized that a longer neck would also allow them to reach more leaves up the trees. Yet only the giraffe grew a longer neck.

3. Cats are carnivores and rabbits are herbivores. Yet they are both in the same branch of evolution: mammals. Are cats carnivores because some day they just happened to kill and eat a mouse, and liked it so much that they consciously decided from then on to always eat meat only? A similar question could be asked of rabbits. It is absolutely clear that cats and rabbits did not consciously decide to become carnivores and herbivores respectively. The only explanation is that cats are a kind of mammals that acquired a DNA mutation that made them carnivores. A similar explanation holds for the rabbits.

4. Scientists know that feathers evolved over many millions of years to become perfectly aerodynamic today. Paleontologists know that the first feathered animals had perfectly

symmetric feathers, that is, the quill shaft was positioned along the center of the feather:

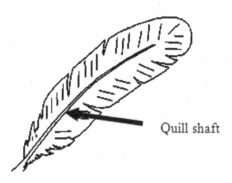

Quill shaft

These feathers had the wrong shape to allow for flight. Scientists know that birds that can fly have asymmetric feathers where the quill shaft isn't centered giving the feather a shape perfectly adapted for sustaining flight. Is it the bird that learned to shape its feathers this favorable way? Of course it didn't—birds know nothing of aerodynamic principles. So clearly this is another case of undirected DNA mutations that led to a perfectly aerodynamic feather—right? The trouble is that the DNA doesn't know anything about aerodynamic principles either. Should we accept outright that somehow by pure luck the correct DNA base sequence of the codes adenine (A), thymine (T), guanine (G), and cytosine (C) appeared into the complex DNA structure?

In the case of the rabbits in item (1) above, the mutations were a lucky coincidence because the mutations resulted in the substance melanin normally present in the hair to fade away. The result was that the hair turned white which conveniently matched the color of the snow.

In the case of the birds, should we believe that some random DNA mutation led to the perfectly aerodynamic feathers? Then the birds realized that they could fly. In this case, I opt for a directed DNA mutation. I suggest that the birds were already trying to fly by climbing up trees with the help of their claws and then glide. The flow of air around the feathers *slowly* led to mutations in the genes dictating the shape of the feather. Contrary to the case of the white rabbits that absolutely needed a *quick* adaptation that was assured by a mutation already present in their DNA, the bird did not require such quick adaptation because it was already adapted to its environment despite that it could only glide. So a very slow directed DNA mutation is plausible in this case while still respecting the rules of evolution. But what's the mechanism that allows such a physical interaction to lead to changes to the DNA?

I propose a mechanism *elevated out* of the physical world as follows. The DNA is a set of codes, information essentially. Information leads to more information thereby evolution. So then the physical interactions are perceived by the living as... information! But the living isn't conscious of that. The consciousness comes from a nonphysical source. Information is *interpretation* of data—in this example, data is air flowing around the feathers. If air flowing by a feather is information then there has to be some sort of mechanism that *interprets* the air flow. Interpretation is necessary because it turns raw data into useful information. Since the DNA is information then an interpretation of a physical input can lead to a change into the DNA.

Many might refute my argument by saying that air flow will change the shape of anything over time. That is true. For instance, air flow will change the shape of a flag perhaps. But we are talking about life here. It's clear that physical interactions are captured as information by a living being – even though the living isn't aware of it. Finally an ability to interpret its environment requires *conscienceness*. In this case, this consciousness doesn't come from the brain of the bird, but from some other source (perhaps each cell in the feathers) then the information became elevated out of the physical world. In Chapter 10, I call that world of consciousness, the "hidden universe". That world consists of system of thoughts, and thus includes the cosmic language and algorithms mentioned earlier.

I propose in this chapter and the next one that living cells possess a level of consciousness. It is this capacity that is elevated into that hidden universe. There has to be some sort of algorithms in that hidden universe that process input from the physical environment and produces a translation in form of DNA mutations. Here's a picture of the process:

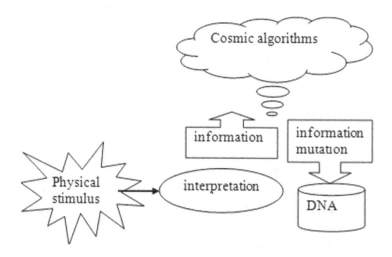

This sort of directed DNA mutations is driven by cosmic algorithms in the universe of consciousness (see Chapter 10). Since that universe is nonphysical, its influence on the physical universe is impossible to detect. Consequently the directed mutations will *appear* to us as undirected random mutations even though they are directed.

5. Most animals possess an ability to sense their environment. For instance, they possess eyes, ears, and a nose to allow them to see, hear, and smell their environment. Scientists will say that these abilities are the result of undirected random DNA mutations. Indeed. Did the animals wish to have the ability to feel their environment, and then the necessary DNA

mutations came about to fulfill their wishes? Of course they didn't. Animals have no conscious control of their DNA. So it has to be that undirected DNA mutations allowed for this ability of sensing the environment. The trouble though is that the DNA knows nothing about the physical environment in which the animal lives in. The DNA is in a completely different world than the one that the species lives in. The DNA is just a molecular structure containing codes. Should we accept that mutations of the DNA appeared, and by some luck, these mutations reflected what the animals needed? How can the DNA produce random mutations that eventually gave animals eyes with lenses that follow optical physical properties so well? How can the DNA produce ears with an acoustic mechanism that follows so well the laws of vibrations when the DNA knows nothing about physical laws?

In the examples four and five, the DNA mutations are—in my opinion—directed, and are not some random undirected mutations. According to the definition offered by scientists, directed mutations are mutations that reflect wishes from the living being— the living being has conscious control of its DNA. I believe that the definition should be much wider. Where's the consciousness, that causes directed mutations, coming from?

In the next section "Molecular Structures and Life," I show that life was created from the hidden universe, the world of consciousness. As I'll show in Chapter 10, this mysterious universe consists of thoughts and consciousness that make their way into our physical world at the quantum level. So that's where the consciousness comes from when directed DNA mutations occur. There exist in that universe cosmic "sorting-out" algorithms for all kinds of things such as the requirements of the creation of life, and perhaps even the rules for evolution.

Scientists will remark that directed DNA mutations have never been observed in laboratories. They *never* will be simply because of the assumed origin of directed mutations as pointed out in the previous paragraph. So, they cannot be detected at all from scientific means. Consequently, even if directed mutations were detected, they would appear as if undirected! Also if directed DNA mutations do exist, then they most certainly are mutations that require a very long time to occur, meaning over numerous generations. The reason is that they are the physical result of *awareness* on the part of the species for the need for a change in the DNA. If such awareness is possible, then nature has three major obstacles that cannot be overcome in an instant:

1. *The awareness itself.* Is the species' level of consciousness sufficient to realize what needs to be changed such as in its physiology? For instance if the climate has been changing for a few generations, do the species understand that they need to grow fur and store more fat to keep warm?

2. *DNA mutations.* Even if the species were to possess sufficient awareness, what mechanism would allow this to cause a mutation into the DNA? How can consciousness affect a DNA sequence? As stated in section "Codes and algorithms for evolution", DNA mutations occur at an unconscious level of the species.

3. *Survival.* Does the species need the desired DNA mutation within a short term or perhaps even right now? If so, then a directed DNA mutation would take too long a time to occur, and so in the meantime, the species' survival would be at risk.

These three obstacles are so great, especially the last two, that directed DNA mutations cannot possible explain evolution. This declaration is in agreement with scientists who claim to have proven that evolution is driven by undirected, random mutations.

Having said this, however great the three obstacles listed above is, it does *not* prove that directed DNA mutations don't exist. Perhaps the awareness that can enable directed DNA mutations isn't at the species' level but rather at a level closer to the DNA level. So perhaps the necessary awareness is rather at the cellular level! You might recall that I mentioned this possibility already before. A cell could alter its own DNA when its environment changes. Is that possible? Is there any evidence?

In the next section "Molecular Structures and Life," I'll use another argument to suggest that numerous directed DNA mutations probably occurred when life first appeared.

There are different types of mutations. The ones that I've considered all along are called *insertion* mutations where new DNA bases are constructed. The DNA underwent a huge numbers of other types of mutations. There exist three other types of mutations—substitution, deletion and and frameshift mutations. Because the number of bases of the DNA of species has remained stable in the past many millions of years suggest that indeed insertion (and deletion) mutations are very rare. Substitution and frameshift mutations occur much more often. These two types of mutations don't really create brand new bases—they rather "shuffle" around DNA codes.

The realization—according to experts—that random mutations (which is mostly caused by substitution and frameshift mutations) are a sufficient factor to explain evolution of life is no reason to conclude that nonphysical factors did not play a role. In the next section, I'll show that the creation of life has a nonphysical cause. Consequently, it's quite plausible that evolution is partially driven by nonphysical factors as well. Biological changes can occur as a result of thoughts. What thoughts could they be? More will be said about the importance of thoughts in this chapter in the section "Exponential "Learning"" and in Chapter 10.

The term "random mutations" is a misnomer. It suggests that evolution is driven by random events. How can random events have led to the evolution of life? The DNA is a long sequence of adenine (A), thymine (T), guanine (G), and cytosine (C) bases. That sequence isn't random. It's a coded message that dictates the look and behavior of species. Since life makes sense, so does that coded sequence. What's random is the combination of the DNA sequence and the physical environment in which the species live. Some environments will favor some DNA sequence S_1 while other environments might favor some other DNA sequence S_2. Combining all this with random mutations, you end up with the wide variety of species inhabiting the earth, hence evolution.

What about the many examples I gave in the section "Origin of Life and DNA Structure" of cooperation between species? Is this learning a result of a change inside their DNA? In some cases this is, but in other cases, it's simply learned cooperation that is communicated from one generation to the next. The examples make it clear that all living things develop some ability to think, some more than others. Even bacteria show an ability to adapt and cooperate. Thus, should we conclude that bacteria can think? Based on my point of view on evolution: yes! Of course this doesn't imply that bacteria can think up complex tasks to do, as they can only think of what's meaningful to them in their microscopic environment. (In other words, don't expect bacteria to grow little hands any time soon so that they can learn to write.) Bacteria live in simple, primitive environments. The feedback they receive from their environment is so simple that they've never felt the need to evolve in the last billion years!

This brings me back to the extremely remote odds of our DNA having formed by chance as was calculated in the section "Origin of Life and DNA Structure." It's rather evident that some *consciousness* had a role to play in the creation of life on earth; especially should scientists find life on another planet (because that would be even more unlikely). However if life really did appear by pure chance, then the astronomical odds of 1 against $10^{1926591898}$ calculated in the section "The Probabilistic Model" would bring me to conclude that life *does not exist anywhere in the universe* except here! If life does exist elsewhere, then the likelihood that life appeared by chance there and here is so slim that we would have to conclude that life had to have been designed by some higher intelligence. Could we be the only planet that supports life? I can't conceive of such a possibility, given the many billions of galaxies out there. Life probably didn't occur entirely by chance here on earth or any other planet out there. This is the belief conveyed in section "A purpose for Intelligent Life" of the next chapter.

Nature thought up a "sorting out" algorithm and feedback similar to the one depicted earlier to greatly reduce the number of arrangements of the A, T, G, and C base molecules to be tried. Did electro-magnetic forces "force out" most of the arrangements? No, because each molecule A, T, G, and C has the same molecular binding with the DNA backbone as mentioned before. So in this case, this electromagnetic force had little to do with the "sort out" thinking. In this case here the "sorting-out" algorithm has to be a *biological* one.

This means that nature at the same time had to assemble all the necessary molecules to make up the proteins that manage DNA, the proteins that repair DNA, and all the other things that make up a cell. Due to the mind-boggling odds calculated, it's clear that nature didn't create the necessary ingredients of a living cell in a linear sequence such as this:

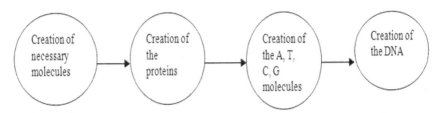

The evolution toward creating a cell must have occurred at a speed that was more or less *exponential,* with one event leading to *many* other events *simultaneously,* which in turn led to many more events. You'll realize later in this section and the next one why I suspect that evolution grew exponentially. These events had to converge with a goal in mind: creating a living cell (with reduction in entropy in mind). It's like mixing two chemicals together: the outcome isn't random, but rather depends on molecular rules. It's the same thing with the evolution toward life. Nature had to learn rather *quickly* and at an accelerated rate how to create the DNA and those proteins. So the sequence of evolution occurred more like the form of a tree:

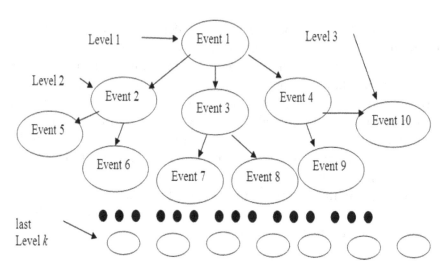

Each bubble represents some event toward the goal of creating DNA. The event of creation 1 led to the events of creation 2, 3, and 4 *simultaneously*, which together led to the events of creations 5 to 10 simultaneously, and so on down the tree. I'm not giving the details about what these events might be due to my lack of knowledge in molecular biology. But it's fair to assume that each event in any level resulted in any one of these:

1. Creation of molecular structures that eventually lead to a necessary protein.

2. Creation of proteins such as enzymes.

3. Creation of molecular structures that eventually lead to a base molecular structure A, T, G, or C or more primitive structures. It's quite conceivable that the first form of life wasn't based on these base molecular structures but on simpler molecular forms.

4. A single DNA base A, T, G, or C.

At the very last level at the bottom of the tree, the entire DNA was finally fully constructed. It's clear that these processes must have taken millions of years to complete. The word *simultaneously* is important because if it were not so, nature would never have had time to create the DNA. This was clearly proven in the section "Origin of Life and DNA Structure" with combinatorial calculations.

Each event occurred in a context of some sort of *feedback* system that I discuss briefly below and also in a context of *cooperation* in which multiple events worked together to produce another event. In the section "Darwin's Theory of Evolution and DNA mutations" I gave plenty of evidence for cooperation between species. Similarly, some type of cooperation had to take place at the biological and microbiological levels toward the development of the DNA. This is what the tree structure above depicts.

What nature essentially had to do in order to construct the DNA in time was to organize a team very much like a company with a structure: president, directors, etc. all the way down to the workers on the field. This suggests strongly that nature is capable of consciousness at the molecular as well as other levels. Could it be that the laws of evolution also took place at the molecular level as well?

This tree-like structure would have led to the creation of the genes of DNA. And guess what—this is precisely the structure scientists have discovered! Although humans have over 20,000 genes, less than a thousand of them, called *control genes*, control the creation and thus determine the functionality of the other genes. Scientists found that the same holds true of all species of animals. So a small mutation of a single control gene may have a significant impact on the features an animal will have such having teeth, tail, or not.

DNA wasn't created in a linear fashion, as there was clearly cooperation and "communication" perhaps even at the molecular level in order to explain the tree-like structure. So the creation of DNA is not entirely the result of blind luck. Consequently, the huge odds I calculated earlier in the section "The Probabilistic Model" are flawed because they don't take into account the cooperation and feedback that clearly took place at all levels in nature from the microscopic at the DNA level all the way to the macroscopic level or cooperation among species. Therefore in the calculations presented below, I found that the odds of life appearing on its own are much higher than I first thought.

Back to the tree structure depicted above. Note that the number of events *multiplies* from one level to the next (1, 3, 9, etc.). This is due to the feedback system depicted in the diagram at the beginning of this section. To explain, suppose a simplified version of the diagram:

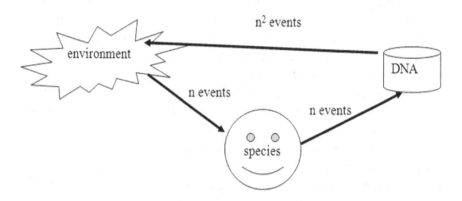

The species senses n new events from the environment that are then fed into the DNA. The DNA reacts to this new information by adding it into itself—essentially adapting the DNA. The DNA has now n^2 events coded. Then the environment changes some more. The species senses again n other new events from the environment that are then fed into the DNA. So the DNA has continually modified its coding to better the chance of the species surviving in a changing environment.

The role of the feedback system is to favor useful information that adapts to the environment and remove or neutralize useless information. Such a feedback system is the only one that can reduce the extremely high odds against the creation of the DNA. It's most natural to propose that DNA developed a few pieces at a time based on this feedback system. DNA clearly did not develop itself in one single attempt. Again the calculations in the section "Origin of Life and DNA Structure" made that clear. As was calculated then, the odds against such an approach are insurmountable.

So, the sequence of events is of order $1, n, n^2, n^3$, etc. This scenario is exactly the same idea as when we plug a number into a calculator, say 5, then press the multiply button, then the equal button (yielding 5^2), then press the multiply button again, then the equal button again, giving 5^3, etc:

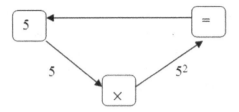

The sequence of numbers generated would be 5, 5^2, 5^3, etc.

How it this analogy useful here? Here is how. First, let's go back to the tree structure of the events shown earlier that lead to the development of the entire DNA:

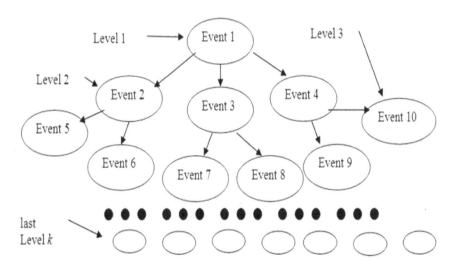

Let

> k be the number of levels needed until the DNA is fully developed for a particular life-form.
>
> n be the number of simultaneous branches from each event of the parent level.
>
> t be the time it takes to go from one level to the next.

The mathematical work that follows for the rest of this section requires knowledge of logarithms, a mathematical tool used in situations of exponential growth. This is the case here.

Using the definitions of the variables above, we can see that after time $T = kt$, the entire DNA is fully constructed. At level 1, one DNA base or a protein is constructed; at level 2, n DNA bases or proteins are constructed; at level 3, n^2 DNA bases or proteins are constructed, etc. The entire number of events in this tree is

$$1 + n^2 + n^3 + n^4 + \ldots + n^k$$

Given the definition presented earlier of what an event represents in the tree, this sum includes necessarily the number of DNA strands and proteins and all associated events toward their creation. Let that sum be D. This sum can easily be proven to be (I'll let you work out that proof as an exercise)

$$\frac{n^{k+1} - 1}{n - 1} = D$$

As the terms n^{k+1} and D are so extremely large compared to n, this equality is approximately

$$n^k \approx D$$

Again we're interested in the time it took to create the DNA (i.e., we seek a value for the term $T = kt$ mentioned earlier). When applying the logarithm base 10 to the formula $n^k \approx D$, we get $k \log n = \log D$. So

$$T = kt = t\left(\frac{\log D}{\log n}\right)$$

The term *log* signifies the logarithm. The *log* gives the exponent value of a number N. So if $N = 10^a$, then *log* $N = a$. This implies that *log* $10 = 1$.

If you're not trained with this tool, it suffices to know the following identities. For any x and y,

$$\log(xy) = \log x + \log y$$
$$\log(x/y) = \log x - \log y$$
$$\log(x^y) = y \log x$$

or grab a mathematics book and learn about it!

Let T_B and T_H be the time it took to develop the first bacterium and the first descendants of humans respectively, then

$$T_B = k_B t = t \left(\frac{\log D_B}{\log n} \right) \quad (1)$$

$$T_H = k_H t = t \left(\frac{\log D_H}{\log n} \right) \quad (2)$$

where D_B and D_H are respectively all the events that took place to develop the first bacterium's DNA and the length of the DNA of the first descendant of humans. Assuming that the variables t and n are the same for bacteria and human DNA (are these fair assumptions?), the two equations put together give

$$\frac{T_B}{\log D_B} = \frac{T_H}{\log D_H} \quad (3)$$

There's no way of knowing the exact values of D_B and D_H, but certainly $D_H > 10^{1926591898}$ (this is the number of human DNAs *biologically* possible calculated earlier) and $D_B > 10^{361236}$ (this is the number of DNAs possible for the simpliest bacteria for which the length of its DNA sequence is known). So

$$\log D_H > 1926591898 \approx 1.9 \times 10^9$$

and

$$\log D_B > 361236 \approx 3.6 \times 10^5$$

and this formula (3) becomes (assuming that the equality still holds)

$$\frac{T_B}{3.6 \times 10^5} = \frac{T_H}{1.9 \times 10^9}$$

Scientists estimate that the first bacteria for which the DNA length could be estimated appeared on the earth when $T_B = 10^9$ years (one billion) and that ancestors to humans appeared some 3×10^6 years ago, so $T_H = 4.5 \times 10^9 - 3 \times 10^6 \approx 4.5 \times 10^9$ years, where 4.5×10^9 is the age of the earth. Unfortunately, when that is plugged into the formula, the equality no longer holds

$$\frac{10^9}{3.6 \times 10^5} \neq \frac{4.5 \times 10^9}{1.9 \times 10^9}$$

This means that either one or both assumptions made earlier are *incorrect*:

1. That the variables t and n are the same for bacteria and human DNA.

2. That replacing the unknowns D_B and D_H by $10^{1926591898}$ and 10^{361236} respectively wouldn't alter the equality of the formula.

The assumption that the variables t and n be the same for bacteria and human is wrong. So we need new variables $t_B \neq t_H$ and $n_B \neq n_H$ such that (using the equations (1) and (2))

$$t_B/\log n_B = 2.8 \times 10^3 \ and \ t_H/\log n_H = 2.4 \ (4)$$

These give the equality

$$t_B/\log n_B = 10^3 \times \left(t_H/\log n_H \right) \ (5)$$

At this point, we have to try different values for the four variables. If you're curious and mathematically inclined, you may wish to give it a try. However, finding these possible values isn't important for the purpose of this section.

Note that the right-hand side of the formula gives a ratio of 2.4, that is, for humans

$$T_H / {\log D_H} > 2.4$$

or

$$T_H > 2.4 \times \log D_H = 2.4 \times (1.9 \times 10^9) = 4.6 \times 10^9 \; years$$

For bacteria, the ratio is around 2.8×10^3—about 1000 times higher than for human DNA. So

$$T_B / {\log D_B} > 2.8 \times 10^3$$

or

$$T_B > (2.8 \times 10^3) \times \log D_B = (2.8 \times 10^3) \times (3.6 \times 10^5)$$
$$= 10^9 \; years$$

These two inequalities suggest that nature probably did have time after all to create life, bacteria *and* humans! In addition, the fact that the corrective value of 2.8×10^3 for bacteria is so much higher than the corrective value of 2.4 for humans suggests two things:

1. That nature required a lot more time (in relation to the complexity of its DNA) to develop the first bacteria than to develop the first human! This is totally counterintuitive given that humans are much more complex creatures than bacteria.

2. That nature had to work *extremely* fast at developing human DNA. Indeed, it took only four times longer to develop the human's DNA even though the DNA is $10^{1926591898} \div 10^{361236}$ $= 10^{1926230662}$ times more complex than the bacteria's DNA!!

Conclusion: nature had to "hit the ground running" to create human DNA. But keep in mind that nature had by then gained quite a bit of learning and knowledge (such as control genes) about constructing DNAs, thereby speeding up the process of evolution toward the creation of humans. This includes the creation of sex, of tissue, or muscles, of moving parts, and of a brain.

Most scientists would reject my findings and say that learning abilities have nothing to do with evolution. As I noted many times before, the accepted theory is that evolution is the result of random DNA mutations. This theory was presented in section "Darwin's Theory of Evolution and DNA Mutations." Scientists might also explain the extremely slow evolution of bacteria by the fact that they evolved during the time that the Earth climate was very hostile, and not very favorable for life. There's certainly some truth to that.

These numbers once again suggest that nature has an ability to think up "sorting-out" algorithms to create life. The calculations of this section make a case toward the hypothesis that nature is able to think. In a couple of sections, I'll come back to these two discoveries (items 1 and 2 above).

In the calculations, I used human's DNA, but I could just as well have used pretty much any animal because their DNA is nearly as long. Moreover according to the web site news.softpedia.com/news/Dinosaurs-Shared-the-Same-DNA-with-Birds-48875.shtml (material used with permission), it's known that dinosaurs were around for over 200 millions of years:

The other group of dinosaurs, ornithischian dinosaurs (herbivorous like horned ceratopsian dinosaurs, armored dinosaurs or duck billed dinosaurs), had longer genomes, more similar to current crocodiles and alligators (about 2.5 or 3 billion base pairs).

What about mammals? The web site www.earthlife.net/mammals/evolution.html says (material used with permission):

The first mammal may never be known, but the Genus Morganucodon and in particular Morganucodon watsoni, a 2-3 cm (1 inch) long weasel-like animal whose fossils were first found in caves in Wales and around Bristol (UK), but later unearthed in China, India, North America, South Africa and Western Europe is a possible contender. It's believed to have lived between 200 MYA and 210 MYA.

So nature had to "hit the ground running" to create dinosaurs as well as mammals and humans. Consequently humans were not favored much more than other living species in the process of evolution. What sets humans apart is, of course, our ability to abstract thinking and complex emotions.

Molecular Structures and Life

These numbers obtained in the previous section suggest that life had to have happened through cooperation at the microbiological level. Cooperation suggests some kind of "sorting-out" ability on the part of nature, hence some intelligence. But intelligence at the *microbiological* level implicates the molecular level too. Cooperation or intelligence at that level seems a *totally* crackpot idea, don't you think?

Perhaps, it isn't. Later in this chapter, I argue in the section "The Thoughts–Energy Dimension" that thinking is first triggered at the quantum level. A thought isn't triggered at the molecular or at the cellular level (although it does manifest itself at that level and above it). Recall that in the section "The Quantum World" of Chapter 5, I calculated that at the quantum level, an atom "looks" as big as our galaxy.

If I were shipped into any galaxy, all that I would see are stars in the night sky. There would be no difference to me whether I am in this galaxy or another one. Similarly, at the quantum level, all atoms look the same. Consequently when a thought is triggered, it doesn't even "know" that the trigger is inside the brain. So perhaps thought may be triggered outside the brain or elsewhere altogether.

Does this mean that a thought may be triggered inside a rock? Can a rock think? Of course, it can't. Even if quantum particles inside the rock could capture a thought, the rock possesses no built-in mechanism to do anything with that thought. Only the living has such a mechanism, and so then only the living can demonstrate thinking abilities. This leads to the question of what separates the living from the inanimate. Scientists have no definite answer.

Based on the ideas concerning consciousness in previous sections, I propose the following answer. Perhaps the *fundamental difference* between the living and the inanimate is that the living has a mechanism to use thoughts for itself. The living uses thoughts to better itself. In other words, the living has some primitive level of awareness of its own existence.

There's a clarification worth making. I stated in section "On Consciousness" that there's some research evidence that cells do indeed have some capacity to memorization and learning. But this capacity comes from *cooperation* among the cells. This is true not just for neurons, but for other cells as well. So it's clear that a single cell has a level of consciousness that's insufficient to allow for further cellular evolution.

Consider this: proteins such as enzymes have a fairly complex molecular structure, yet even though they're necessary participants in the function of a living cell, they're not living beings. Likewise, DNA isn't a living being even though it's at the center of the existence of all living things. Proteins and DNA definitely have no consciousness! DNA is believed to be the most complex molecular structure on earth, yet it's not a living being. All living beings are made of complex molecular structures, but complexity isn't in itself a sufficient criterion for being considered a living being. Clearly the DNA in animals is more evolved than in bacteria. But the DNA didn't evolve of its own will. Or did it? This is the big question. Did a conscious desire to evolve come from beyond the level of that DNA structure? We'll ponder about this later on in this section.

In the section "The Quantum World" of Chapter 5, I calculated that at the quantum level, a unicellular organism "looks" nearly as big as the entire universe. So you can well imagine how complex a single living cell is. Anyone who has taken an introductory course on microbiology will agree. A living cell is a "universe" of molecular structures interacting with one another. At the biological level, consciousness is beyond a single molecular structure although molecular structures are part of the mix.

In light of this, there appears to be two criteria that have to be met for something to be a living being:

1. The being has to be more than a just a very complex molecular structure: it has to be made of individual molecular structures that *cooperate* toward a common goal.

2. This "universe" of molecular structures, as a whole, has to demonstrate a level of *consciousness*. I listed in the section "On Consciousness" a few criteria for consciousness, some of which are:

    a. The will to replicate itself.

    b. The will to feed itself.

    c. The ability to interact with its environment, such as interacting with other cells, and making sense of its environment.

A living cell certainly meets those criteria.

From these two criterias, it can be deduced that cooperation among entities does *not* imply that there is necessarily some consciousness or life going on. Take for instance the water cycle in the atmosphere that transforms water into clouds from heat from the sun, and also involves winds to produce weather patterns. There's a lot of cooperation going on, but clearly, there's no consciousness or life involved whatsoever.

Did life on the earth start off in the form of unicellular organism? The answer is certainly no! Scientists know that the simplest living cells are bacteria. But even these are extremely complex. It's not reasonable to suggest that life first appeared in the form of bacteria in a single miraculous "one shot" directly from the inanimate. There was no need for such a miracle. In this section we'll develop a perfectly logical reason for the appearance of life.

We have to understand that the emergence some three billion years ago of unicellular organisms is itself a result of evolution. Evolution didn't start when the first unicellular organisms appeared. It has to be that evolution started even before these living cells appeared! Consequently—as unbelievable as this seems—the complex cells made of a DNA, RNA and the rest came later through a slow evolutionary process governed by the laws of natural selection. It's therefore possible that the first living cell didn't even have DNA or RNA molecular structures! No scientist would dare propose this, but then, despite all that is known about the molecular working of life, scientists remain at a loss to explain what started life. The issue remains open for debate.

Instead, at some point in the distant past, a few relatively simple molecular structures must have been created (possibly by chance) that were able to cooperate, thereby forming an entity. This entity somehow was able to sustain and replicate itself and develop other functions. This cooperation didn't likely appear by chance alone though. Indeed, the emergence of such functions eventually elevated the entity out of the physical world into a world of consciousness. Once consciousness enters the scenario of life, even at its most basic form, chance alone cannot explain evolution. My opinion here seems to contradict what scientists claim to have proven: evolution is driven solely by *random* DNA mutations.

I'm not deterred though, and I maintain that there's more to evolution than random mutations. I expressed my point of view in the section "Codes and Algorithms for Evolution." The theory of evolution doesn't explain how life came about in the first place. What sort of evolution led to the creation of the DNA? Or what led to the creation of the thousand or so control genes that all complex living beings share from insects all the way to humans?

But how did that entity, the first life form, obtain that consciousness? It certainly cannot be from any of its molecular components (RNA, DNA, enzymes, or proteins) because these aren't life. Clearly, life is more than the sum of its parts. In other worlds, life is nonphysical, and so, it doesn't come from the physical world. How did it plant itself into complex molecular structures that react with one another? Answer: life was captured from "thin air" at the quantum level. Why do I say this? Because I show in the section "The Thoughts–Energy Dimensions" later in this chapter that thoughts show up at the quantum level. Also I find in the section "Our Virtual Universe" of Chapter 10 that the quantum level is the entry door into a universe of thoughts and consciousness I call it the "hidden universe." This universe isn't just here on earth, but everywhere in the physical universe. So it just happened that three billion years ago, here on earth a system of molecular structures that started out as just reacting chemically captured primitive thoughts in a meaningful way. These particular thoughts were about life, and bingo, life came to be!

Intuitively, we know that a minimal level of complexity was required for this ability to capture or formulate thoughts. Is it possible that the first living cells appeared at such a low level of complexity? It is possible. One thing is sure; the first living cell was most certainly *a lot* less complex than bacteria. Consider these questions for instance:

1. Did the first living cells even have a well-defined membrane to separate them from the outside? Yes, most probably. In fact this membrane probably formed solely from chemical forces (i.e. no consciousness involved) *before* the molecular structures system developped into a living being. The reason is that without a protective membrane, the molecular structures system would have been exposed to the environment. The likely consequence is that the system would have been vulnerable, and possibly would have been dismantled. Entropy would have won. Consequently, this molecular system would possibly have never been able to sustain itself—and so life would never have come to be from it.

    This threatening situation would have been detrimental to a trigger of life. Clearly, a living cell could not have had a high enough consciousness to realize that not having a protective membrane would be a threat to its survival. Therefore that membrane formed before the emergence of the living cell.

2. Did the first living cells even have a nucleus? Just like the brain—the center of a nervous system—is the result of a long evolutionary process, the nucleus is undoubtedly also the result of some evolutionary process. So the first living cells possibly didn't even have a nucleus. Consequently, they probably didn't even have DNA molecular structure in place. Moreover, DNA is far too complex to have been created by pure chance. DNA is itself a result of evolution. Many scientists will not agree, but they have to admit that little is known about how the DNA came to be.

Here's another argument that leads to suggest that the first living cells didn't have a DNA. The DNA is a code. A code is the result of interpretations. Interactions require information processing. Information is meaningful data. But only the living, through its consciousness, can capture meaning from an environment. Consequently the living came before the DNA. But, can living cells capture information without having a DNA in the first place? If not, then we have to conclude that the DNA was created out of the supernatural. Could the original DNA be the result of some supernatural signals? I rather doubt it. So, it has to be that that the first living cells didn't have a DNA. They most likely acquired a simpler molecular structure first, perhaps the RNA (ribonucleic acid), although even that structure is quite complex.

So, the first living cells possibly didn't have a nucleus. Note that the emergence of the nucleus was bound to happen eventually due to the spherical geometry of a cell which tends to concentrate things toward its center. But then did the nucleus formed as a result of awareness on the part of the living cell, or did it form as a result of chemical reactions and geometry? When explaining evolution, scientists lay out in great detail how random mutations altered the DNA of species which eventually led to the great variety of species on earth. But how do scientists explain the appearance of the DNA molecule itself? I'll cover that question after the third item below.

Here's yet another reason to suggest that the first living cell probably didn't have a DNA or a nucleus. Experts say that the DNA appeared some two billions years ago at the earliest. Moreover the length of the DNA was in some cases already nearly as long as they are today. Does this leave any room for the random formation of the DNA itself? It's somewhat doubtful. So, in the beginning of life, numerous *directed* DNA mutations *have to* have occurred. I theorize that these mutations created the control genes from which all other genes came from. These are genes that (a) tell that the species will have a head, eyes, legs, etc, and (b) tell what other genes should be there.

How can random DNA mutations possibly have produced such *vitally meaningful* information, and so early in the development of the DNA? Is it plausible that random mutations have gotten it right so early?

Let's do some calculations for a possible clue. I obtained permission to reprint from the Web site www.genomenewsnetwork.org (Genome News Network) this information:

> In the yeast study, the researchers identified 203 DNA sequences that regulate genes (the human genome may have ten times this number).

These DNA sequences that regulate genes make up the control genes. According to this study, our DNA contains about 2000 sequences that regulate genes. So, there are $4^{2000} \approx 10^{1204}$ possible sequences. The combinatorial calculations in section "Origin of Life and the DNA Structure" were for the case of the chance of creating the entire human DNA anywhere in the universe. Let's adjust the results for the creation of the control genes only. I remind you that we had calculated that there were 10^{104} trials of creating the control genes anywhere in the universe since the Big Bang. At that rate, it possibly could have taken nature $10^{1204}/10^{104} = 10^{1100}$ years to create the correct control genes. Given that the universe is only about 10^{10} years old, nature had a much better chance to win the lottery many times over before coming up with the correct control genes. Is it still plausible that nature's creation of the human control genes was entirely random?

If you understood well the exponential tree-structure model of the previous section, you know that it allowed me to conclude that nature did have time to create the DNA by way of cooperations. So, why didn't I use that model here in my calculations for control genes? The reason is that these genes were the first ones. So, there was possibly little around to cooperate with. Suppose that I'm wrong, that there was cooperation involved to create those control genes. So, then nature had plenty of time to create them. Did that cooperation occur before or after life appeared? In an argument in the second paragraph of this item (2) that we're in, I concluded that life appeared first. Let's now look at it from the point of view of mutations. There are two possibilities:

a. The mutations were undirected. Since this type of mutations don't involve any sort of consciousness, it's possible that the molecular cooperation that took place didn't necessitate life. If that's the case, then the DNA appeared at the same time as life itself.

b. The mutations were directed. Since this type of mutations require some level of consciousness then there has to have been life there already. Consequently the DNA appeared after life's creation.

Scientists don't know for sure, but most are in favor of the scenario (a). But this choice leads to the following dilemma. Since DNA is composed of meaningful codes, it must have come as a result of numerous interpretations. Yet this scenario (a) says that life didn't exist yet. Therefore there was no consciousness present and so, no meaning could have come out of anything hence no interpretations could have taken place. So, the DNA's meaningful sequence was the product of pure luck. I can't conceive of such a lucky "toss" of millions of DNA sequences on the part of nature.

Should we conclude that the DNA appeared out of nothing? Or perhaps the DNA molecular structure came from a simplier structure out of a cocktail of molecular reactions that was not life yet. There's such a structure: the RNA (ribonucleic acid) and this is the subject covered after item 3 below.

3. Can we talk of evolution when life had not appeared yet, when all there was were chemical reactions? Can we consider as evolution the formation of increasingly complex molecular structures? According to Merriam-Webster's dictionary, there are many definitions to evolution such as "being a process of continuous change from a lower, simpler, or worse to a higher, more complex, or better state." This suits well life. Then elsewhere, it defines evolution as "ways the action or an instance of forming and giving something off." This suits well inorganic chemical reactions. So evolution doesn't apply to life only. Evolution started even before the first form of life appeared!

While being ignorant of the details of the evolution of life, some of my views and deductions expressed in this section so far are confirmed by scientific evidence! I've searched the Internet, Wikipedia in particular, for answers as to the origin of the DNA molecule. Biologists agree that bacterial cells cannot form from non-living chemicals in *one* step. If life arose from non-living chemicals, there must have been some intermediate form of pre-cellular life.

Of the various theories of pre-cellular life, in the early stages of evolution of life on Earth, the most popular view today is the "RNA hypothesis" first proposed by scientists Leslie Orgel and Francis

Crick in 1968. There exist other theories of course, but I feel that this is the most probable one.

The hypothesis states that the earliest living cells emerged as a result of chemical evolution billions of years ago in a process called *abiogenesis* (generation of life from non-living matter). The term is used to refer to theories about the chemical origin of life, most probably through a number of intermediate steps, such as non-living but self-replicating (referred to as *biopoiesis*).

The earliest form of life was the prokaryote, unicellular bacteria possessing a cell membrane and probably a ribosome (consisting of complexes of RNA and protein making up 95% of RNA content within a cell), but lacking a nucleus. This matches what I speculated in my questions (1) and (2) in my list of three questions presented earlier. These bacteria thrived in aquatic environments, and ruled the Earth in its early history some 2.5 billion years ago. Like all modern cells, it used DNA as its genetic code, RNA for information transfer and protein synthesis, and enzymes for catalyzing reactions.

RNA (ribonucleic acid) is a nucleic acid consisting of nucleotide monomers, which act as a messenger between the DNA (deoxyribonucleic acid) as the carrier of genetic information, and ribosomes, which is responsible for making proteins from amino acids. Single rather than double-stranded, RNA is a precursor to the the original DNA and acts as the intermediate leading from DNA to protein.

Based on this knowledge, in 1968, biologists Leslie Orgel and Francis Crick proposed that that RNA must have been the first genetic molecule. According to them, evolution based on RNA replication preceded the appearance of proteins. They also believed that RNA, besides acting as a template, might also act as an enzyme (catalyst), and in so doing, catalyze its own self-replication.

Because it can replicate on its own, the RNA performed the task of both the DNA and proteins. So, RNA is believed to have been capable of initiating life on its own in the early history of Earth! Consequently, in the early stages of life's evolution, all the enzymes may have been RNAs after all and not proteins. Recall that enzymes are catalysts. These serve as stimulants that perhaps gave the "spark" needed for life.

My view is that self-replication requires a sort of cooperation among chemical molecules. However the cooperation wasn't in this case the product of some consciousness on the part of the RNA because the RNA isn't life. Could the RNA molecular compound be considered life? I doubt it because life requires some level of consciousness. The self-replication capability of the RNA could have happened by a chance assembly of molecules that possessed just the right properties to allow it to take in energy and self-replicate or reproduce. An ability to reproduce definitely suggests life. But I don't believe that it's a sufficient requirement. For instance consider again the example used earlier of the water cycle in the atmosphere that produces clouds. This can be seen as a sort of reproduction where energy from the sun evaporates water from the seas thereby continually reproducing clouds and wheather patterns. Yet this water cycle is clearly not life.

The self-replicating RNA molecules were believed to be common 3.8 billion years ago! Yet, as mentioned a few paragraphs ago, it's believed that the first bacteria appeared 2.5 billion years ago. There's a gap of over one billion years. This confirms my hunch that life must have appeared first in a form much simplier than a bacterium.

As for the DNA, it couldn't initiate life on its own because even the shortest DNA strand needs protein to help it replicate. This is the "chicken and egg" problem (genes require enzymes; enzymes require genes). Which came first, the chicken or the egg? DNA holds the recipe for protein construction. Yet, that information cannot be retrieved or copied without the assistance of proteins.

Which large molecule, then, appeared first in getting life kick-started: proteins (the chicken) or DNA (the egg)? The answer has already been provided in the previous paragraphs. A simple solution to the "chicken-and-egg" riddle, in my view might be the "RNA hypothesis" containing RNA molecules that serve to catalyze the synthesis of themselves, and the first step of evolution proceeds then by RNA molecules performing the catalytic activities necessary to assemble themselves.

The first self-replicating RNA that emerged from non-living matter carried out functions now executed by RNA, DNA and proteins.

According to scientists, a number of additional clues seemed to support the idea that RNA appeared before proteins and DNA in the evolution of life. Many small molecules, called "cofactors", play an important role in enzyme-catalyzed reactions. These cofactors often carry an attached RNA nucleotide with no obvious functions. These structures have been designated as "molecular fossils", relics descended from the time when RNA alone, without DNA or proteins, ruled the biochemical world. The big question is: *how did the first self-replicating RNA arise?*

Many scientists are opposed to this RNA hypothesis for many reasons:

1. The instability of RNA when exposed to ultraviolet light.

2. The known range of its catalytic activities is rather narrow.

3. The difficulty of activating nucleotides (molecules that comprise the structural units of the RNA).

4. The lack of available phosphate required to constitute the backbone of the RNA.

5. The chemically fragile RNA and the difficult to synthesize abiotically.

6. The origin of an RNA synthesis apparatus is unclear.

Note that the DNA's history suffered similar obstacles, yet life came to be nonetheless.

Even if the pre-cellular life is presumed to exist in the early epoch of the Earth's existence, there's a problem getting from there to proteins, genes, and cells. The random production of protein isn't an explanation. Although its emergence from non-living matter is hard to conceive, scientists believe that pre-cellular life appeared almost spontaneously! This matches my belief expressed earlier that life was captured from "thin air" at the quantum level. Think about it. A molecular system either has life or is lifeless. There was never such a thing as a molecular system that had half a life, or one-third a life. There was never a transition to life: it had life or it was lifeless. It has life or it's lifeless. There isn't a "in between" state. Consequently the creation of life was spontaneous just as the creation of the universe was so too.

Even here, science remains clueless about the timing of spontaneous infusion of life into pre-cellular entity. Researchers are also unclear how even some of the shortest amino acid chains, called peptides, formed prior to the arrival of living organisms.

It was said earlier that the difference between the inanimate and the living is that the living possess some level of consciousness. The difference is so clear cut that whatever has a consciousness is necessarily living. Moreover since the world of consciousness resides in that "hidden universe" mentioned ealier in this section, then life resides in that universe as well. The physical mechanism of life resides in this physical universe, but life itself doesn't! *Life is elevated out of this physical universe.*

So then could it be that evolution is itself part of that world of consciousness? No, because, by definition, evolution deals with the dynamics of life forms in the *physical* world. The relation seems to go both ways though: wherever life is found, there's evidence of evolution too. Consequently evolution is likely partially driven by consciousness. This assertion cannot however be scientifically proven. Also this assertion doesn't mean that life forms appeared out of nowhere into this world.

Because that hidden universe is absolutely everywhere in the universe, the same rules of evolution therefore apply everywhere in this physical universe. From the first day that life appeared, evolution was already at work.

So, imagine this: the seeds for life were and are already present in the "hidden universe." But life itself wasn't present yet in our physical universe because the necessary molecules were not yet present in a favorable manner. As explained at length in previous paragraphs, even with all the necessary molecules, it remains unclear how life started. We face here the same dilemma encountered in Chapter 7 whereby we had to conclude that the universe was created from a system of thoughts. The same thing happened with life: it was *created* by a system of thoughts!

So, the appearance of life was a sort of Big Bang. However, there's one important difference: there was no Universe at the time of Big Bang. As stated in Chapter 7, the Big Bang was the result of thoughts that generated an immense amount of energy. In the case of life, the process was somewhat the other way around. The universe already existed. Somehow the correct combination of inanimate molecular structures captured from the "hidden universe" a thought of *awareness*. At that moment, evolution took root into the physical world and went on from there.

I suggest that the thought of evolution comes from the "hidden universe." You might ask: why then is life not blossoming everywhere in the universe? Answer: the physical environment in most planets of the universe is harsh and not favorable for life. It very much seems that although the "hidden universe" tries to plant life against all odds, the physical universe doesn't want life to appear. But once life is planted, the forces of evolution will do all they can to make life flourish.

But evolution isn't perfect for two reasons: (a) the logical process of evolution is imperfect, and (b) evolution has to deal with a physical environment that isn't perfect. From day one, there were undoubtedly numerous trials and errors. Here are a few lessons that evolution must have learned along the way:

1. "You realize the need for something when it does work."

2. "You realize that you don't need something when it doesn't work."

3. "You realize that you don't need something when you never use it."

4. "You realize the need for something when you lose it."

These principles relate to features that a species already has or had in the past. These principles suggest a consciousness on the part of the species although this can't be proven. But none of these principles account for a species that acquires a brand *new* feature. So in addition to having a consciousness, a living cell either has an intelligence of its own to *invent* some new feature or it's in communication with some entity that has intelligence.

The first living structure was a relatively simple one, and things built up from there with a few random events at times, but sometimes, the buildup required a "sorting-out" ability type of cooperation, meaning that bad biomolecular designs were rejected for favorable designs. Is that possible that natural selection didn't only occur at the biological level but at the molecular level as well? There appears to be no way around that because otherwise DNA, enzymes, and proteins would have developed by chance, which I've proved in this chapter to be *impossible*.

Hold it! Some people including scientists might reject my argument that a conscious cooperation took place at the molecular level on the basis that as the universe became more complex on its own, as time went by this inevitably led to yet even more complex structures. Consequently, the exponential tree-structure that I depicted in the previous section would have taken place even if no consciousness of any kind existed.

My reply is that my tree-structure and the calculations of exponential growth associated with it do prove that cooperation had to take place. Whether this cooperation was the result of some level of consciousness or not is debatable. Or is it? In my opinion, this debate is already ended. Based on what I argued in previous paragraphs, we can't deny that there's this fundamental difference between the living and the inanimate: the capacity to hold thoughts and consciousness. Life manifests its presence in this physical universe, but life itself isn't of the physical world: it originates from that hidden universe of consciousness. You might recall that I mentioned this already before.

What puzzles me though is that the exponential growth of DNA seems to have stopped many millions of years ago. Our DNA isn't much longer than the DNA of the crocodile, which has not evolved for 300 million years! Yet we are obviously more evolved than crocodiles because of our intelligence. Perhaps it's not only the length of the DNA that I should have taken into account in my calculations, but also how the DNA dictates the functions of an organism. Our DNA has advanced functions that a crocodile's DNA doesn't.

One view is the following one. The DNA may be viewed as a hard disk containing data about species inside a computer, and the computer software would be all sorts of applications (species' behavior and the environment) that use the rules of evolution to operate. When life first appeared on the earth, the computer was brand new and the hard disk was nearly empty. So the DNA was small. As the computer used its applications, these applications learned new tricks all the time. These tricks were stored (insertion mutations) inside the hard disk, thereby increasing the size of the DNAs. The data inside the disk grew in size rather quickly because a lot had to be learned—hence the exponential learning curve and growth of the DNAs. As time went by, the applications had less tricks to learn because over the millions of years, the same combinations of species-environments came back once in a while.

For instance, the numerous ice ages that went by served as a stimulus for the appearance of new species. But at each ice age, the species more or less came up with the same tricks for adaptation because these adaptations were already learned from previous ice ages. Therefore the tricks were already inside their DNAs, dormant. So, undirected DNA mutations took place when needed at the time that a specific environment came back. At times tricks learned are no longer needed, so the applications (i.e. species) either deleted the related information (deletion mutations) from the hard disk or left them there becoming "junk DNA".

Although many different types of species evolved over the millions of years, this didn't translate into longer DNAs. I suggest then that insertion and deletion of DNA strands have become rare for millions of years already, and that other types of mutations such as frameshift and substitution mutations have become more common. These are mutations that "shuffle" DNA codes around resulting perhaps in new species or new features acquired by species.

Because the DNAs haven't grown any longer for a very long time, should we conclude that life on this earth has learned just about all there is to learn? What circumstances might trigger the evolution of longer DNAs?

Exponential "Learning"

By now we've seen that "sorting-out" algorithms improved learning, and vice versa. This cosmic language and algorithms must have been the result of God's thinking, right? I'm not so sure for we've also seen in this chapter and earlier ones that there are different levels of consciousness:

1. At the level of quantum physics, nature has to interpret the distance between emitted energy of a particle and its arrival as time, very much like we interpret motion on a watch as time. Nature *invented* the mechanism of time, not God. Note that God created the thought of time—not the time itself. This invention appears to have been a necessary condition for motion to be possible. But then it could instead be dark energy—the fabric of space—that is the mechanism enabling motion. Recall that this was my conclusion in the section "Motion Finally Explained!" of Chapter 8. Also I suggested in an earlier chapter that God's thoughts are at this quantum level.

2. At the subatomic and atomic level, nature had to realize that entropy at that level could be reduced by inventing electric charges, thereby creating atoms, and allowing for the creation of molecules and matter. As the electric field is at the quantum level, is it reasonable to believe that God's thoughts create that field?

3. At the molecular level, nature had to create carbon-based structures such as methane and ammonia. This initial environment eventually led to life and the DNA structure a bil-

lion years later. That structure contains codes that have a meaning at the biological level. The DNA isn't just a complex molecule that floats around aimlessly. The previous section discussed at length consciousness at the microbiological level. If the DNA molecule conveys a "coded purpose", perhaps other molecules do too such as water, methane, or carbon.

4. At the cellular level, nature had to think up algorithms to enable complex molecular reactions needed to operate a unicellular organism. One of those algorithm involved enzymes and other complex structures such as DNA needed for the creation of cells. The enzymes are proteins acting as a catalyst encouraging reactions in the cell but without interfering with them. In other words, enzyme is a neutral partner, and so it does not add "polarity". This is important because otherwise, the enzymes would become a "third player" or a third pole and so would cause an increase in entropy (recall the discussion about the relationship between polarity and entropy in the section "Entropy Versus Organisms").

    Here's what Wikipedia has to say about enzymes:

    > Enzymes are proteins that catalyze (i.e. accelerate) chemical reactions. In enzymatic reactions, the molecules at the beginning of the process are called substrates, and the enzyme converts them into different molecules, the products. Almost all processes in a biological cell need enzymes in order to occur at significant rates. Since enzymes are extremely selective for their substrates and speed up only a few reactions from among many possibilities, the set of enzymes made in a cell determines which metabolic pathways occur in that cell. Like all catalysts, enzymes work by lowering the activation energy for a reaction, thus dramatically accelerating millions of times faster the rate of the reaction. As with all catalysts, enzymes are not consumed by the reactions they catalyze, nor do they alter the equilibrium of these reactions.

So enzymes were an important partner in the *rapid* evolution of the species without which it's doubtful that humans would be around yet.

Also nature had to figure out how to create neurons to form a brain. Note that neurons are specialized cells that "store" thoughts and enable thinking. How could a thinking organism be created out of cells that weren't able to think at all? As the saying goes, you get nothing from nothing. It seems that thinking was already involved before neurons were evolved. We'll come back to this idea in the next chapter.

5. At the biological level, mature had to realize that cell splitting was too slow a process to reduce entropy, and so it invented sex.

6. At the celestial level, nature had to realize that matter could collect under gravitation pull and create bodies, solar systems, and galaxies. This was discussed at great length in Chapter 6.

7. At the logical level, nature had to realize that evolution is better assured with cooperation between species and feedbacks.

The discoveries in section "Codes and Algorithms for Evolution" that nature:

1. Required a lot more time (in relation to the complexity of the DNA) to develop the first bacteria than to develop the first human even though humans are much more complex creatures than bacteria.

2. Had to work extremely fast at developing the human's DNA considering that it's $10^{1926230662}$ times more complex than the bacteria's DNA!

suggest strongly that nature experienced an *exponential learning curve*. It took it one billion years to figure out how to develop the simplest bacteria, but only another three billion years to develop mammals and humans. Yet a human's molecular biology is *exponentially* more complex than that of bacteria. Thus can pure luck explain alone this exponential learning curve?

I wish to provide you with a very brief overview of the history of life and you should see emerge an evidence for exponential growth of life and also of earth's evolution. Most of the information in the list below was obtained from Wikipedia.

1. The early Earth was lifeless and inhospitable with its atmosphere dominated by materials from the solar nebula, especially light gases such as hydrogen and helium. During the first period of Earth's existence called the Hadean epoch (4.5 to 3.8 billion years ago), the earth is believed to have undergone a period of heavy meteoric bombardment (leftovers from the formation of the solar system) for about 700 million years. Many scientists now theorize that life may have come from outerspace in the shape of an asteroid or a comet. Scientists have found evidence that these bodies contain a lot of water and organic compounds.

   Note that organic compounds are essential to life but aren't life themselves. So life appeared on earth, not inside the comets. But, even this, is now put into question. Scientists have found evidence of some type of bacteria that hibernate, totally frozen, for millions of years, and then come back to life when warmed up. Perhaps there exist bacteria inside asteroids and comets.

2. The planet is believed to have cooled quickly with formation of solid crust within 150 million years, and formation of clouds in about 200 million years. The subsequent rains gave rise to the oceans within 750 million years of Earth's formation, making it a habitable planet for the first time in its his-

tory. That's about 3.8 billion years ago. Could life have taken root under the oceans that early?

3. It's generally believed that until 2.4 billion years ago, Earth's atmosphere was devoid of oxygen. Volcanic activity was intense, and without an ozone layer to hinder its entry, ultraviolet radiation flooded the surface. So no life could possibly form in such an environment mostly due to the radiation that would have destroyed all organic material.

4. However, despite the hostility of the elements, life eventually took root due to these three environmental developments:

   a. At the beginning of life, the Earth was a hotter planet due to increased greenhouse effect of a carbon dioxide (CO_2) in the atmosphere. Nonetheless bacteria can live in some extremely hostile places. The first bacteria are believed to be "archaea", unicellular microbes that can survive in the harshest of environment on the planet—considered extremely inhospitable for habitation by humans and other creatures. They're believed to be among the first life on the Earth 3.8 billion years ago. They're presumably the first version of life on our planet when its atmosphere was devoid of oxygen, and comprised largely of ammonia, methane, water vapour and carbon dioxide. It's believed that relentless lightning in the young atmosphere helped form some of those gases. Could bacteria live without ozygen?

   This early atmosphere didn't contain oxygen until about 2.4 billion years ago. That didn't stop bacteria from thriving because some bacteria don't use oxygen at all. In South Africa, samples of rock from gold mines at 2.8 kilometers depth below the surface show the presence of thriving thermophiles that derive all its energy from the decay of radioactive rocks rather than from sunlight! There are microbes known to depend exclusively on geologically produced hydrogen and sulphur compounds for nourishment, and lives in conditions similar to those of early Earth.

b. As mentioned earlier, within 750 million years of Earth's formation, water was already filling seas and oceans. Water provides an excellent environment for the formation of complicated carbon-based molecules that could eventually lead to emergence of life. Steam escaped from the crust while more gases were released by volcanoes, creating the second atmosphere in Earth's early history. Microbes have been found to thrive in other extreme environments such as extreme acidic, alkaline, and saline as well. Bacteria having the capability to thrive at temperatures between 80 - 121°C have been found around volcanic vents at the bottom of the oceans. These hydrothermal systems are deemed to be our closet link to the very first organisms to have evolved on the Earth.

Moreover geologists have recently discovered 1.43 billion year-old fossils of deep-sea microbes, providing more evidence that life may have originated on the bottom of the ocean.

The seas were ideal for the emergence of bacteria and more complex life forms because these organisms were were free of potential violent lightning storms or lethal gases. Moreover ultraviolet rays don't penetrate water because water scatters and absorbs the rays. Therefore the seas protected the DNA of all living organisms from harmful ultraviolet rays that would otherwise cause all sort of deadly mutations of its structure.

c. It's the existence of bacteria in the seas that allowed the atmosphere to contain the nitrogen and oxygen of today. Otherwise, carbon dioxide may very well have remained trapped in the oceans.

5. Once the atmosphere was equiped with nitrogen, oxygen and an ozone layer, Eukaryotic cells (the first unicellular organisms with a complex internal structure) developed about 1.5 billion years ago in a stable environment rich in oxygen.

   These unicellular organisms were not structured like bacteria. Why did nature come up with a different type of cell? I suggest that it's simply because the earth's environment had changed. As nature was still in its stage of thinking up ways to develop cells, this led to a different type of cell being created.

   The oldest fossils of land fungi and plants date to 480 million to 460 million years ago, though molecular evidence suggests the fungi may have colonized the land as early as 1 billion years ago, and the plants 700 million years ago! In comparison, modern humans—Homo sapiens—are believed to have originated somewhere around 200,000 years ago or earlier; the oldest fossil dates back to around 160,000 years ago.

   So, it took life's evolution 3.5 billion years to go from the first bacteria to plants! But those weren't the luscious plants of today. They more like to small weeds that demanded little ozygen. Then it took less than 500 million years to go from small plants to mammals! This is the exponential learning that I talking about earlier. But I'm getting ahead of the story.
   (Section "Molecular Structures and Life" covers in more details the early days of the DNA and life.)

6. One reason that it took so long for life to evolve is that for about 2.8 billions years, the earth's environment was too violent and poisonous for the development of anything much bigger than a bacterium. As the earth cooled down, its environment became less violent and more hospitable, giving a chance for the development of more complex organisms.

This peaceful environment encouraged the development of more complex organisms and DNAs.

Note that the complex organisms were made of tissues. The first ones appeared about one billion years ago. Although tissues are a collection of cells, these cells are different from bacteria. Bacteria are free to move around, but not the cells of a tissue. These types of cells are physically attached by tiny adhesive proteins and sugar that act like glue. This implies that nature had to think up another type of cell—the multicellular organisms.

What event brought those on?

I learned during my high school years that unicellular organisms appeared before cells that "glued" together, making up multicellular organisms (biological tissues). There are two views as to what brought about multicellalar organisms:

a. If unicellular organisms joined, there had to have been an evolutionary purpose for forming more complex organisms. I suggest that there was perhaps some consciousness at play here either at the cellular level or at the DNA level or at some other level in the hidden universe that, I theorize in Chapter 10, is a universe that contains consciousness of everything from our physical universe. A sort of cooperation may have occurred with unicellular organisms. Cells also had to be aware of their surroundings in order for them to cooperate in joining together. You may recall that I already mentioned earlier in this chapter that *cooperation* was a key to the success of evolution of life.

b. The glue along the membrane of cells came because of a random mutation. So there was no evolutionary consciousness involved.

I must admit that the second explanation, view (b), is more readily believable! But not so fast: tissues of plants, animals, fungi, and so on, are of different structures. Therefore it's believed by biologists that multicellular organisms evolved *independently*—they have different evolutionary paths. So then, it becomes unlikely that some independantly random DNA mutation occurred for *each* of these type of tissues. What's the chance of that? So then, perhaps view (a) is plausible. My approach is to never discount straight away the capacity of consciousness in the evolution of life.

The evolution toward tissue was important for the next step to evolution: structured organisms.

Let's now fast forward in time this presentation of the evolution of life.

7. Note that it's certain that complex life forms first appeared in the sea, for the waves of the water made it easy for organisms to move around without muscles (which weren't yet invented by nature). Organisms made of tissue eventually led to the evolution of different types of tissues such as muscles. Could it be that organisms realized that it's easier to move around and control motion with muscles? Or is it just some freak accidental DNA mutation that is behind this evolutionary step? So, now these organisms could move around in the sea. The era of animals was born, and it started 600 million years ago.

    These muscles had to be activated by something—hence, the evolution toward nerve cells. But I'm getting ahead of myself. In the evolutionary process, what appeared first, muscles or nerves? We know that muscles are activated by nerves, and therefore, nerves most likely evolved before muscles. What evolution led to nerve cells? It's possibly the motion of water slowly led tissues to create specialized tissues (nerves) that could react to water's motion? Essentially, organisms became touch-sensitive. Indeed, nerves give us the sense of touch. These sort of animals appeared some 30 millions years later, which is 570 million years ago.

Note that at this point, nature had not yet invented the brain. But could complex organisms move around without a brain? Yes, and this still happens to this day. Some sea creatures such as the starfish have no brain, and yet have a nervous system that allows them to move around.

In the ladder of evolution, the brain made its appearance a very long time after muscles and nerves: from 100 millions later for the fish to 450 million years later for mammals.

This is an awful long time just to create a brain, but it makes senses. Evolution always goes from simpler structures to more complex ones. Nothing is more complex than a brain! In fact, neuroscientists believe that our brain is more complex than the entire universe. It makes no sense that the brain would have developed first, then nerves and muscles. Also it isn't even plausible that the brain, nerves, and muscles evolved at the same time. What always drives evolution is adaptation and improvement. If brain, nerves, and muscles all appeared at the same time, what could the brain have been an improvement of?

Creatures of the seas slowly evolved complex nervous systems that became centralized: the brain was born about 550 million years ago.

How could that come about?

I theorize later on that cells have an ability to "think" primitive thoughts. What a crackpot idea, right? But later in this chapter and in the next chapter I give arguments in support of that view. Of course, I'm not suggesting that cells can think up complex ideas such as what humans are capable of. I propose that at some point in the evolutionary process, better "thinking" cells evolved and invented physical connections with each other to create a thinking organism. This organism is called a brain. But it's not just the brain that seems to have this ability. As I explained in section "On Consciousness", scientists have discovered that immune cells build "bridges" with its neighbours to communicate the identity of an invader.

Let me be clear: there's *no* reason to believe that the appearance of a brain wasn't part of evolution. A brain wasn't created out of nothing; it's part of evolution just like everything else. There's no reason to believe that the neurons in the brain aren't part of that evolution. As the brain is able to think, it *had to have* evolved from something that was also able to think. We'll continue regarding this issue of brain cells evolution in the section "Our Virtual Universe" of Chapter 10 where I'll provide a solid argument in favor of consciousness at the level of cells.

It's clear that the evolution of the brain *accelerated* evolution of the species. The brain allowed the species to make complex decisions to better its chance for survival.

8. As organisms became more complex, their level of consciousness also rose up—especially once equipped with a brain. This led some of them to think up new ways to live and thrive. For instance, some of them may have realized that crawling near the seashore was a better way to get food.

9. Eventually after thousands of years, this learning changed their DNA such that their bodies now had legs and they crawled out of the sea. This was 360 million years ago.

There it is: a quick overview of earth's and life's evolution.

You probably noticed the exponential growth in life's evolution. Note that the evolution of the brain was much slower than other parts of the body of any species. If you look it up on Wikipedia at the web address en.wikipedia.org/wiki/Timeline_of_evolution, you'll find that it generally took 30 to 50 millions years to evolve from one group of animal to another. But it took nearly 400 million years to develop the brain to the level of that of mammals, and I'm not even including humans who had to wait nearly another 200 million years to evolve!

The brain is so incredibly complex, yet it evolved relatively quickly considering that it took 2.5 billion years to go from the simpliest bacteria to the complex unicellular organisms. Yet their complexity is insignificant compared to that of the brain. Given this timeline, can the theory of evolution satisfactorily explain the brain's evolution?

Scientists believe that bacteria appeared on earth first, suggesting that they may be more primitive than animal cells. This doesn't imply that our body doesn't profit from bacteria. In fact, there are approximately ten times as many bacterial cells as human cells in the human body. Many of those bacterial cells protect us against viruses or are used to help digestion.

A Thinking Nature

Nature is capable of thoughts—what a ridiculous idea, right? Not so as I explain in this section and the next one.

I've often mentioned nature in these scenarios of development. Keep in mind that these involve different levels of consciousness, of learning and thoughts. So nature has different levels of consciousness. But learning requires a thinking process. But where's the brain? Unicellular organisms, plants, and some types of fish have no brain. However I've thrown hints in previous sections that there was thinking involved in their evolutionary process. So, just perhaps, thinking can occur without a brain housing it! Is this a crackpot idea? I don't think so. In a paragraph below, I give a simple example to suggest that thinking may take place without a brain.

I've suggested before that thinking on the part of nature was in part involved in order to create complex life-forms or even simple bacteria. The fact that mammals developed only three billion years after the first bacteria suggests that nature drew from past experience with simple life-forms to then develop more complex ones.

The whole story of life looks very much like the story of the evolution of manmade technology. The development of technology is impossible without first some scientific discoveries. Some of those discoveries occurred by luck, but there has always been an *intent*. The beginning of the history of technology led to very simple tools, which can be viewed as parallel to the development of simple enzymes or proteins by nature. Then those simple tools were found to be useful to create other more complex tools.

For centuries the evolution of technology was linear for at least two reasons:

1. Technology was not at the center of human activity as it didn't seem all that important.

2. People lived in small villages, disconnected from one another—there was very little *cooperation*.

But in the middle of the nineteenth century, technology started to take off like a rocket. Numerous inventions were made. Part of the intent came from the Industrial Revolution. Human thought up a *purpose* for technology and the result was an *exponential* growth in technological invention. For instance, microcomputer power doubles every eighteen months. There was little randomness about the succession of events that led to the technology we enjoy today. The same holds true with the evolution of life. Just as technological evolution was possible by thoughts, I suggest that the same applies to nature. Nature's thoughts led to life.

Let me offer another example that suggests that thinking may take place outside a brain. I gave numerous examples of cooperation between species in the section "Darwin's Theory of Evolution and DNA mutations" and explained thereafter that cooperation allowed for an exponential learning curve from nature. Cooperation implies that various species have to connect, to communicate somehow. This connectivity may also involve the nonliving physical environment such as a river, the conditions of the air, etc. The living conditions of a single species may depend on numerous subtle connections with other species or inanimate things. Cooperation therefore fosters a network of interconnected species and inanimate things. This interconnection stimulates a level of consciousness in all species involved.

Now compare the scenario above with the working of the brain. I know very little about the inner working of the brain, but I know the following:

1. That a thought generated by the brain involves tiny electrical signals sent between thousands (if not millions) of neurons. I see this as *cooperation* between neurons to form a thought.

2. That a single neuron is totally unaware that it's part of the formulation of a thought beyond itself. A neuron has *no way* of knowing that thinking is taking place. The neuron is only aware of its immediate microscopic surroundings, very much like a bacterium. However, neurons are more evolved than bacteria in that they send weak electric signals to neighboring neurons.

3. That a single neuron is insufficient to form a thought, but thousands of them may be enough.

4. That it's not just the number of neurons that is relevant but also how they cooperate—how they are interconnected. This same sort of scenario occurs in nature.

Now compare this list with nature:

1. There's evidence of cooperation between various species in the ecosystem.

2. A single living being involved in this cooperation is only aware of its own immediate surroundings, and is not aware

of the "big picture" of the ecosystem. So a living being, even a human, can't be aware whether the ecosystem is engaged in thinking.

3. A single living being isn't sufficient to make an ecosystem function.

4. The way that living beings are interconnected dictates the complexity of the ecosystem.

You should be able to see a parallel between the brain and an ecosystem part of nature. Based on this, why not suggest that in just the same way, nature can form thoughts made evident by cooperation between various players in the ecosystem. Certainly, nature doesn't have a mouth and so it can't speak to us or tell us what its thoughts are. But this is no reason to dismiss the possibility that nature might be capable of thinking.

Some believe that if we can't detect it, then it doesn't exist, but I don't buy that logic. My view is very similar to what's called the *Gaia hypothesis* that suggests that the whole earth has a consciousness. Barely a handful of scientists believe in this hypothesis, but that doesn't deter me.

Most people would surely judge my idea as complete lunacy. But how do we know that our brains think? We evidently know simply because, well … we have thoughts. A person next to me knows that I think from the way that I behave such as by what I say or do. It's all extremely obvious because we have our five senses to help us communicate. The trouble is that nature doesn't have any means of letting us know that it thinks. A biologist would be better suited to take this idea further than me, so I'll leave it at that, and let you ponder.

My idea seems silly as we don't feel like humanity or all of life is participating in some great consciousness, and we aren't aware of any greater awareness. But this certainty shatters rather quickly when we realize that a single neuron isn't conscious either. A single neuron is unaware of the part it plays in giving rise to a conscious us, but it does so nevertheless.

Another reason to believe that nature is able to think is when we consider our relationship with God. How can God have an influence on nature? God can only communicate with nature via thoughts. If nature were unable to think, God would have no means of being involved even indirectly with nature. For those of us who believe that God is the Creator of nature, God has to be able to communicate with it, hence the need for intelligence on the part of nature.

You might reason that if nature is able to think, then why did it take billions of years for nature to "think up" DNA and all the other things? Good question, considering that it took humans only a couple of centuries to invent all kinds of very complex synthetic molecular structures. My reply is twofold:

1. DNA is immensely more complex than any molecular structures invented by man.

2. The brain is both an extremely efficient captor and processor of thoughts. Nature, on the contrary, is an extremely slow thinker. I'll show in the next section and in the next chapter that thoughts are captured at the quantum level. So I suggest that nature captures thoughts as quickly as our brain. So why is nature so slow at thinking? Simple—unlike the brain, nature doesn't have neurons! The bacteria and unicellular organisms that nature "thought up" a few billion years ago are no match to neurons. The network of neurons of the brain is essentially a translator of quantum thoughts into a biological form. This is what allows us to think complex thoughts. The ecosystem doesn't possess such an advantage.

A consequence of the different levels of consciousness is that there are different levels of consciousness inside a human body! The human being as a whole is at the highest level of consciousness, meaning that a person can imagine higher thoughts than the cells inside the body. This doesn't imply that the person is fully conscious about what his cells "think." Another example is that a human's level of consciousness is higher than that of a spider, yet the human has little consciousness about what the spider lives through.

So, different levels of consciousness imply different levels of intelligence and different types of knowledge. The intelligence is directly related to the environment in which the being lives. Just like there are different levels of consciousness, the same applies to intelligence:

1. Higher levels of intelligence are like ideas and imagination.

2. A medium level of intelligence might include the use of words and language. This includes mathematics.

3. A lower level of intelligence would include more mechanical processes such as chemical and physical reactions.

For instance, the act of walking is a lower level of intelligence. Studying a map of the city of New York is a medium level of intelligence. Walking through New York using this map is a high level of intelligence.

Because consciousnesss is at different levels, one level has to convey information to the level above in a language that it understands. The problem is that the level above is more intelligent than the level below. How can each level communicate? What's the process that acts as the intermediate between thoughts and physical events?

The Thoughts–Energy Dimensions

In previous sections, I've used analogies to explain a few aspects of entropy such as entropy being the reverse of organisms. But there's more: entropy is the reverse of thoughts! To help see what I mean, consider this simplistic yet valid example. Suppose a child takes the letters of the alphabet randomly from a bag: *M, A, D, L, N, D, B, O, I*. Then I look at them and put them in this order: *A, B, O, L, D, M, I, N, D*: A BOLD MIND. By using a sorting-out algorithm, I made sense out of disorder, thereby reducing entropy. This happened as a result of a thought.

So thoughts *reduce* entropy. This makes sense. For instance, a new company is created from thoughts. Its organization is governed by thoughts. If no thoughts were involved, then the company would deteriorate rapidly and its entropy would increase. Recall that entropy is a measure of the amount of energy no longer available to do work with. If a company is totally disorganized, very little useful work will be accomplished indeed.

There's nothing in physics that accounts for human communication such as the words *A BOLD MIND*. Thoughts are beyond the physical world, so thoughts are not "things." Yet as the simple example above shows, thoughts appear to influence things in the physical world because in this case, thoughts reduce entropy. Finally, entropy is related to energy so there appears to be a connection between thoughts and energy. This section will demonstrate that it is so!

While a thought may lead to a manifestation of energy, that energy is *not* an exact manifestation of the thought. Consider these examples:

1. Suppose we think of a triangle. We can draw a triangle on the sand. Our human minds can deduce that for a perfect triangle, the sum of the angles of the three corners should equal 180 degrees. Although neither humans nor even nature can produce a perfect triangle, as even a triangle shaped by the water molecule H_2O isn't a perfect and static triangle as there's a very small level of uncertainty about its shape. Yet our minds can imagine a perfect triangle.

2. In the section "Solar Systems in Equilibrium" of Chapter 6, we determined that the speed of a celestial body around the sun is given by the formula

$$v = \sqrt{Gm_s/r}$$

This speed allows the body to maintain a state of orbital equilibrium. Just like a human is unable to draw a perfect triangle, but yet is able to deduce its various properties with total precision, nature is unable to follow *exactly* the elliptic orbit that respects this formula which expresses the ideal situation. The physical reality is that a celestial body has to perform a balancing act all the time to maintain its orbit—there's essentially a feedback system here at work. So the celestial body actually follows a slightly wavy path along its orbit such as this:

Of course, this picture exaggerates greatly the waving motion. In reality that motion is so small as to be negligible. It's somewhat like when an acrobat walks on a tightrope and has to rely on feedback from his feet as the tightrope sways back and forth like a wave.

3. The position or speed of a quantum particle can't be determined by nature with absolute certainty. This fact has been mentioned a few times already where it has been called the Heisenberg uncertainty principle. A precise thought can be formulated for the behavior of an electron (e.g., by using equations), but nature can't make good use of it. Nature has to think about what matter is; otherwise matter is little more than invisible electromagnetic waves. Matter isn't a thing but rather a concept, the result of thoughts. Because of the Heisenberg uncertainty principle, these thoughts result in nature having an imprecise "feel" of matter. This is my philosophical take on that principle.

Just like humans, nature isn't perfect. How shocking! You might recall that I already arrived at that conclusion in section "Another Formula for Fuzziness" in Chapter 5 and in section "Does God Exist?" of Chapter 7.

Note that it's a very good thing that nature isn't perfect because if nature were perfect, meaning that it could execute all mathematical thoughts perfectly, then nature wouldn't exist at all! You shouldn't be surprised by this statement because I already proved in Chapter 7 that perfection can't be attained by the system of thoughts that created nature. Perfection is *bad* for Nature.

It's no coincidence that I used mathematical equations in my examples in order to talk about thoughts. Once scientists started to use mathematics to explain nature, it started to make sense. There was no longer a god of the wind, a god of the thunder, etc. Mathematics could explain natural phenomena. In fact mathematics can explain so many things about nature that we might wonder if mathematics isn't the language of God. It could very well be so based on the findings of the two mathematicians below.

In the nineteenth century, mathematician George Boole discovered Boolean algebra, the algebra of logic, making it a new branch of mathematics. Wikipedia states that Boole "did not regard logic as a branch of mathematics (...) but he pointed out such a deep analogy between the symbols of algebra and those which can be made, in his opinion, to represent logical forms and syllogisms, that we can hardly help saying that (especially his) formal logic is mathematics restricted to the two quantities, 0 and 1."

In the early twentieth century, mathematician Bertrand Russell in his book *Principia Mathematica* presented the view that mathematics is in some important sense reducible to logic on which all of mathematics can be built—essentially proving that all of mathematics is equivalent to Boolean algebra. In other words, logic isn't a discipline of mathematics: logic *is* mathematics.

I just suggested above that mathematics seems to be the language of God. So then logical thinking is what God uses. But is it really God that uses logic? In Chapter 7, I proposed that God is beyond logic, more precisely beyond any logical system. It's the *Creator* of the universe that uses logic. As stated in that chapter, it can't be proven that God is the Creator. Nonetheless, so as not to shock anyone, let us attribute this task of creation to God because we are more comfortable with that concept than with some unknown Creator.

Mathematics is logic. Logic involves a set of thoughts and rules. So there has to exist out there a world of thoughts. It seems that our thoughts are from another dimension. Then the thought is communicated to nature somehow, and nature carries out the thought in form of energy. Two questions come to mind:

1. How does nature carry out the thought? Suppose that the thought is an equation such as the one mentioned a few paragraphs ago for the motion of a celestial body. Does nature execute the equation by performing countless calculations, just like a computer? If this were so, then nature would have to perform calculations for every single moment of time as the body orbits the sun. This is impossible, and I'll provide further reasons in section "On Virtual Reality" in Chapter 10.

    I suggest that nature doesn't calculate anything at all! Nature acts as a result of thoughts and feedbacks. An analogy would be a basketball player throwing a ball. The player doesn't perform any calculation to assure that the ball falls into the hoop—a successful throw is the result of thoughts that came from experience. It's the same with nature. Nature learned to do things such as orbiting a body around another or creating life. Albert Einstein said it best: "God does not care about our mathematical difficulties. He integrates empirically." It's the same with nature.

2. How does nature capture a thought? I'll answer this question more completely later, mostly in the section "Our Virtual Universe" of Chapter 10. For now, it suffices to realize that in order to capture a thought, nature has to *understand* it! So we have to deduce that nature has the ability to *think* and to *learn*. You might recall that I already mentioned this possibility earlier in this chapter. This idea seems totally wild and crazy, but I have good arguments to support it coming up later.

    To see that capturing a thought requires an understanding of it, take for instance an ordinary three-year-old child whose music teacher is attempting to teach him how to play Beethoven's Ninth Symphony. You can be sure that nothing will enter the child's brain because he doesn't possess the required understanding of music. The thoughts conveyed by the teacher won't translate into any meaningful action from the child.

Just the same, we'll see later in this section that nature does indeed capture thoughts simply from the obvious realization that nature makes sense. Does this imply that nature has a consciousness? Earlier, I gave six criteria for an entity to acquire in order to demonstrate a level of consciousness. Depending on the level of nature (such as the biological level), nature does indeed possess a capacity for consciousness. At the quantum level, I doubt very much that there's any consciousness although there may exist very rudimentary thoughts. (I give my reasons a few paragraphs below.) Note that there's a difference between a thought-capable organism and a consciousness-capable organism. The latter has to be *aware* that it thinks! For instance, when a dog barks at an approching person, its action is triggered by the thought of protecting its territory. There's thinking involved. But is the dog aware of its thoughts or just executing them?

Since the advent of computers, it became clear that any information can be coded in binary form of electronic currents of 0s and 1s. Mathematicians have proven that such a machine is equivalent to a logical system. So then, logic can express any thoughts and anything in the universe can be expressed using a logical system. That logical system would be part of a thought dimension. I mentioned in the second paragraph of this section that thoughts appear to influence things in the physical. Consequently, nature reflects that logical system, and so nature makes sense. Contradictions are therefore not possible in nature, even though nature isn't perfect.

In various places in this chapter and previous ones, I've suggested that our thoughts might be physically "stored" or felt at the quantum level. In this section, I further suggest that thoughts affect the physical world. These are rather far-fetched ideas, don't you think? No—the paragraphs below will show that such connections between thoughts and the physical world do exist! Actually they have to exist; otherwise the universe would simply not be able to function.

When a computerized axial tomography (CAT) scan is made of a brain, the colors clearly indicate various regions of the brain that are activated depending on what the person thinks. This suggests strongly that thoughts are stored in the neurons. But these neurons are simply evidence of the manifestation of the thoughts. The neural network is involved in the process of thinking, but the neurons are not the thoughts. As an analogy, consider a CD (compact disk) of songs. Does the storage medium constitute the songs? No—the CD acts as storage for the songs. But the songs themselves are just thoughts elevated out of the physical world.

The colors from the CAT scan of a brain show evidence of activity among neurons. The question is this: is it the activity that creates the thought, or is it the thought that creates this activity? If it's the activity of the neurons that creates the thought, then what causes this activity in the first place? If "something" caused this activity, then what created that "something"? You see here that this leads to endless questions, just like the one we encountered in Chapter 7 when we wondered about what created the universe.

So then should we conclude that *nothing* creates the thought? Should we accept that the thought creates itself? We face here the same dilemma encountered in Chapter 7 when we came to the impossible situation that the universe supposedly created itself. If you recall, we had to conclude in that chapter that a system of thoughts created the universe—and thoughts created the physical world. I propose that the same applies with the brain: a nonphysical entity, a thought, triggers physical signals in the brain! It *can't* be any other way.

Furthermore a thought manifests itself at the quantum level as these two arguments demonstrate:

1. It's now clear that it's the thought that causes the activity in the brain. Consequently, the thought can't have a physical cause, but it can cause a physical effect such as in the brain. I can't help but find a parallel here to Chapter 7 where I demonstrated that the Big Bang is the result of thoughts. Our

brain experiences "little Big Bangs," little moments of creation, many times a day—in fact every time we have a thought!

2. Moreover, because a thought is nonphysical, it doesn't occupy space. Consequently, when it shows up in space (in the form of energy), it shows up in a single *dimensionless point*. Don't you find that this scenario looks similar to what Stephen Hawking's "singularity theorem" describes (recall the section "The Singularity Theorem" of Chapter 8)? There's a parallel here with the cause of the Big Bang created from a system of thoughts as suggested in the section "Does God Exist?" of Chapter 7.

Finally, because the singularity point and the Big Bang that followed were created first at the quantum level, I propose that thoughts manifest themselves first at the quantum level as well. In fact this is certainly what happens because as a thought doesn't have dimensions, it has to manifest itself into this world as a singularity point which obviously occurs at the quantum level. Consequently, the thoughts in our brain are captured by our neurons first at the quantum level.

Think about this: it's impossible to pinpoint a thought in the brain as we can only see its physical manifestation. Sure, brain researchers know that certain parts of the brain are activated when a person has a thought. Is it the activation of those parts of the brain that create the thought? If this were so, we'd have to ask what caused that activation of the brain. This leads to endless questions as I explained before. Clearly it's the reverse that happens: the thought activates the brain.

There's another way to see that it isn't the neuron that forms the thought. How do brain cells recognize items as complicated as a chair, the number seven, a dog, Elvis Presley? Are single cells like transistors in a computer or pixels on a computer screen that only when combined with the output of thousands or millions of other cells form the complex pattern that means Elvis Presley? Or can a single neuron learn to recognize that face? Of course, it can't.

Most neuroscientists adhere to the pixel view of neurons, arguing that individual neurons can't possibly be clever enough to make sense of a concept as subtle as the face of a person; after all, even the world's fastest supercomputers have difficulty performing pattern-recognition.

These two points prove that thoughts in the brain manifest themselves at the quantum level!

Also at the quantum level, nature can't tell the difference between a brain and a molecule or anything else. So if a thought manifests itself in the brain at the quantum level, we can be sure that the thought also has the capacity to manifest itself at the quantum level *anywhere* in the universe simply because the quantum world is everywhere in the universe. Hold this thought in mind because I'll come back to it in the section "Discovered Thoughts" of Chapter 10 to offer some daring possibilities.

I made a connection between the physical manifestation of thoughts and the Big Bang. Physicists know fairly well what happened 10^{-40} seconds *after* the Big Bang, but are at a loss to explain what started it all. You might recall that I provided a philosophical proof in the section "What Created the Universe?" of Chapter 7 that the Big Bang was triggered by a system of thoughts. The paragraph below provides a second way at arriving at that same conclusion, although it's more symbolic and simplistic.

Remember back at the example I presented in the section "On Consciousness" of this chapter of the letter N formed with quantum dots. I said that the dots are things, but the letter N is a state of organization, not a thing. So, as far as the *physical* world is concerned, that letter is *nothing*. So, by creating that letter, I created nothing out of something, the dots! Well, as far as nature is concerned, the beginning of the universe started the other way around: something was started out of *nothing*. Suppose that there's nothing on the table. I think of the letter N and I draw it on the table. The drawing was done with something physical, but the source (i.e., the *purpose*) is nonphysical. To a two-dimensional world, the letter N was constructed out of nothing—but in fact it was created out of a thought.

Using this analogy, I suggest that the universe was created by a thought! Moreover, a few paragraphs ago, I said that mathematics and logic are one and the same. As mathematics can explain the universe, logic could very well be the language of the universe—the cosmic language. Finally, thoughts can be expressed in logical terms, so the universe is all about thoughts.

Although the brain needs energy to store a thought, the creation of that thought is elevated out of this physical world. So a thought possesses at least these two properties:

1. It's energyless. It doesn't require energy to exist. But then, in Chapter 8, I proposed that dormant light existed prior to the Big Bang. That light is said to be energyless, but more precisely, it was determined that it's energy that hadn't "woken up" yet. Could thoughts be composed of dormant lights as well? Then when a thought is triggered into this world, it "explodes" into energy such as inside the brain. I proposed that this dormant light resides into that hidden universe that I talk about in the next chapter.

2. It's spaceless. It does't require space to exist.

I propose that thoughts reside in some nonphysical thoughts dimension. As I made a connection between the thought and the energy that it requires manifesting itself in the physical world, I propose now a thoughts–energy dimension. We'll continue this idea in Chapter 10.

Thoughts are surely elevated out of the physical world. In fact, in Chapter 10 I suggest that they reside in some "hidden universe." In that chapter, I also suggest that this hidden universe is closely intertwined with the physical universe. The hidden universe is all around and even inside us! So my thoughts, although in the hidden universe will still "appear" in my brain as if physically there.

It should become clearer to you that the universe is made up of two parts: the physical, and the nonphysical. The nonphysical consists of thoughts while the physical is the *manifestation* of those thoughts. As time is a nonphysical dimension, it resides in the world of thoughts! Now because thoughts aren't in the physical universe, they don't require energy (more precisely, they don't *spend* energy). Consequently, they are ageless in the sense that once a thought is formed, it never dies. This makes sense. Writings by Socrates from the old Greek society may look old on tablets or paper, but his thoughts themselves don't age and thus still reside on that thoughts dimension to this day.

Because thoughts reside outside the physical universe, then perhaps they reside in a thoughts dimension. We can sort of feel the presence of a thoughts dimension when we're deeply concentrated, thinking of something. We become so focused as to no longer even feel time going by. We are in the *now*. Our body moves along in time and in space, but our current thoughts remain in the now:

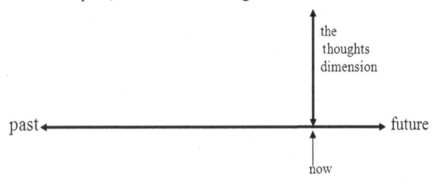

We don't think of the past or future while in that state of mind. This suggests that perhaps the thoughts dimension doesn't include the thought of time after all. We'll see in Chapter 10 that this might be so. Because time is a thought, as I've come to realize, I've decided to bring back into this book the time dimension even though it doesn't exist as far as God is concerned. But time might very well exist as far as nature is concerned, and surely exists as far as we are concerned.

Based on the very simple formula $\Delta E \Delta t \geq g$ that I derived in the section "The Beginning of Nature" of Chapter 8, or equivalently, Planck's formula $\Delta E \Delta t \geq h$, I conclude that nature could not have created the universe. Consequently the creation of the universe is *supernatural*. I don't suggest that God created the universe—I simply say that some supernatural entity created it. (This is also what I concluded in Chapter 7.) So there appears to be two kinds of universes living "together": the natural and the supernatural. How do they communicate? Using the time dimension is one possibility—or more precisely the thoughts dimension. This is the second reason that I had to bring back the time dimension, with the first reason being that nature had to create it to allow for quantum motion. I don't think that God has anything to do with time as a dimension or even nature, but we humans use time. I'll say more on this in the next chapter in the section "Our Virtual Universe."

Creationism and Intelligent Design

Evolutionists believe that life has evolved on its own rather than through divine intervention. In the section "Molecular Structures and Life," I showed on the contrary that life appeared on earth as one single act of *creation*. This act came as a result of a sort of 'bing bang' coming from some singularity point from the "hidden universe" (which is a universe of thoughts and consciousness as explained in the previous sections) into this physical universe.

People of religious faith might call that hidden universe, God, but there's no way to know that. This initial act of creation is definitely one point that creationists have in their favor. Some time thereafter, evolution took root and species evolved naturally. As stated in the section "Molecular Structures and Life," this evolution is based in part on certain levels of consciousness from that hidden universe but in large part through random events. Of course, by consciousness, I'm reffering to consciousness possessed by living species.

It could be argued that God is the Creator of the universe, and that God created the very first life-form on earth, and even that God invented the rules of evolution. However, none of this can be proven by scientific methods. In Chapter 7 and this chapter, I showed that a "hidden universe" elevated out of the physical universe did create the universe and the first form of life. The rest can't be considered out-right as acts of creation.

Some creationists believe that God created all living beings spontaneously. These creationists think they are honoring God's intelligent design, but actually they're doing the opposite! The way these creationists think can be compared to Joe Smith going to the hardware store to buy a preconstructed shed (a kind of spontaneous creation), and then bragging about what he has bought as if he'd constructed it himself. It's a false sense of pride. These creationists demonstrate their lack of faith in God's laws of nature to be able to let evolution make life-forms and humans flourish over time. This is the beauty of God's work—*evolution*.

As I theorize in this chapter, nature appears to have an inherent intelligence that might possibly play a role behind evolution. All living things possess consciousness, and the more complex biologically a living being is, the more evolved its consciousness tends to be.

According to Wikipedia, *intelligent design* is the assertion that certain features of the universe and of living things are best explained by an intelligent cause, not an undirected process such as natural selection. I somewhat disagree on the last part of that statement. As this chapter explained, natural selection isn't entirely an undirected and random process. Many people equate intelligent design to Creationism, but on the contrary, I believe that intelligent design should be associated with the process of evolution. Creationism is about creation "out of nothing." Only three things were created this way: the universe, the first life form and some "brilliant" thoughts—that is, thoughts that seem to come from nowhere. Evolution isn't about creation but about life forms changing over time. For instance fishes in the ocean were not created: they evolved from some other life forms.

Have you ever come across the book titled "The Blind Watchmaker: Why the Evidence of Evolution Reveals a Universe without Design" by world-famous author Richard Dawkins? It's a fabulous book that claims to prove that evolution is without design and that there's no intelligence involved whatsoever. This book seems to prove most of my ideas in this chapter to be wrong. Arguments advanced in the book include:

1. That evolution is without design. The argument is that to design something, there has to be a plan, and a plan is for the future. There has to exist some awareness on the part of nature for the future. There's no such evidence of planning from living beings. Humans are able to plan for the future, but there no evidence that this ability eventually leads to DNA mutations and evolution. DNA mutations are therefore undirected, and this leads to the second point.

2. That the DNA mutations occur randomly. This point has been made numerous times in this chapter. Species have no direct control over their DNA. This means that if a mutation occurred, it was a random event. It also says that, no matter how intelligent the species are, that intelligence cannot control the DNA, and so there's no intelligent design.

My response to the first point is that it's true that evolution isn't driven by some desire to design anything. Evolution is driven by reactions of species to their environment and interactions with it. These reactions and interactions are the basis for the rules of evolution discovered by Darwin and for the four lessons listed near the end of section "Molecular Structures and Life." These rules and lessons favor good designs over bad designs. It sure feels like some intelligence is implicated. Richard Dawkins claims that this is an illusion.

This leads me to a response to the second point. To prove that evolution is without intelligent design, scientists have to prove that DNA mutations are never directed, that is, *all* mutations occur randomly. Have they proven that? To prove that evolution *is* with intelligent design, scientists have to prove that at least some of the DNA mutations are directed. Unfortunately, as I explained in section "Codes and Algorithms for Evolution", this is impossible to prove because directed mutations are caused by consciousness, something that doesn't reside in our physical world and therefore cannot be measured scientifically. So, just like the existence of God, intelligent design will remain in the realm of faith. This doesn't imply that it doesn't exist though. In section "Quantum Physics and Thoughts" of the next chapter, I present an argument that suggest that in some cases, thinking was involved in the evolution of the brain—if with the brain, then perhaps with other type of cells as well...

On the one hand, it remains true that most DNA mutations occur randomly. The consequence is that probably most DNA mutations form a variety of species with features that are not at all tuned into the environment that they live in. These species therefore die off rather quickly. Only species with favorable DNA mutations will thrive. This is a kind of 'sort-out' algorithm that weeds out the failed designs. Oddly enough, evolution is largely driven by failures. It doesn't feel that way because we only see successful species.

On the other hand, consciousness of species is part of evolution. Nobody will deny that species have over time developed increased awareness, none so more than humans. The question is: what has evolved? Is it the consciousness or is it the capacity of species to capture that consciousness that is already there in that hidden universe of conscience that I introduce in the next chapter? In that chapter, I theorize that consciousness in all its highest magnitude has always existed. Life simply is the physical means of "pulling" it into this physical universe.

Another notion discovered in this chapter is that life isn't of this physical world. It belongs to, what I call in the next chapter, the hidden universe of consciousness. People view creationism and evolution as two contradictory theories. This chapter has attempted to show that they are complementary theories. Evolution is about the *how* of life; creation is about the *why*. To put it in computer technology terms, evolution is to hardware what creation is to software.

Chapter 10: Is There an Afterlife?

Some five billion years from now, our planet will inevitably disappear. Humankind will go down as well—if not much earlier than that as some scientists assert! There's no way around that fate. What about our own lives? We are not just some piece of rock. Life reaches out beyond the physical as we saw in the previous chapter. So it's natural to ask "Is there an afterlife?" Actually a much better question is "Is there life *before* conception?" Here are a few arguments.

What are Souls?

Clearly there's no scientific way of determining what a soul is simply because its definition is entirely based on what humans believe it to be! Whether souls exist depends on who you ask. Here are a few plausible characteristics of a soul:

1. By definition, a soul survives physical death. Therefore, a soul is immaterial. The main motivation for the belief in souls is that they live on beyond death.

2. By definition, a soul has an identity. My soul has to be about *me*, obviously. During life, the body is associated with its soul. The identity of the soul clearly can't be based on the body because the body doesn't survive death. The identity has to be defined by some core thoughts that survive death. These core thoughts would never change during life or after death. Therefore they identify the person during life and after death.

3. A soul is a set of thoughts. This follows from item (2).

4. A soul has spiritual properties. Could the core thoughts define the spirit of the person? By definition, the spirit of a person includes fundamental emotions that determine that per-

son's character. As emotions are possible only via a surge of energy, they manifest themselves in the physical world. So do emotions survive death?

5. A soul grasps the notions of right and wrong.

6. A soul isn't associated with its body, not even with the brain!

    a. The first part of this characteristic is easy to understand. There are only a finite number of human genes. Thus, there are only a finite number of unique bodies that can be conceived. This means the possibility exists that two humans could be born with exactly the same genetic makeup—essentially two *identical* bodies. Well, we already know that to be true, because twins exist. Yet each twin has his or her own soul.

    b. The second part of this characteristic follows from the fact that the brain is physical. However, thinking takes place inside the brain. As a soul consists of a set of thoughts, should it not be associated with the brain? No, because thoughts are immaterial. They manifest themselves inside the brain in the form of energy, but they're not inside the brain! (More on that idea in the section "Our Virtual Universe" coming up). For instance, it's known that a person suffering from Alzheimer's disease loses a sense of identity and other capacities. Has that person lost his or her soul, or is it that the soul has lost its medium to manifest itself? I believe that it's the latter.

Now here are a few arguments related to some of these characteristics.

I have no recollection that I lived before my present life. I know that I never existed before. So if God could do without me for billions of years, why would it be so important for Him that I be "around" after my death? The counterargument to this is that the fact that I have no recollection of a previous life is no proof that I never existed before.

Now suppose that I did exist before my current life. What does the "I" mean then? Is my current "I" the same as the one from supposedly a previous life? Does the "I" mean my soul and body, just as it is now? When we die, our bodies decay and disappear altogether over time. So it certainly can't be the body that will go from one life to the next: it can only be the soul. The "I" must refer to my soul only. So maybe my soul has existed before—there's no way to know. Our soul surely contains our spirit. Do they also contain our knowledge? Do they contain our thoughts? Can that be carried from a previous life to this one?

If my soul survives the death of my body, then perhaps my soul could go from one body when it dies to another body that is conceived. It wouldn't be the same body, but the same soul. Would that really be me in both cases? It certainly wouldn't feel like it. How can I be "me" in two different bodies in two different lives? Part of the "me" is my identity. The trouble is that my identity is influenced by my parents and my social environment in general so that I couldn't possibly remain "me" in another life. Therefore, I can't exist twice as *flesh*.

Indeed, if we are to understand what a soul is, we have to remove all notions related to the physical world and stick to only the nonphysical. But then what's left in my soul that ensures that it remains *different* from another soul? What characteristics of spirituality ensure that my spirituality is *unique* from any other human that ever existed and will ever exist in the entire universe?

To have a soul requires consciousness. But is that enough? In the previous chapter, I concluded that *any* lifeform has some level of consciousness, but it seems totally ridiculous to suggest that a bacterium has a soul. So, possessing some level of consciousness is definitively not enough to have a soul. In the section "On Consciousness" of the previous chapter, I listed six criteria that characterize a being with a consciousness:

1. It shows evidence of making sense of its environment and interacting with it.

2. It demonstrates a will to live such as wishing to consume food.

3. It demonstrates a will to replicate (or reproduce).

4. It evolves.

5. It's able to produce thoughts.

6. It demonstrates some level of intelligence.

These criteria are not enough for a being to have a soul. Based on the characteristics of a soul given at the beginning of this section, the list lacks the requirements that:

1. A being has to know that he or she has an *identity*.

2. A being grasps the difference between *right* and *wrong*.

3. A being has other characteristics such as the notion of the divine and an afterlife.

Clearly humans possess these characteristics. What about animals? Why would God favor humans over animals? After all, we are all a product of nature. It's clear to me that some animals have some conception of right and wrong. When my cat wants out of my arms, he doesn't bite me to hurt me; he just opens his mouth and pretends to bite me. This is his signal that he wants out. Why does he not bite me to hurt me? Because he knows that it would be going too far to simply send a signal. In other words, cats do show some control. This is evidence of some level of consciousness and knowing right from wrong.

Note that the requirement of grasping right from wrong means that the being understands that there are right and wrong things going on out there. The requirement doesn't suggest that the being knows what's right and what's wrong. In other words, the being may not have the capacity to distinguish which action is right and which action is wrong. Humans can grasp the fact that there is right and wrong. However, influences such as cultures and upbringing may easily blur what's considered right from what's considered wrong.

What about identity? Do animals have a sense of identity? A zoologist is in a better position to know, but my hunch is that they do. So then do animals have a soul?

There has to be more to a soul than that. Is it necessary to have a sense of the divine in order to know right from wrong? It depends at what level of abstraction. This is what really separates humans from animals: *abstraction and the sense that something intangible higher than us exists.*

But there's more. A soul embodies the *why* of a thing's existence, its meaning and purpose. It's a thing's "inner identity, it's raison d'être." Animals definitely don't have those capacities. Or maybe they do—we have no way of knowing. Does this mean that a person suffering from a deep depression who has lost all purpose in life has also lost his or her soul? No, it would just mean that the person's brain has temporarily lost its connection with the soul.

Just like the "soul" of a musical composition is the composer's vision that energizes and gives life to the notes played in a musical composition, the actual notes are like the body expressing the vision and feeling of the soul within them.

Based on the characteristics of a soul given earlier, it appears clear that souls exist during our lives. This isn't surprising as we first assumed that souls exist, and then attached characteristics to them. But do souls still exist after death? This is the *big question*. Here are two arguments that sadly suggest an answer in the negative:

1. Science now knows how a baby is conceived: sperm S from the father meets egg E from the mother, and baby SE is born. Congratulations, proud parents! In the semen, there are millions of sperms S_1, S_2, S_3, etc. Every S_i has a shot at meeting egg E, but only *one* of them will be the winner. Every other combination $S_j E$ will *never* happen. As every combination $S_j E$ represents a *unique* human being, then it follows that there are millions of human beings who will *never* exist— and this argument applies to all couples on this planet and other planets with life in the entire universe every single day.

So there are billions of billions of humans who will never exist.

If God can do without all those potential humans, then why would He care that I exist in any form after my death? It's egocentric to believe that life was created only on this earth. It's equally egocentric to think that I should exist in some form after my death. I don't intend to trivialize our lives as we are more than a lump of rocks. Life is precious—all forms of life. Human life is even more precious because we have the intelligence to grasp its significance, and most of all, we get only one shot at it! So live well.

2. Another argument that puts into question life after death is simply the purpose of it. Why should it be so necessary that I exist again—in any form—in another life? If none of me, neither my body nor my soul, remains after my death, then none of me will be there to be aware of it. So my entire, complete death can't affect me, but it would affect the people still living! This is one reason that the question of life after death is so important to all.

Some might dismiss the arguments above by suggesting that God doesn't really care about bodies, He cares about souls. Perhaps God already has created all possible souls at the time of creation of the universe (or perhaps before that) and when a baby is conceived, its brain receives one of those souls. This idea could have some merit because the soul is independent of the body. So even if billions of combinations of S_jE will never exist, it might be fine because these combinations represent a body, not a soul. Those souls would simply remain bodiless forever.

But there's a problem with that theory. According to Wikipedia, a soul is a "spirit or essence of a person, usually thought to consist of one's thoughts and personality". Another Web site says that the soul is "the true identity of a human being, the immortal self that surpasses death and lives on into the other world." I've seen many other definitions fairly similar to these. Two possibilities:

1. If those definitions of a soul are true, then a soul requires a body. Consequently, God can't possibly have created all souls at the time of the creation of the universe. A soul is created only when a human is conceived. Therefore, billions of billions of souls will never exist because the human being with the sperm–egg combination S_jE will never exist. So why would it be so important that my soul still exist after my death? There would be no reason for it as far as nature is concerned, and perhaps even as far as God is concerned.

2. If those definitions of a soul are false, then a soul isn't a set of thoughts and it doesn't have a personality. What would a soul be then? Could we even say that souls exist? If souls exist, then they have no description.

Either way, the definition of a soul is unclear, and thus its existence after death even more unclear. Consequently, the existence of souls resides in the domain of beliefs. Just as in the section "Does God Exist?" of Chapter 7 I indicated that it's impossible to prove that God exists, it equally appears impossible to prove that souls exist, let alone after death. I've seen on the Internet people who claim to have scientific proof of the existence of souls, but this is bogus. Humanity will never be able to prove that souls exist just as we'll never be able to prove that God exists. Both will forever remain in the realm of faith.

Or perhaps they won't. Is there a method to prove the existence of a soul? As a soul doesn't have any features coming from nature, it seems to make no sense that nature could be used to prove the existence of souls. Thus should there be any proofs, they'll be indirect ones.

How about the handwriting style of person? It's extremely difficult to mimic the handwriting of someone else. Even twins have different handwriting. What factors determine handwriting? Personality, isn't? By definition, personality is a part of the soul. Now is handwriting influenced by the environment? Is it influenced by the way the person was raised? It's probable. How about being influenced by the genes? This is not so sure—I have ten siblings, and although we were all raised in the same environment, none of us have similar handwriting.

So, handwriting seems to be independent of physical factors and would be a good candidate for determining if a soul has lived through more than one life. If two persons of distant generations who are not even related have nearly identical handwriting, could that be proof of souls existing after death (and before conception)? This idea deserves to be studied by scientists.

It's most difficult to prove the existence of the nonmaterial. We do nonetheless know that thoughts exist. I'm clearly engaged in thinking as I am writing this paragraph. A thought is clearly not something material. It's conceivable that my thoughts might remain somewhere after my death. In the next section, I'll call that somewhere the "hidden universe." If you read the previous chapter, you already know a few things about that mystical hidden universe. What about emotions? My view is that emotions are thoughts expressed with energy in the physical world. Consequently, our emotions won't survive our death, but the thoughts associated with them would.

Scientists quite often dismiss the idea that there might be any "supernatural" entities. But the "supernatural" may not be "super" at all—it may just be something natural beyond what humans can understand. This might be the case with souls and other nonmaterial things. This reminds me of Arthur C. Clarke's famous comment that "any sufficiently advanced technology is indistinguishable from magic." This can be phrased another way: any society *not* sufficiently advanced technologically will be unable to distinguish magic from the natural. Maybe quantum mechanical waves and fields are the physical portal to the spiritual world. If so, then science may already be studying spirit without really knowing it. This idea goes well with the words I expressed in the previous chapter, especially in the last six sections starting with the section "Codes and Algorithms for Evolution".

Where Do Souls Reside?

The impossibility of proving the existence of souls is no reason to reject the possibility of their existence. Let's assume that souls exist. In section "The Thoughts-Energy Dimensions" of the previous chapter, I explained that thoughts are stored in a nonphysical dimension. What about souls? By definition, souls include some kind of thoughts. So a lot of what I said previously about thoughts may apply here too. It's within reason to think that a soul is immaterial, and that it has no energy. Let's look at this from two different views:

1. If a soul did have energy, it would exist in space because as you recall in Chapter 1, we said that where there's energy, there's space, and vice versa. If a soul is nonmaterial, it doesn't contain energy. Consequently, it can't occupy any space and can't reside in the space dimensions. Where is a soul located then?

    The only option left is that it exists in the thoughts dimension that I theorize the existence of in section "The Thoughts-Energy Dimensions" of the previous chapter. But I'll instead use the time dimension because it ties well with the ideas expressed in other chapters. Indeed, I showed in the section "The Time Dimension is Nonphysical" of Chapter 4 that if an entity occupies no space and emits no energy, it's confined to existing in the time dimension only. Now this is interesting because the Bible says that God is the keeper of all souls, and in section "The "God's Light" and the Expansion of the Universe" of Chapter 8 I explained that God indirectly created (through nature's mechanism) the time dimension when God's Light was released from the Big Bang. This suggests a possible link between souls and God. Hmm! There might be more to that dimension than what meets the eye.

    Anything sitting on the time axis (time dimension) occupies no space and therefore doesn't contain any energy or matter. Consequently, all things that sit on the time axis can't be seen or felt in any way by anything that lives in the space dimension. Souls, if they exist, might reside on the time dimension, which confirms the previous paragraph. Souls would live on the time dimension during our life and after death. My soul would be *exactly where I am* because we remember from Chapter 3 that the time axis is everywhere in the universe. When I die, my soul will be all that is left of me: I'll no longer be in the space dimension.

2. The first view above uses the finding in Chapter 1 in which I said that where there's no energy, there's no space. Then I showed in Chapter 4 that if an entity occupies no space and emits no energy, it's confined to existing only in the time dimension. This is all good stuff, but how correct is it? Could it be instead possible for an entity to exist in space and yet not emit energy?

   Yes, it's possible, and this finding was discovered in the sections "The Energy Dimension" and "Motion Finally Explained!" of Chapter 8 in which we discovered the existence of dark energy. Recall that this is energy that constitutes the fabric of space. Moreover, dark energy contains energy—but it does *not* emit it. So anything hiding in the fabric of space would seem as though it doesn't contain energy. Ah! Just like souls.

   Perhaps souls reside in the fabric of space, dark energy, and not in the time dimension. Because space is everywhere, just like the time dimension is everywhere, my soul can easily be right where my body is, hidden in the fabric of space. So what I said in the first view above applies here, except that the time dimension is replaced by the dark energy which is like an energy dimension as proposed in Chapter 8.

I find both views correct and they actually complement each other. The first view deals with the nonphysical aspect of souls whereas the second view deals with their physical aspects.

But hold it! Based on the definition of souls from the previous section, souls don't have any physical properties. But there's no contradiction if we assume that the fabric of space would just be a "storage area" for a soul—just like our brain is a storage area for our thoughts. This is like a CD being used to store songs while the songs themselves are nonphysical entities. The nonphysical time and the physical energy dimensions go hand in hand. As will be seen in a few paragraphs, the time dimension is the *portal* to the nonmaterial world while the energy dimension is the *receptor* of the nonmaterial world.

In my view, the scenarios (1) and (2) developed above are simply another way to describe a singularity point described in section "The Singularity Theorem" of Chapter 8. Indeed it is said there that the singularity theorem describes "what happened the moment before the Big Bang. So it's no wonder that the laws of physics broke down because these laws weren't even created yet!"

Here's my reasoning. Since a singularity point has no volume, it lacks the space dimensions. However it might still possess the time dimension and dimensionless energy. Recall my explanation given in Chapter 8, whereby energy transfers requires space, but energy itself doesn't. So, it's possible that a soul might always contain energy, but it is "dormant" energy inside a singularity point! Then when it wants to enter our physical world, it unleashes its energy and appears in our physical world. So the appearance of a ghost could be a sort of little "big bang."

When a person dies, the reverse happens. The energy associated with the soul is channelled along the time dimension into a singularity point, and becomes dormant. I mentioned many times by now that the time dimension is associated with light. Now, isn't this a coincidence? Many near-death survivors claim to have been channelled inside a tunnel during the near-death episode and then to have seen a very bright and pure *light* that invited them into the other world, that I call the "hidden universe" of Consciousness.

I might be onto something here...

The role attached to the time dimension is fascinating, but shouldn't souls be instead in the thoughts dimension that I speculated about in the previous chapter? Indeed, souls are composed of thoughts, perhaps among other things.

In the section "The Thoughts–Energy Dimension" of Chapter 9 I stated that time is a thought with the reason being given in the section "Motion in Space–Time: The Plot Thickens!" of Chapter 5. Also Einstein's Special Theory of Relativity proved that time is a dimension. The argument below will suggest that time is a *thought* dimension of its own.

After we die, if anything of us lives on, it can't be anything more than our thoughts (and soul) because the universe's thoughts don't have a sense of anything outside that. So it makes sense that our body decays after death. This goes well with what the Bible says that after we die, our soul goes back to God. As a soul is a form of thought, then our soul goes back to God. However, in the next section, we'll conclude that after death, our souls will return to some hidden universe rather than to God—although we'll see that this hidden universe has a very close tie with God.

The time dimension could be seen as the portal to the thoughts dimension that holds everything that is nonphysical such as souls. So souls wouldn't reside on the time dimension after all but rather in a thoughts dimension. The thoughts dimensions would be two: one for thoughts and one for consciousness. Why two other dimensions? Because thoughts can be represented as a sequence of codes very much like in a computer as was explained in the section "The Thoughts–Energy Dimension." That is one dimension. Consciousness is the result of thoughts, so it's elevated out of thoughts and hence into its own dimension:

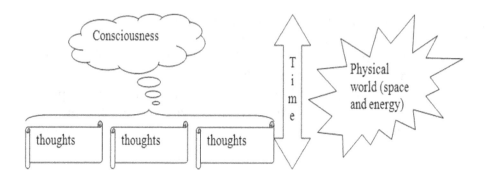

The diagram shows that the time dimension is the link between the non-physical and physical world. As we saw in Chapter 7, a system of thoughts was present *before* the creation of time. So there was no such link before the Big Bang. Also, if the universe comes to an end (an end that will never come according to my view in the section "The Expansion of the Universe and Its Fate" of Chapter 8), the time dimension will inevitably disappear because it's associated with energy (the energy dimension as mentioned earlier).

What will happen to the thoughts dimension when time is gone? The figure above seems incorrect in tying the thoughts dimension with time. But no, the figure is correct. Thoughts can't possibly reside on the time dimension—and they don't, just like energy and matter don't reside on the time dimension either. The present argument and the following suggest that time is indeed a dimension for three reasons:

1. In our physical world, thoughts do seem to be tied with time. Indeed, I said earlier that the storage area of souls might reside in the fabric of space. In the section "The Time Dimension is Nonphysical" of Chapter 4, I drew a parallel between the time dimension and the fabric of space.

2. We'll see later in the section "Discovered Thoughts" that thoughts are timeless. Although I said in the section "Motion in Space–Time: The Plot Thickens!" of Chapter 5 that time is a thought, its dimension can't reside on the side of the thoughts dimensions because whatever reside there is timeless. But time cannot be timeless, of course. So it would be a

total contradiction to place time in the thoughts dimension. In addition, because time is nonphysical, it can't reside in the physical part of the universe. So the only place where the time dimension can be is *between* the nonphysical, nonmaterial universe and the physical universe.

3. In Chapter 7, I deduced that the universe was triggered by the thought of time on the part of the Creator. Then nature created the mechanism that led to the creation of the time dimension. It seems that everything gravitates around time. In his space–time theory, Einstein attributed the wrong role to the time dimension. Time has nothing to do with gravitation: time is warped by gravitation, but time doesn't cause gravity.

You might see a contradiction in the picture above. It seems to link the time dimension with the thoughts dimension, yet I stated more than once that thoughts are timeless. Indeed, they are timeless in the sense that they're not affected by time simply because they possess no energy. A thought never gets old by itself. We may view a thought as old, but it's us getting old, not the thought!

There might be some truth to souls being able to store themselves in the fabric of space because it might explain the appearance of ghosts. I've never seen a ghost, but some people firmly claim to have seen one. Let's suppose for a minute that ghosts exist. How do they suddenly appear out of thin air? Well, if souls may reside inside the fabric of space, then perhaps ghosts may too. Ghosts would reside in dark energy that somehow at times comes out of the fabric of space, which makes them appear in form of light.

People who claim to have seen a ghost often experience either:

1. A sudden chill in a well-defined small region in the room.

2. A sudden electromagnetic fluctuation in that region.

3. A breeze.

I advance here the idea that these experiences are consistent with ghosts residing in dark energy. The reason is simple. For a ghost to appear in some form or energy—a light or even a human shape—that energy can't come from the dark energy because that type of energy can't radiate. So the ghost has to *draw* energy from the environment around the room precisely where the ghost hides. The consequence is that the room temperature will decrease slightly at the ghost's *precise* location. It's like if we placed a big block of ice in a room. The ice will draw energy from around the ice.

As for the electromagnetic fluctuation, it would be caused by the energy being drawn to the small region in the room. The dark energy would act like a small magnet. The breeze sometimes described during the apparition of a ghost could be explained by the fact that a lowering of temperature changes the air pressure causing air to move slightly. This is the same phenomenon as in the atmosphere when a high pressure system meets a low pressure system.

When we die, the popular belief is that we end up on the "other side" with God. This is also what the Bible says one way or another. However, based on what I theorize about souls, it seems that we won't be with God, or at least not as close as is believed. Two reasons:

1. God is much more than us, even considering our souls. We won't be on the same level. God isn't a soul. God isn't an entity anyway we think of it. It's a bit like my soul being inside a house, and God being the owner and caretaker of the house. I made it clear in Chapter 7 that God is beyond any system, and beyond everything. So it's clear that we won't be alongside God!

2. The physical storage of our soul supposedly resides in the fabric of space. That's not where God is, as God is above all that. Even should our soul reside in the thoughts dimension, then we'll still be in the universe, but in the nonphysical part of it. We'll probably be closer to our still living loved ones than to God. In fact, our soul would be *very* close to our

loved ones, just confined in a hidden dimension, invisible. We'll pursue this idea some more in the section below.

There's life after death. Of course nobody can know what it will be like. But it's clear that it'll be a life of thoughts alone for two reasons:

1. We won't identify with our bodies because our bodies will have disappeared and will no longer be relevant.

2. We'll also lose our own identity as a person. Why do I say this? Recall from the previous section the argument that perhaps God already has created all possible souls at the time of creation of the universe (or perhaps before that), and that when a baby is conceived, its brain receives one of those souls. But then what about all the babies who will never be conceived because billions of sperms will never conceive? Their souls will never receive any identity. Clearly a soul receives an identity only when a baby is conceived, and this is what's confirmed by the definition given earlier that a soul requires a body.

   The consequence of the above argument is that—as unbelievable as this seems—when we die our soul will lose its identity! This means that after death, we lose the sense of association to all aspects of our identity while we were living. Our culture, nationality, color, and personality will become irrelevant. We'll loose our ego. However, because thoughts are never lost, we'll still retain in the "hidden universe" all information related to our identity. It's just that the identity itself will become irrelevant.

   So imagine this: after your death, suppose that someone from the living is able to connect with you and ask you questions about what your name was or what nationality you were. You would be able to answer, but not because you understand your identity, but rather because your thoughts remain after death. So you just retrieve the facts.

Understanding requires a *thinking process*, and this requires energy present in the *physical* universe only. The "hidden universe" requires the physical universe for execution of its thoughts. So when a ghost appears in our physical universe, it might do so because it wants to think, to communicate either with a living being or with another ghost!

What will it feel like after death? I'll say some more about this in section "Heaven", but for now, I suggest that it might feel like when you are daydreaming very deeply, so deeply that you no longer feel your own presence in space and time or your own identity. In such a state, you are just aware of your thoughts. Were you ever in such a state? Of course, this is just my theory. Humans will never know the truth about this.

Our Virtual Universe

Remember in the section "Who is the Observer?" of Chapter 5, we learned humans see and feel physical phenomena as nature lets us perceive them. But how does Nature (at the quantum level) feel something? Answer: it's the fabric of space that "feels" energy. You might recall that I've already stated that a few times before.

In the section "The Beginning of Nature" of Chapter 8, I concluded that nature is part of the universe, and so it's not the universe itself. With my suggestion in item 2 of the section "Who is the Observer?" that our reality is limited to what nature sees, this implies that we do *not* see the real universe! There are three reasons for this:

1. Nature masks the real universe from us because nature has to interpret what it feels. One example of this is nature's interpretation of time as discussed in the section "Motion in Space–Time: The Plot Thickens!" of Chapter 5. Other examples were given elsewhere, mostly in the previous chapter.

2. Then our five senses feel what we see, hear, smell, touch, and taste. The brain has to interpret what it sees and feels— and it can only see and feel nature's interpretation of the real universe.

3. The brain then has to make an *abstraction* of what it feels with our five senses. For instance, when we touch a brick, our brain attaches a meaning to it that is different than if we touched a simple rock. Another example is when I see a dress and pants. In some cultures the first one is strictly associated with a female while the other one is associated with a male. And so on …

So humans get a triple dose of interpretation. The first two levels of interpretation are beyond our control whereas the third is based on our upbringing, collective and personal experiences, etc.

Nature and our brains let us see one universe, but that implies there's another one *deeply* hidden! What's the constitution of that hidden universe? You can be sure that it's quite different from what we perceive in our daily lives. The following two points taken from the section "The Thoughts–Energy Dimension" of Chapter 9 propose an answer:

1. Noticing that the study of physics is filled with mathematical equations, it's clear that mathematics explains the physical world. Mathematicians such as Boole and Russell discovered that mathematics is in some important sense reducible to logic on which all of mathematics can be built—essentially proving that all of mathematics is equivalent to Boolean algebra. In other words, logic isn't a discipline of mathematics: logic is mathematics. Logic enables the formulation of thoughts.

2. Thoughts manifest themselves in a physical form in the universe.

These two points bring me to conclude that the hidden second universe is a mathematical one, one made of thoughts. So the thoughts dimension, whose existence I theorized in the section "The Thoughts–Energy Dimension" of the previous chapter and used in the previous section, might actually be much more than a dimension: it might be the *real* (but hidden) *universe*! So what feels real isn't, and what feels unreal is real. Clearly the real nature and universe are not the ones we feel and see. The implication of this is that we live in a *virtual universe*! Here's a picture that shows the interaction between the hidden true universe and our virtual universe:

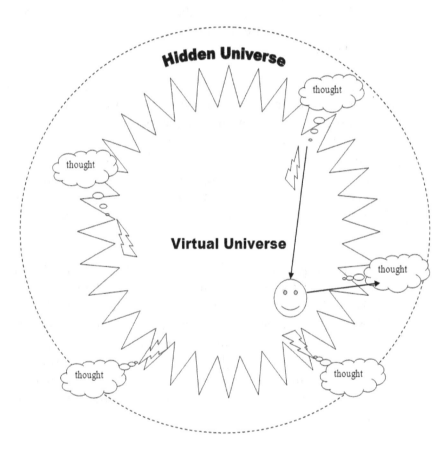

The lightning symbols represent the physical manifestation by nature of a thought. Recall that in the section "The Thoughts–Energy Dimensions" of the previous chapter we proved that thoughts manifest themselves in the form of energy in our universe, the virtual universe. The face represents a human. Note that human thoughts reside in the hidden universe (the arrow going outward), and also that humans can discover thoughts (the arrow going inward) from the hidden universe as well. We'll see the reasoning behind these observations later. Note that the hidden universe isn't in some isolated corner of the universe: it all over our virtual universe. So then our thoughts are always where we are. The reason, as was deduced in the section described above is that the hidden universe is spaceless.

The way that the picture above is depicted, it seems to suggest that the hidden universe wraps around the virtual universe. But that isn't so. The hidden universe is closely intertwined with our virtual universe. It's all around us, and even inside us! The hidden universe has that same property as the fabric of space in that it's everywhere, including inside us, except that the hidden universe is spaceless and energyless.

Our hidden universe consists of pure thoughts and possibly souls. People of religious beliefs might think of the hidden universe as heaven where our soul resides after death so the thoughts and souls of our departed loved ones are around us. That is a reassuring thought.

The hidden universe is intertwined with our virtual physical world. This gives us the feeling that thoughts seem to be "out there." Moreover, I showed that the brain's thoughts reside in the hidden universe. As thoughts are nonphysical, can I really say that my thoughts are inside my brain? They surely manifest themselves physically inside the brain, but they can't be in there simply because the hidden universe doesn't reside in space!

(Please note my concept of a virtual universe shouldn't be confused with the virtual reality you may have heard about. I clarify this in the section "On Virtual Reality" below.)

My proposition that we live in a virtual rather than real universe seems very far-fetched, but then there's no means of determining whether it's true. Indeed, in a virtual universe, everything in it will seem real to its inhabitants. But read on and you'll see that this hidden universe truly exists!

As we concluded in Chapter 7, the universe was created by a system of thoughts. The question is: do all thoughts manifest themselves in a physical way in our virtual universe? For instance, do souls manifest themselves in our virtual universe? We already dealt with this in the section "Where Do Souls Reside?": souls manifest themselves through the fabric of space. Note that the time dimension is actually part of the hidden universe because it's nonphysical. However, as it's also useful in the virtual universe, we may visualize that dimension as a portal between the two universes. Everything that goes from one universe to the other would have to go through the time dimension.

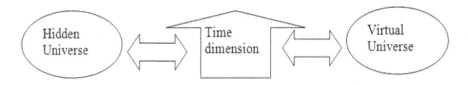

You might recall that I had mentioned in section "Molecular Structures and Life" of the previous chapter that life originates from this hidden universe. So all thoughts related to life reside there too. Perhaps the rules of evolution reside there as well. *All* thoughts, no matter the kind, reside in the hidden universe.

In our *virtual* universe, there's *energy*. In the *hidden* universe, there are *thoughts*. But do you think it's totally absurd to suggest that *thoughts* in the hidden universe translate into *energy* in our virtual universe? It's not absurd if you understood well the arguments made in the section "The Thoughts–Energy Dimension" of the previous chapter.

A thought in the hidden universe can and *does* translate into a burst of energy in our virtual universe. Note that a thought doesn't generate new energy. If it did, that would violate the First Law of Thermodynamics which states that energy can't be created or destroyed. To manifest itself in the physical world, a thought has to *draw* from energy already present in space. Do you find a similarity here with the manifestation of a soul or ghost in space discussed in the previous section? Based on what I said about souls, I theorize that a thought enters the brain as follows:

1. It traverses the hidden universe and goes through the time dimension, the portal between the two universes.

2. From that portal, it stores itself in the fabric of space. We found in the section "The Time Dimension is Nonphysical" of Chapter 4 and other places that there exists a parallel between time and the fabric of space.

3. Once in space, it draws energy from the brain. How does the thought find its way into the proper area of the brain? I propose that "entangled" quantum particles within neurons always remain connected to the thought that created that entanglement in the first place. Physicists have proven that entanglements of quantum particles do exist. I say more about this idea of entanglement later on, so read on, please.

In our virtual universe, energy transfers at no faster than the speed of light. At what speed can thoughts travel in the hidden universe? In that universe, there are no space dimensions and there's no energy; therefore, speed has no meaning in there. Consequently, everything in that universe happens *instantaneously*! The result in the introduction of the following section might prove me right.

Quantum Physics and Thoughts

According to quantum theory, it has been *mathematically* proven that two quantum particles at the same origin that are "entangled" and then thrown many light-years away from each other remain *synchronized*, meaning that the behavior of one particle is *instantaneously* replicated by the other! Two particles are entangled when they are coupled so that they must swivel in opposite directions. Forcing one to spin clockwise will set the other spinning counterclockwise, so that no matter how far they are separated in space, they are fatefully intertwined. This implies that communication is possible at a speed much faster than the speed of light. Although physicists understand the mathematics involved, they admit they don't understand how this can be physically possible.

Moreover, researchers have been successful in synchronizing more than two quantum particles. When a quantum particle changed state, all the others changed state as well and *instantaneously*, no matter how far apart they are from each other. As an analogy, consider a church choir where every singer sings the same song. They are synchronized, and this is possible only if they sing the same words, and so consequently *think* the same thing.

I theorize therefore that "entangled" quantum particles are of the *same thought*. I theorize also that the more complex the thought is the more "entangled" quantum particles are required. Based on this and my model of a hidden universe, that synchronous property may be explained this way. There are two factors at work here:

1. As long as these quantum particles are of the same thought, we'll remain entangled and therefore synchronized.

2. Because in the hidden universe, there's no distance, both particles will remain synchronized no matter how far apart they are. One particle could be at one end of the universe and the other at the other end, yet they'll remain synchronized. Consequently, in theory anyway, two living beings could be sharing the same thought no matter how far apart they might be as long as their brains have quantum particles that are entangled with the same thought. However, in practice this is highly unlikely.

Actually this synchronization is possibly what happens in the brain as well. When a thought is formed, researchers have found that it's not just a single tiny point of the brain that's activated. A region—and sometimes many regions of the brain—are activated at the *same* time. As a thought manifests itself at the quantum level, it has to be that the thought activates quantum particles in many places *synchro-*

nously. Essentially these quantum particles are "entangled" by the thought.

Physicists call this the principle of *quantum nonlocality*, and it was that principle that the two "entangled" quantum particles of a few paragraphs ago were obeying. The principle of nonlocality states that objects may experience the same event even though they are not in the same local environment. This doesn't make sense in our macroscopic environment as a person's thoughts at the other side of the earth have no impact on my thoughts, but at the quantum level, it's a different world altogether.

Could it be that the brain is structured in such a way that quantum particles are "entangled" in the process of forming a thought? Depending on the thought, numerous quantum particles would be entangled. But there's one problem: a few paragraphs ago, I stated that only *two* quantum particles may be entangled. However, I found on the Web site physicsworld.com/cws/article/news/2871 (material used with permission):

> When two or more particles are "entangled," the wavefunction describing them cannot be factorized into a product of single-particle wavefunctions. This means that a measurement on one particle will immediately influence the state of the other particles in the entangled system. A group of physicists in the US has now "entangled" four particles for the first time.

There may be some truth to this quantum entanglement in the brain because it's well-known that the structure of the brain is made of over 100 billions neurons physically connected in *tangled* webs of connections. These webs would facilitate our brain in translating thoughts from the hidden universe into physical manifestations inside the brain.

It has to be that through evolution; somehow the structure of the brain and neurons became in tune with making sense of the quantum level and bringing it up to the level of neurons. A sort of translation would exist from the quantum level up to the cellular level. As a result, biological connectors evolved between neurons. This evolutionary process is clear evidence that cells and neurons that evolved from them possess a level of consciousness. It *has* to be so!

I propose that neurons are unique cells unlike the other billions of cells in our bodies. One evening, my son was watching a science TV program in which the narrator said that all cells throughout our body die out and are replaced by new cells a few hundred times in a lifetime. So although my skin never seems to change in appearance, its cells are replaced numerous times throughout my life! The narrator also noted that this replacement takes place everywhere in the body *except* inside the brain where the neurons are never replaced.

Right then, I wondered why this is so. What's unique about neurons that made them evolve into cells that have to last the entire *lifetime* of the living being? The answer came to me right away. A few paragraphs ago, I theorized that quantum particle entanglement is the way nature uses to store a thought inside the brain. I suspect that this entanglement takes places inside the neurons. If neurons died out periodically and were replaced by new ones, this entanglement would *disappear* during the process of replacement. As a result the thought related to that entanglement inside the brain would *disappear* too! Note that the thought itself that resides in the hidden universe wouldn't vanish, but its connection with the brain would. If the neurons lost this quantum entanglement, every time that we learned something, we would forget it a few weeks later, when the neurons were being replaced. Our brain would be totally dysfunctional.

You might attempt to refute my idea by saying that other cells in the body are replaced periodically such as the cells of the skin, yet the skin always retains its physical properties. So why would new neurons not retain their thoughts?

The answer is that the cells replicate at the *molecular* level. The molecular level is elevated at a higher consciousness level than the quantum level where thoughts are captured. Through evolution, the molecular level learned to participate in the replication of a cell. However, at the quantum level I suggest that replication is impossible because there isn't enough consciousness at that level. So quantum evolution was never able to figure out how to replicate. It *has* to be that through evolution neurons realized this limitation and figured out a way to solve it by simply ensuring that neurons never replicate!

Think about what happened some 300 million years ago. The first neurons that appeared probably replicated themselves just like all other cells. But the replication was causing the entangled quantum particles within the neurons to be lost, and consequently the neurons kept losing their collective thoughts. One of two things happened:

1. Eventually the neurons themselves realized this loss and figured out that the only way to not lose information in the brain was to never replicate, or

2. The brain realized that it kept losing its information, and "instructed" the neurons to never replicate.

If the first scenario is what happened then we have to conclude that neurons have a collective consciousness (i.e., that neurons are aware they are part of a system of thoughts). If so, then neurons possess a level of consciousness higher than that of other types of cells.

In the section "Exponential "Learning"" of the previous chapter, I presented a short list of events that helped the process of evolution, starting with bacteria and leading all the way to us. I suggested that as the brain is able to think, it had to have evolved initially from some cells that were also able to think. But what was the evolutionary process that transformed cells that replicate into neurons that don't replicate? Neurons didn't appear out of nothing. Evolutionary *thinking* must have taken place.

Remember that entanglement of quantum particles is evidence of a thought. As I suggested in the previous chapter that cells have an ability to acquire thoughts, there must have existed entanglement of quantum particles inside the cell, but when the cell died, the thought was lost because the entanglement was lost too and the new cell had to reacquire thoughts. With this kind of replication, the level of consciousness of the cells could never rise. I propose that at some moment, cells must have figured out that the only way to ensure that their thoughts were not lost was to simply not replicate. There isn't doubt this evolutionary decision is evidence that cells are *conscious* that they possess thoughts!

By now it should become clear to you that living cells possess a certain level of awareness!

I found on the Internet numerous Web sites relating to research done by scientists who are finding that the brain does seem to work based on quantum events. However, there's no consensus. Some researchers claim that the brain is more complex than the working of the entire universe! I believe this without any hesitation.

You might recall that in the section "The Quantum World" of Chapter 5, I calculated that to the quantum level, a single cell is as large as the whole universe. Imagine now the hundred billions of neurons cells making up the brain. Some scientists propose that the brain may change its surrounding environment at the quantum level. This is also what I propose later in this chapter. I'm sure that the brain holds many surprises yet to be discovered.

This possibility of "entangled" quantum communication defies our senses. We must keep in mind that whatever happens at the quantum level doesn't translate in the same way into our macroscopic world. This synchronization phenomenon couldn't possibly exist at our macroscopic level, not even at the atomic level (except possibly in the brain where neurons might synchronize to manifest a thought). This impossibility exists because in order for two atoms to be synchronized, *every* single one of their *thousands* of quantum particles would have to be synchronized at the same time. This is statistically so remote as to be impossible. This provides further evidence that thoughts are captured at the quantum level.

In various places in this book, I stated that nature is only sensitive to energy in the form of waves, and that thoughts manifest themselves in the form of that energy. In Chapter 7, I found that God (or the Creator) is a system of thoughts. Consequently, God is only sensitive to thoughts, and His thoughts find their way into our physical world at the quantum level. The same holds true of our own thoughts—even though it certainly doesn't feel that way. I don't mean to say that we think only of quantum things, as of course that's not so.

At the quantum level, only a few things exist, including electric forces. These have only two possible values: positive or negative. This is equivalent to 0s and 1s. Referring back to the previous paragraphs, so then the quantum level has the right tools to represent thoughts and store them. Why shouldn't thoughts be storable in a physical medium higher than the quantum level? There are at least these two possibilities:

1. With a living possessing a brain, thoughts are captured by neurons at the quantum level, and then the network of neurons enables the brain to capture thoughts at a level high enough to enable the formulation of complex thoughts.

2. With a living being without a brain, thoughts would be captured "raw" by the cells. Deprived of specialized neuron cells, that living being could only capture very simple thoughts. This would explain why cells don't demonstrate intelligence—they are more like robots that obey chemical reactions. But how did evolution make possible such complex chemical factories if not through some evolutionary intelligence from nature? This question was discussed in the previous chapter, especially in the section "Molecular Structures and Life."

Note that thoughts at the quantum level are the most simplistic ones, especially if there's no quantum entanglement. This suggests that at the Big Bang, only the simplest thoughts were manifested. In Chapter 7, I suggested that the thought of time is what triggered the Big Bang. Could time be the simplest thought coming from the Creator? But why would that thought have triggered the entire energy of the universe?

Recall that the creation of time immediately led to the creation of light. Light is made of electromagnetic waves. Is electromagnetism the manifestation of a more complex thought? In order to form more complex thoughts, quantum events have to be "entangled" somehow. This is what happens with the brain, and perhaps some of that happens at lower levels such as at the molecular level. I can't help but notice that light is made of *tangled-up* electromagnetic waves. Could it be that these tangled-up quantum events be the result of a thought from the Creator?

I showed in the section "The Thoughts–Energy Dimension" of the previous chapter that thoughts in the hidden universe translate into energy in the virtual universe. Can the reverse happen too? Consider this: the act of thinking requires time and energy, so then *thinking* can only take place into the *physical* universe. But the thoughts generated by that thinking process reside in the hidden universe. Thus, physical events have to find their way into the hidden universe in the form of thoughts.

How is that possible? As thoughts manifest themselves in the virtual universe in the form of "tangled-up" quantum particles, then perhaps the reverse is also true: the thinking process that takes place in our virtual universe inside the brain creates "tangled-up" quantum particles that translate into thought in the hidden universe. Here is the picture:

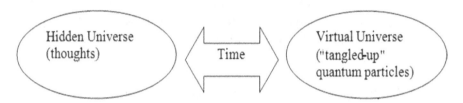

If you recall from Chapter 7, the Creator was referred to as the system of thoughts that created the universe. Well, the hidden universe is made of the thoughts dimensions; hence it's a system of thoughts. Therefore it has to be that the hidden universe was created *before* the physical universe.

There is another method at arriving at the same conclusion. The section "Discovered Thoughts" below, I suggest that thoughts are timeless. As thoughts are timeless, it follows that the hidden universe is timeless. So, unlike the physical virtual universe, it had no beginning. Consequently the hidden universe was there even *before* the creation of the physical universe! This follows from the fact that the system of thoughts referred to as the Creator in Chapter 7 *is* the hidden universe. So we live with our Creator every day! As our thoughts reside in the hidden universe, we are part of the Creator. The big question is: what are those thoughts in the hidden universe that triggered the appearance of the physical universe?

Physicists study how the universe works. They should also study how the universe *thinks* because the universe is a thought much more than a structure. This has become clear by now, don't you think?

Discovered Thoughts

In the section "Our Virtual Universe", I deduced that thoughts trigger physical manifestations inside the brain. But what creates the thought in the first place? It certainly could be another thought, or a logical series of thoughts driven by a thinking process. But this series of thoughts can't go on forever, as at some point, there has to be an initial thought. That thought was triggered by *nothing* at all! It came all by itself. Isn't that what geniuses experience when they have a flash of great insight? They are often at a loss to explain where the thought came from. There appears to be no other way around this: some thoughts trigger themselves without any logical process behind them. Actually, this isn't mysterious as people who demonstrate the ability to "think outside the box" do this very feat.

As was stated in the section "The Thoughts–Energy Dimensions" of the previous chapter, thoughts in the brain manifest themselves at the quantum level. As at the quantum level, nature can't tell the difference between a brain, a bacteria, a molecule, or anything else, you can be sure that a thought may also have the capacity to manifest itself at the quantum level *anywhere* in the universe simply because the quantum world is everywhere in the universe.

If a thought may be captured by the brain, perhaps it can be captured by other things as well. And who knows? Maybe that happens all the time! Our bodies and brains aren't equipped to capture thoughts "out of thin air" (actually the brain does have that ability, but only for our own thoughts). So there are two possibilities:

1. Might my own thoughts be captured by something else, living or inanimate? Let's consider both cases:

a. *Living things such as bacteria.* I recall seeing results of a research that showed that plants sense emotions. Here is what I gathered (with permission) from the Web site www.care2.com/greenliving/psychic-reactions-from-house-plants.html, that received this information with permission from the Web site www.thefoursgates.com:

> Cleve Backster, the scientist working with the police, decided one day to attach the electrodes of a lie detector to the leaf of a dracaena plant to see if the device was sensitive enough to pick up reactions from a non-human subject. After the device was attached to the leaf he thought maybe the reaction would be stronger if he burned the leaf. As soon as he had this thought there was a dramatic peak in the polygraph chart, a trace signature that Backster recognized as fear.

> Intrigued, Backster continued his research and the results were always the same: The plants always reacted to his intention before any action was taken. Backster concluded that not only are plants as sensitive as human beings, but they are able to read emotions and intentions, because there's a form of psychic connection, or affinity, between plants and people.

> Backster's other results show that plants have memory, emotions, and very humanlike reactions, as well as psychic abilities. In other experiments, Backster demonstrated the love or empathy between a plant and its owner. One day he accidentally cut his finger and noticed that a plant being monitored was demonstrating a stress reaction of its own, as if it was experiencing Backster's pain and shock at the sight of his blood.

How to explain a plant's ability to sense emotions? Well, emotions are certainly triggered by thoughts, but they also accompanied by a burst of energy, negative or positive. So I propose the equation: *emotion = thought + energy*. As I showed in this book that nature is sensitive to energy and thoughts, then nature is sensitive to emotions.

Emotions are based on instinctive thoughts such as fear, joy, hunger, sadness, etc. So the thoughts associated with emotions tend to be *primitive* ones. Thus, thoughts related to emotions are much simpler than other thoughts. Consequently, the "entanglement" of quantum particles associated with those emotional thoughts is likely to be simple too. So it's perhaps possible for some quantum particles within the plant to entangle synchronously and then capture those simple thoughts.

If a plant can feel emotion, then it's possibly able to make sense of it. So if a plant can feel sadness from a person, then the plant can possibly possess the ability to develop sadness on its own! This seems totally crackpot, but don't be so sure.

Note that a plant would have no chance of capturing complex human thoughts. When a flower shop clerk does his bookkeeping calculations to balance the books at the end of the work day, you can be sure that the plants don't capture any of that!

b. *Inanimate things like a chair.* This is surely complete lunacy to propose that a chair could capture a thought! Indeed, but let's think about it anyway as we have no way to know whether a chair can capture thoughts or not. Keep in mind:

i. Thoughts are captured at the quantum level, not at the cellular level. They might have an impact at the cellular level, though, such as in the brain.

ii. If they can capture thoughts at all, clearly quantum particles from the chair could only capture extremely primitive ones because a chair doesn't have a brain and isn't a living thing. So a chair has an extremely simple structure compared to that of a living organism. As my thoughts are complex, those quantum particles from the chair could only capture a small "piece" of my thought, if at all, and could do nothing with it anyway.

iii. Numerous quantum particles from the chair would have to be "entangled" in order to capture even a small "piece" of my thought. This entanglement is extremely unlikely to happen.

Finally, in the section "Molecular Structures and Life" of the previous chapter, I stated that only living things have some consciousness. As a chair is inanimate, it surely can't realize that a few of its quantum particles capture a thought. Consequently, a chair can't think, and even if it could capture thoughts, it couldn't do anything with it. Of course, you already know this, but it's good to formulate a proof.

In principle, my thoughts might affect the surrounding environment. However, certainly, my thinking alone of lifting an object will not trigger the desired effect. Keep in mind:

o The influence of thoughts over the physical world might be as subtle as that a butterfly flying by will have over the weather a few weeks from now. There's no way of knowing what that influence might be. Thus, the physical alteration that my thoughts might cause would be very subtle, and the effect might appear to have no connection at all with the thought that caused it.

o As a thought isn't defined, it's impossible to measure its influence on the physical world (unless the thought is spoken out loud). Although I don't know the mechanism behind this relationship between thought and the physical, it may be analogous to a song on a CD. The CD is physical, the song manifests itself in form of waves (sounds), but the song itself is a set of thoughts elevated out of the physical medium.

2. Might my brain capture thoughts from somewhere else? In item (1), we investigated whether my thoughts might affect the environment around me. Now, let's look at the reverse. In the first paragraph of this section, I deduced that thoughts may trigger themselves. If so, then how do I know that my thoughts are my own? How do I know if a thought that comes to my mind is a created thought of my own or a discovered thought that entered my mind from the nonphysical part of the universe? Here are two reasons that limit our brain's ability to know if a thought can be captured:

a. My brain isn't equipped with "sensors" to allow me to detect the difference (i.e., to know who is the owner of the thought).

b. My brain might not have the necessary neural complexity to capture a "foreign" thought. For instance, I could try to teach a child Einstein's gravitation field equations, and I assure that his or her young brain will capture absolutely nothing. So even though the hidden universe holds all the mathematical thoughts imaginable by the Creator, most of these thoughts aren't captured by anyone on earth due our limited intelligence.

I'm not suggesting that all our thoughts are from discovered ones from "out there." For instance, it's obvious that when I plan on having a nice cup of coffee that this thought isn't a discovered one—it's *my own* thought. Nevertheless, it remains that the thought is triggered from the hidden universe. So in the hidden universe, there's a little compartment in which my thoughts reside. The religious pictures of Jesus or saints with auras above their heads may also suggest this.

Because there's no space in the hidden universe, that universe seems more like a pool of thoughts. Where are my thoughts in that pool? Are they mixed in with thoughts of other humans? It's not at all clear how a thought comes to manifest itself inside my brain rather than in someone else's brain. How much control do I have in the creation of my own thoughts? Are they *my* thoughts?

In the previous section, I suggested that "entanglement" of quantum particles form a thought. If two persons have the exact same thought, this should translate into the exact same "entanglement" of quantum particles, no? If so, then should we conclude that they share the same thought located in that pool of thought?

So I'm not convinced that humans create *all* of their thoughts with their brain. My theory is that our brain *discovers* some thoughts from that pool of thoughts and doesn't create them all. This idea sounds totally crazy. But consider this: what do people say when they have a thought? They say "a thought *came* to my mind" suggesting that the thought was discovered by the mind; hence the thought came from *outside* the mind. People don't say "my mind *created* a thought."

In the circular picture presented in the section "Our Virtual Universe," nature is omitted because nature is everywhere in the virtual universe. This implies that nature's thoughts are everywhere in the hidden universe. I gave hints in various sections of the previous chapter that nature may possess an ability to think and learn.

This gives me an original idea: Suppose that intelligent life already exists somewhere else in this virtual universe. It's nature that brought about that life, so then could nature have used the knowledge it acquired to bring about life *faster* (because nature already knew how to create life) here on earth? Can nature learn from past experiences it acquired somewhere else in the universe? This is no more a crackpot idea than those people who claim that aliens brought life here on earth.

Should there however be a civilization in some distant galaxy that has the technology to capture thoughts, then, in principle, it could detect our presence here on earth *instantly*! Who knows? Maybe this Earth is already known by some distant aliens. But they are much too far away to come visit us. Perhaps that civilization is sending their thoughts to us. We have no way to tell.

But even if our brains had a sensor to capture thoughts, the thoughts from that civilization might be far too advanced and intelligent for our brains to make sense of it. In other words, these thoughts might reach us, but when it comes time for our brains to manifest these thoughts, our brain's neural network might not be advanced enough to configure and "store" them. It's a bit like if we try to watch a high resolution television program from a DVD on a TV screen that is too old to understand the signals from that high resolution TV. The high resolution of the DVD will be ignored by the TV.

Thoughts are timeless and spaceless, and they have another property: they can't be removed from the pool of thoughts. Indeed, when I have a thought, I can't remove it from my brain. Even when I think that I've forgotten it, it's actually still around. Consequently, the pool of thoughts includes all thoughts all the way to the beginning of time! But there's more: because there's no time in the hidden universe, the pool of thoughts contains all the thoughts *before* the beginning of all time (i.e., from even before the creation of the universe). Our brain might be capturing thoughts from the Creator Himself because it's the Creator that holds that system of thoughts. Does that mean that we are also very close to God? This is impossible to answer because God is beyond description.

Would an event be of pure chance or rather evidence of two minds connected by discovered thoughts "out there"? I should add that probably only logical and true thoughts are discovered, for they are around floating in the universe. Would the universe "store" incorrect thoughts? If so, then how did they become incorrect? Because it's certainly not God that would has created incorrect thoughts.

When a series of events seem to be coincidental, people—especially scientists and mathematicians—tend to calculate the probability of such a series happening. If the probability is extremely remote, they get stuck and don't know what to say. Scientists will thus tend to refrain from concluding that there's a nonphysical connection or explanation. Their reaction isn't logical because these same scientists (most of them) believe in the space–time theory of the curvature of space caused by gravity. But don't they realize that this theory is based on a dimension that is nonphysical? If they so firmly believe in this nonphysical time dimension, then what's stopping them from believing that perhaps there exist other nonphysical dimensions in the universe?

If it's true that the hidden universe truly exists (I believe it does), that it stores thoughts, and that everything happens instantaneously, it opens the door to all kinds of possibilities such as telepathy, psychic phenomena, communication with the dead, ghosts, etc.

This brings me to the practice of *meditation*. I've never tried it. People who have say that it takes them into a state of total relaxation. Apparently the goal of a session of meditation is to spend time without thought, that is, the attempt is to remove all thoughts from occurring. If this state of mind is possible, then I can see how it could be relaxing! Moreover some say that meditation is a means of being closer to the Divine.

How does meditation relate to the subject of this chapter? Being without thought implies being in total internal silence. As you've learned in this chapter, thoughts don't reside in this physical universe but rather reside in the hidden universe—the universe of Consciousness. As you've seen in Chapter 7, this universe contains all thoughts that the "logical system" which drives it can contain. Therefore it would include some, but not all of our Creator's thoughts.

Now if my brain is in total silence and is no longer thinking, then there's a problem. The brain's neurons are meant to continually connect. This means that it should be impossible to *not* think. Since in a state of meditation, absence of thoughts is supposedly possible, what might happen then is that some other thoughts from the universe of Consciousness might literally enter the brain of the person in that state of meditation! These thoughts might be directly from our Creator, or could be thoughts from… someone else here on earth or from a far away galaxy!

This connection of thoughts from far away strangers is however easier said than done because for a thought from the world of Consciousness to connect with the brain of the meditator, it has to meet two requirements that have already been explained in this section and reproduced below for your convenience:

1. The tought has to enter the brain at the quantum level where some precise and correct quantum events have to take place so that the thought can take root. This is almost impossible to accomplish except evidently with our own thoughts. However, in theory, it's possible. So, yes, meditation might allow you to be closer to the Divine.

2. The neural network has to be complex enough to "understand" the thought. Are we smart enough to understand thoughts coming directly from the Divine?

I suggest that the hidden universe consists of thoughts from all origins including the afterlife. Just perhaps there are a tiny number of people in the world whose brain and neurons are in tune at the quantum level with thoughts from other people. This is perhaps possible through meditation. This seems crazy, but there's no way to prove or disprove this possibility. The reason that there exist only a handful of supposedly true psychic people might be that our brain learns to connect with its own thoughts only. A normal brain isn't structured to connect with thoughts from elsewhere because our evolution never saw any need for that potential. However:

1. If a hidden universe of thoughts (as suggested in the section "Our Virtual Universe" of Chapter 10) truly exists, and

2. If Thoughts are stored at the quantum level,

then thoughts perhaps are capable of traveling around and being captured in very rare situations. Although this seems crazy, it can't be disproven.

But what about people who are afflicted by a mental disease that affects their thinking? Are their thoughts gone? No, it's just that the physical connections with the thoughts are broken. The thoughts are safe because they reside in the hidden universe that can't be affected by physical tragedies.

On Virtual Reality

You may already have stumbled upon various essays or Web sites about virtual reality. The movie *The Matrix* fantasized about that possibility. Understand that virtual reality is *not* the same idea as the virtual universe I propose in this chapter. Virtual reality is about the creation of a virtual world using powerful computers that compose a world that feel so real that anyone inside it would be unable to tell that he or she is inside anything other than a real world.

My idea of a virtual universe doesn't run on computers but rather it "runs" on thoughts.

A virtual reality world supposes the existence of powerful computers that make billions of calculations a second to account for everything in the universe. I can't conceive of such a possibility as there are numerous difficulties with this theory of a virtual reality driven by computers:

1. Where's that computer? And what created that computer? What was the creator's intent? Was it to replace God with a computer?

2. The processing power of that computer would require more energy than the entire energy of the universe. That is quite obvious because the slightest physical quantum event would require complex calculations, thereby requiring lots of energy.

3. The complexity of the program run by that computer would be beyond the reach of any level of intelligence. That program would have to contain all the mathematical equations that describe everything from electromagnetic forces in atoms to human behavior. The trouble is that *all* those equations are highly nonlinear and unsolvable in many cases. This means that they can't be resolved *precisely* or even approximately even by the most precise computer. This lack of precision would cause malfunctions that would be detectable by us (e.g., inexplicably it would start raining even though the sky is clear).

4. As said above, the behavior of everything would be determined by calculations made by that computer.

   a. This would cause delays in reactions. As a consequence, the speed of light (for instance) would vary among other anomalies. All physicists know that the speed of light is a solid constant and doesn't vary in the slightest.

   b. Because no computer is perfect (even God couldn't create a perfect computer simply because there's no such a thing as perfection—this is mathematically provable), that computer inevitably would have glitches and "bugs". This would translate into occasional physical observations that are contradictory. For instance, all in a sudden the earth would rotate faster due to a bug, or my hands would have an extra finger one day, then the next day it might be gone.

5. The actions of humans would be determined by that computer. My brain would be controlled by that computer. The brain alone is so complex that no computer will ever be able to duplicate it, let alone the whole universe.

6. The computer would have to be made of circuitries in which signals travel millions of times faster than the speed of light.

The reason is simply that the computer would have to calculate the path of light or energy of every single photon in the universe!

7. For the computer to work, it would have to know the difference between objects (for instance, what's an airplane from what's a cloud?). So it would have to simulate their behavior just like in a computer game. Do you think that nature cares about the difference between a cloud and an airplane? To nature, it's just energy stuff.

It's much easier to believe in a universe that's run by a system of thoughts.

People find nonphysical, supernatural stuff out of reach or hard to believe, but I'm the other way around. The physical world is hard to believe as it's all an illusion. Thoughts are real, though. We think that everything around us is real because from birth, we've developed thoughts based on our senses. These thoughts allow us to imagine a "reality," but nature doesn't care about our perception of things.

I propose a hidden universe that's not driven by computers, but by intangible thoughts. That seems impossible, right? Don't be so sure. Take walking, for instance. I don't need to perform calculations in order to walk even though walking is a fairly complex motion involving many messages from the brain and muscles. In order to walk, I simply need to *think* of it.

I saw on television an experiment where it was proven that a person can have a thought of some action, and the brain will send the same signals whether the person has actually performed the action or not! This fact is evident in dreams. Sometimes dreams feel very real. Indeed, scientists have discovered that the brain can't tell the difference between a dream and actuality.

The virtual universe that we live in doesn't perform any calculations. When the moon orbits the earth, no calculations are performed anywhere. Instead the path of the moon is assured by the feedback system I discussed in the section "Solar Systems in Equilibrium" of Chapter 6. This system can be translated into mathematical formulas and precise computations for our purposes, but nature isn't aware of that.

Finally, our brain has a direct connection with the hidden universe. That universe contains all mathematical formulas imaginable, yet those equations don't translate into physical manifestations that reflect precisely those equations. This is because nature uses feedback systems to figure out behaviors of everything in our virtual universe. This was discussed in the section "The Thoughts–Energy Dimensions" of Chapter 9. Nature calculates nothing and doesn't use mathematical equations!

God and the Devil

Through a logical argument in the section "The Thoughts-Energy Dimensions" of the previous chapter, I showed that thoughts in our brain manifest themselves at the quantum level. In the section "Does God Exist?" of Chapter 7, I argued that God does exist simply because of our brain's capability to conceive the notion of perfection. Clearly then God manifests Himself at the quantum level too. Therefore, God reveals His personality in our brain. In the Bible where it says that God made us in His image, according to my arguments, this is clearly true.

If God exists, then should we believe that the devil exists too? Suppose so. If God manifests Himself at the quantum level, then the devil would also manifest itself at the quantum level because any thought, good or bad, takes place at that level.

The trouble now is this: how can nature distinguish between which quantum thought comes from God and which comes from the devil? Actually this question isn't at all relevant because the devil isn't at the same level or of the same "nature" as God.

The image we learn as kids that there's a God in heaven and a devil in hell is somewhat misleading. It gives the impression that both God and the devil are some kind of spirits fighting one another. No, God isn't a spirit—God is beyond everything, both material and spiritual. Spirituality may lead us to connect with God, but God isn't a spirit. Is the devil a spirit? No, because a spirit has a consciousness, by definition. The devil has no consciousness.

So, if the devil isn't a spirit, it's therefore ... nothing! The devil is nothingness. The devil is the abiss, and has no consciousness, which means the devil is void of purpose. Indeed, there's nothing worse for a being than to lose one's purpose for existence.

So, what's the devil? It's simply the *absence* of God. Once a brain is totally disconnected with God, the brain has lost its purpose. However, because the brain is still able to think, the thoughts that it generates become totally meaningless. This is why people who are possessed by the devil behave in a totally unconscious, irrational, and beast-like manner.

But then if the void of consciousness is evidence of the absence of God, then insects are devilish, and living cells are devilish too because they apparently have no consciousness (this is my premise). Should we infer then that insects and cells and bacteria are possessed by the devil? If so, then our human cells would be the product of the devil. But I said in the first paragraph of this section that we are (all of nature in fact) made in the image of God. Devilish entities obviously can't produce something in the image of God.

We're at an impasse here. Where did I go wrong in my logic? Nowhere! Therefore, I have to conclude that it's my premise (that insects and cells and bacteria have no consciousness) that is wrong! Consequently, it has to be that living cells, bacteria and insects absolutely *do* have some level of primitive consciousness. You might recall that I already came to that conclusion in various sections of the previous chapter and in another argument in the section "Our Virtual Universe" in this chapter. So, if handful different arguments led me to the same conclusion of consciousness at the cellular level, then there must be some truth into that conclusion.

This is what intelligent design (ID) is about: all living things having a consciousness. The more complex biologically a living being is, the more evolved its consciousness will be. Contrary to what many people believe, ID isn't *solely* about creationism. There shouldn't be anything shameful for us to admit that we humans are part of that intelligent design (i.e., that our existence is the result of evolution just like everything else). We are simply more evolved and closer to God.

Heaven

I suggested earlier that Heaven might reside in the hidden universe, or might be part of it. As children, the image of Heaven that we were taught is that it's a place of beautiful sceneries, of flowers, of complete happiness. But Heaven (by definition) has no physical properties, so you can forget about flowers, perfect blue sky, and so on.

There exists only one thing in Heaven: peace of mind. But you don't have to wait to die to reach peace. It's not necessary to search for peace either. It's within. Your original state is one of peace. External situations may pull you away from your peace. Internal feelings can also pull you away. Tiredness, for example, leads to irritability.

When we die, we return to a state of peace. The very instant before death, our brain sorts things out quickly. We make peace with the sorrow that we may have caused, and make peace with people who may have caused us sorrow. Understand that peace isn't the same as happiness. Happiness is an expression of emotions. Peace brings a stillness of the mind. Also only our thoughts will remain.

While in our physical world, peace of mind may bring happiness, it can't be so in Heaven because happiness is about emotions, not about the mind. Emotions are like a battery. It's got a positive and a negative pole—positive versus negative emotions. Whenever you have polarity, you end up with energy (which is how light forms). Recall the formula that I proposed in section "Discovered Thoughts:" emotion = energy + thoughts. There's no energy in Heaven. The thoughts *of* emotions may reside in Heaven, but not the emotions themselves. The Bible says that in Heaven, there's the ultimate happiness. That can't be the case for two reasons:

1. Emotions manifest themselves in the physical world,

2. Should happiness happen in Heaven then it's a state possessed by a thought. Should a thought have a state of happiness ever slightly different than another thought, then polarity is formed. This implies that one thought becomes happier

than another one. But this contradicts the idea that in Heaven, there's the ultimate happiness.

Consequently, happiness doesn't exist in Heaven. More precisely, emotions are impossible in Heaven.

A mind is in peace when it has "sorted things out." People who die with a troubled mind will have emotional issues, but these will not follow them in Heaven. These emotions would be like residual energy that might be "floating around" in space. So then troubled people may die and still make it in Heaven with a peace of mind. When we die, our troubles don't follow us in heaven. Isn't that what the Bible says?

I propose the equation:

peace of mind = positive emotions + negative emotions = zero = no residual energy in space

A person that died and left residual energy in space might manifest himself or herself as a ghost. When a person dies suddenly, the person may not have had time to reach a peace of mind, so emotional energy is left behind. This is why ghosts tend to have emotions. They are never neutral. Ghosts sap energy from space such as sucking energy from a battery.

If some ghosts have residual issues, then they may intend to resolve those issues by way of thinking. Since thinking requires energy, the ghost would have to manifest itself into this physical world to draw on energy present. So, if you see a ghost, he or she might just be trying to figure things out, and might not mean any harm to you.

Many billions of people have died since the beginning of Humanity. Yet ghost appearances are extremely rare. Based on the equation above, I conclude then that most people die with a peace of mind no matter what they lived through in their life.

Most people will agree that when a loved one passes away, the emotion of loss is about the same whether the person was a young child or an adult. I know, from having gone through this, that the loss of an unborn baby through a miscarriage also brings the *same* feeling of loss. Life is life no matter the age. In section "Molecular Structures and Life" of the previous chapter, I concluded that the seed of life is already present in the "hidden universe" which is the universe of Consciousness. Since that world is void of energy, there's never a change of state of consciousness. It's the stillness of the mind. This explains in my view why we all feel the same loss when a loved one passes away. To some, the loss is more intense than for others, but it's still the same *awareness* of loss.

As an analogy, it's like when you look at a red light. No matter the intensity of the light, it remains red. It's not the brain that dictates what we feel. It's rather the inner consciousness that dictates to the brain what to feel. The emotion is physical but its cause— awareness—isn't.

Life isn't of this physical universe. Once the baby is conceived, it receives life and is instantly connected to the world of Consciousness. Because the seed of life resides in that world of Consciousness, life is unaffected by a death in our physical world. Consequently when someone dies, he or she is still alive in the hidden universe.

You might ask: if our brain's thoughts are dictated by that world of Consciousness, then how come that I cannot sense the thoughts from other people? After all, their brain is also connected to that world. The explanation is that the brain resides in the physical world. The thoughts from the world of Consciousness are communicated to the brain at the quantum level by a series of quantum events (the notion of quantum entanglements mentioned earlier). These events are unique to each brain, and so cannot be replicated into another brain. In some rare cases, their might be some capacity of replication—this might be the power that psychics people possess.

After we pass away, our thoughts are no longer limited by quantum events. The consequence is that thoughts are instantly connected, and this for *all* people that ever existed, and this for the entire universe! The world of Consciousness is like a single huge brain. As a result of the limitless connectivity of thoughts, after we pass away, *all knowledge of the universe will be revealed to us.* I invite you to believe!

A Purpose for Intelligent Life?

Most scientists nowadays don't believe that the universe was created with the intent to create intelligent life. I beg to differ. My opinion in this section is influenced by these arguments:

1. My belief expressed in the section "What Created the Universe?" of Chapter 7 in which I stated that God created the universe for a purpose. What's that purpose? I suggested that souls are located in the time dimension or the thoughts dimension. When I die, will my soul go back in that pool? Does a soul exist forever? If such a pool of souls does exist, then it must imply that God has created it for a purpose.

   Surely God wouldn't create something for no reason. As souls are meant for humans (or more generally, any being with a sufficient level of conscience), this implies that God had to ensure that humans exist somewhere in the universe so that some of those souls are attached to a living body as needed. In this context, *human* refers to any living form that has a brain with the capacity to receive a soul.

2. The section "Our Virtual Universe" of this chapter in which I found our thoughts are in the same hidden universe as the original system of thoughts that created the universe. By creating intelligent beings, the universe has gone full circle. Indeed, thoughts appear at the quantum level in the brain, the same level as they did at the time of the Big Bang!

3. As argued in the previous section, God's thought is in our brain. So it's by studying the brain that we can answer the question of what created the universe! Neurologists know that the brain is more complex than the whole universe itself because there are more neural connections in the brain then there are atoms in the universe.

4. In section "Codes and Algorithms for Evolution" of the previous chapter, I argued that the DNA structure is the product

of some cosmic algorithm. As we all know, the evolution of the DNA eventually led to intelligent life. Just like the appearance of the DNA was no accident, the emergence of intelligent beings was no accident either. These things were bound to come to be somewhere in the universe.

5. In section "Molecular Structures and Life" of the previous chapter, I concluded that the seed for evolution of life was already present in the "hidden universe." Since the hidden universe is spaceless—not bound by space—the seeds of life therefore try to penetrate our physical universe, everywhere! Since the hidden universe is all about consciousness, there's therefore a conscious intent in the cosmos to bring about life into the universe.

These five arguments bring me to believe that the intent for the creation of the universe was for the evolution of intelligent life, that is, life intelligent enough to understand the making of the universe. The answer to the creation of the universe is inside our brain! What a daring idea that seems totally crazy. But we found in the section "The Thoughts–Energy Dimensions" of the previous chapter that there was a similarity between the creation of the Big Bang and the creation of thoughts. That similarity is a singularity point that I introduced in Chapter 8.

If the above arguments are correct, then there has to exist a *law of certainty for intelligent life* that says that in a given universe, the probability of humans existing on some planet somewhere is always 100 percent.

My view here is very similar to the *anthropic principle* that suggests that the universe has been created with the intent that intelligent life exists on some planet so that the living beings there will have the consciousness and intelligence to discover and appreciate nature's makeup. Moreover, in Chapter 7, we used Gödel's incompleteness theorem to prove that the universe, the hidden universe of the previous section, is incomplete in that it cannot answer all questions.

Therefore, there might be intent on the part of God to bring to the universe beings intelligent enough to be aware of that theorem, and, from that, to realize that perfection can't be attained—hence the need to believe in some ultimate truth that we define as God. God created the universe with the intent of the universe to reflect back onto its Creator.

Moreover, in the section "Codes and Algorithms for Evolution" of the previous chapter, I calculated that nature did have time to create life and humans. This doesn't mean that there needs to be many planets with such life. Just one would do—such as the earth. If that law (of certainty for intelligent life) exists, then we have to conclude that the universe was created with soul-capable creatures in God's mind. Isn't that what the Bible says? It's not that the universe was created for humans, but rather that God had to create the universe in order for *us to be part of it.*

It was extremely unlikely for life to appear on Earth. Yet here it is. Because of the extreme odds against life to happen occurred by chance, some conclude that it was a miracle, an Act of God. I disagree. The force in this case is the will to bring consciousness into the physical world. Perhaps the consciousness comes from God, but there's no way to know.

Consciousness is a force from the other universe that I call the "hidden universe". That universe is everywhere around our physical universe, and it applies the same force everywhere in the cosmos. Consequently it's believable that life may very well exist in other galaxies. As shown in section "The Toughts-Energy Dimensions" of the previous chapter, that hidden universe of consciousness pierces our physical universe at the quantum level. But it cannot do much without some favorable physical event that came by *accident* right here on Earth a few billions of years ago.

This event has to be some complex quantum events that are triggered by the right kind of atoms and molecules. Once this event took place here on earth, the 'flood' gates of consciousness were open for life to plant itself. Note that once life appeared here on earth, it flourished *very* rapidly, as scientists have discovered. Random events cannot alone explain this rapid evolution of life. This was demonstrated in the previous chapter.

Many scientists seem to believe that consciousness is nothing or has no effect in the physical world. Not so, I believe. Quite the opposite, consciousness is nearly everything. Although the existence of a universe of consciousness (the hidden universe) cannot scientifically be proven, most will agree that humans do possess consciousness. Where did that come from? Did that come from nothing? That's impossible. You get nothing from nothing, as the saying goes. So if we have consciousness, then inevitably the source of life possesses consciousness as well.

Scientists cannot detect any evidence for this view of a universe of consciousness. But that doesn't imply that it doesn't exist. Many generations ago, it was believed that people in a coma lost all awareness. Doctors now know that they retain a certain level of awareness. What else out there has consciousness, and yet we believe otherwise?

Some will attempt to refute my view by asking: how do I know that consciousness exists? OK—suppose that it doesn't. Then consciousness is an illusion, meaning that it doesn't exist. One of two possibilities: (a) I'm aware of that—consciousness doesn't exist. But then this implies a level of consciousness on my part. This contradicts the premise that consciousness doesn't exist; (b) I'm not aware that consciousness doesn't exist. But then this means that I believe that I possess consciousness. Whether consciousness is there or not, the fact that I believe that I possess it puts it into existence!

Either way, the implication is that consciousness does exist.

Some might say that if the forces of consciousness are everywhere in the Cosmos, then life should flourish everywhere. No, because the correct combinations of molecules have to be present too, as well as the correct temperature, and numerous other factors. It took a lot of luck for life to appear here on Earth, but once it was planted, luck alone doesn't explain evolution. You certainly recall that I formulated this explanation many already before—it's never a bad strategy to repeat expressing true statements!

Closing Comments

So there it is: my view on those ten questions pertaining to quantum physics, cosmology, and life. My intent was to approach the questions in a logical manner, as best as I could. Here are some of the highlights of what I discovered from my own intuition and imagination:

- The universe had a beginning.
- The universe was created by a system of thoughts, suggesting it's all about thoughts.
- The universe began from a single point and expanded from there.
  - A Creator, not necessarily God, created that single point.
  - God didn't cause the Big Bang but rather created the energyless single point that the Big Bang came out of.
  - The fabric of space is the likely "observer" of the universe.
- The universe isn't infinite, but will last forever.
- Space has four physical dimensions: the three dimensions of space and energy.
- Space contains energy and matter, which is a different form of energy.
- Motion occurs in quantum steps, and it's made possible by nature's inability to instantly detect motion (uncertainty principle).
- Motion is made possible by quantum particles moving in space as waves, then "resting" as particles, then moving as waves again, and so on.
- Time is a thought created by the Creator.
- Time isn't a real (physical) dimension—it's a virtual dimension imagined by nature! God isn't the keeper of time.
- Time is continuous which implies that space is also continuous.
- The fourth physical dimension isn't time, but rather density of energy in space (dark energy).

- To nature, the only thing that matters is energy (and thoughts).
- To God or the Creator, the only things that matter are thoughts.
- Light is the energy of the universe.
- Light is at the base of everything physical in the universe:
  o Creation of space, hence its expansion.
  o Creation of time.
  o Creation of matter.
  o Creation of all forces of nature.
  o Creation of life.
- Light is the clock of the universe.
- There's no physical reality to space–time, so gravity isn't *physically* explainable by space–time. Consequently, there's no physical reality to the curvature of space in the way explained by space-time.
- Gravitation is caused by a difference in space density in a region of space.
- The mechanism of gravitation is carried out at the quantum level.
- The universe expands with acceleration, but hasn't always done so.
- Although life is made of things from the physical world, life might be the result of a law of the cosmos with intent to create it somewhere in the universe. But probabilistic models and arguments can't determine that with any certainly.
- The universe equipped itself with rules that reduce entropy, suggesting that the universe was created with a purpose, perhaps a purpose to support life.
- Life was an act of creation, and evolution took over from then on.
- Every life form possesses some level of consciousness.
- Evolution is mainly driven by random events with which life forms interact and interpret them.
- Nature, in part through its intelligence, had time to create life on earth, including humans.

- Nature is one part of the universe; there's also a nonphysical part of the universe.
- The supernatural (nonphysical) world exists and resides in a hidden universe.
- Thoughts are part of the hidden universe.
- Our soul or thoughts remain in the hidden universe during life and after death: there is a kind of afterlife.

I tried to make a connection between the natural and the supernatural without making any reference to religions. I figured that if I believe in God and the supernatural, I may as well try to figure out how all of that comes into the picture of the universe.

My intent in this book was to reflect on how the universe *thinks* and not on how it works. What has driven me to sacrifice a lot of my free time to composing this book? I'm not a physicist, scientist, or philosopher, so my motivation did *not* come from an eagerness to impress anyone or to find the correct and true answers to those ten questions. It was the intellectual challenge that drove me—an attempt to come up with something out of nothing, like pulling a rabbit out of a hat. But did I instead pull out a donkey instead? You be the judge.

Or maybe Jill Bolte Taylor should be the judge. She's a neuroanatomist who suffered a stroke in her left brain ten years ago. Many months before I sent my manuscript in to my publisher, she was interviewed on the very popular *Oprah Winfrey Show*. Ms. Taylor recounted the events of the moments while she was having the stroke and during the medical treatment that followed.

She said she lost the ability to speak, although she was still able to think clearly. She also said that she no longer was able to understand spoken words. She no longer was able to recognize people's faces either. She couldn't even recognize her own mother. Despite these disabilities, she could still sense people's feelings and emotions even though she couldn't make out a single word coming from them! Finally, when she looked at her arms or legs, all she saw were fuzzy shapes without any well-defined contour. She had the same sensation when she touched her limbs. She said that her limbs felt like a lump of fuzzy energy that was merging with the energy in the environment around her. She'd lost the notions of hardness or softness of a surface and of boundaries.

It's clear that her brain had lost the ability to interpret her physical surroundings as well as language. Finally, she felt that time had stopped: she could only feel the present moment, just like a newborn baby. She'd lost the notion of time.

This fascinating story proves that my views presented in this book are correct. Our reality is based on the brain's interpretation of what we see, touch, or feel. When her left brain failed, the brain lost what it had learned about the physical aspects of nature. As a consequence, her brain couldn't form any interpretation of the world. Information came into her mind "raw."

Her observations agree with what I said about our reality being an illusion in the sections "Reality Versus Abstraction" of Chapter 2, "The Thoughts–Energy Dimensions" of Chapter 9, and "Our Virtual Universe" of Chapter 10. Shapes and the constitution of matter had become to her abstract concepts. She also had lost the sense of time, suggesting that time is an abstract concept as well. This is precisely what I suggested in the section "Reality Versus Abstraction" in Chapter 2 and other places.

Why are these things abstract? It's because they are clearly interpretations from the brain. As nature doesn't have a brain, eyes, or ears, you can be sure that nature's idea of physical surroundings is quite different from ours. I said in various places in this book such as Chapter 1 that Nature only feels energy and in the section "The Thoughts–Energy Dimension" of Chapter 9 that the universe is one of thoughts. This is precisely what Ms. Taylor experienced. She could only feel people's energy and emotion. Emotions are the energy associated with thoughts.

As stated in the section "Discovered Thoughts" of Chapter 10, emotion = thought + energy. So to some extent, she could feel thoughts as well as emotions even though she'd lost the ability to communicate! Her brain was able to discover people's emotions. All this goes so well with my suggestion that the universe is all about energy and thoughts which reside in what I call the "hidden universe." The rest are details invented by nature and our brains.

Appendix 1

(This appendix goes with Chapter 4: Why is Time Relative?)

Comparing the Two Formulas for Time

Early in Chapter 4, I expressed the notion of time using frames. This was our first interpretation of time. So just for fun, let's take the formula obtained in the section "The Correct Tick Formula" using the second interpretation, $T^2 = t^2(c^2 - V^2)$, in which time is expressed in seconds instead of frames, then let us manipulate it and see how it changes when it's expressed in frames units. You'll be surprised at the result!

As explained in the section "From the Planck Units to a Time Tick Formula" of Chapter 4, there are $1/P_T$ frames per second, and there are $1/P_S$ steps per meter that the object travels through. So the object moves at a rate of $V \times (1/P_S)$ steps per second, and there are 1 P_T frames of time per second. So then T seconds is replaced by T/P_T frames, t seconds is replaced by t/P_T frames, c meters per second is replaced by $c \times (1/P_S)$ steps per second or per $1/P_T$ frames, and V meters per second is replaced by $V \times (1/P_S)$ steps per second or per $1/P_T$ frames. So the equivalences are:

1. T seconds $\leftrightarrow$ T/P_T frames

2. t seconds $\leftrightarrow$ t/P_T frames

3. c meters/second $\leftrightarrow$ $c \times (1/P_S)$ steps per second or
   $c \times (1/P_S) / (1/P_T)$ or
   $c \times (P_T/P_S)$ or
   $c \times (1/c)$ or   1 step/frame

4. V meters/second $\leftrightarrow$ $V \times (1/P_S)$ steps per second or
   $V \times (1/P_S) / (1/P_T)$ or
   $V \times (P_T/P_S)$ or
   $V \times (1/c)$ or V/c steps/frame

So, if we plug those equivalences into the formula $T^2 = t^2(c^2 - V^2)$, we obtain

$$(T/P_T)^2 = (t/P_T)^2(1 - (V/c)^2)$$

So, the "tick" of time of the particle as seen by the observer is

$$number\ of\ frames = (T/P_T) = (t/P_T)\sqrt{1 - (V/c)^2}$$

or

$$number\ of\ frames = (T/P_T) = (\frac{t}{cP_T})\sqrt{c^2 - V^2}$$

Since $P_S/P_T = c$, then $cP_T = P_S$ and

$$number\ of\ frames = (T/P_T) = (\frac{t}{P_S})\sqrt{c^2 - V^2}$$

or taking the V outside the square root

$$number\ of\ frames = (\frac{tV}{P_S})\sqrt{(c/V)^2 - 1}$$

The term tV is the distance traveled by the particle, and when divided by P_S, it gives the number of steps. So

$$number\ of\ frames = number\ of\ steps\sqrt{(c/V)^2 - 1}$$

Finally, dividing by "number of steps," we get

$$number\ of\ frames\ per\ step = \sqrt{(c/V)^2 - 1}$$

So the formula I derived in the section "A Better Tick Formula" of Chapter 4 lacks the square root, but otherwise it's correct.

The algebraic manipulation above shows a stunning result. The *imaginary* interpretation of the observer taking snapshots of the par-

ticle (this is the first interpretation of time) and the *reality* that it's
the delay between emitted energy and the detection of the particle
that defines the time (this is the second interpretation of time) actually both give almost the same formula. Yet they are contradictory
interpretations. The one with snapshots implies that as the observer
takes "pictures" of the particle, there are fewer pictures taken as the
particle speeds up because the observer has a harder time locating
the particle, so more steps of the particle are missed. But with the
other correct physical reason *none* of the steps are missed by the
observer!

You might recall that this imaginary interpretation of the observer taking snapshots of the particle led me in the section "Space–
Time Fuzziness" of Chapter 2 to the fuzziness formula $\Delta s \Delta v$.

Appendix 2

(This appendix goes with Chapter 5: What Makes Motion Possible?).

An Absolute Frame of Reference

Nature detects the quantum particle (more precisely, its quantum energy) only when its wave is at its top amplitude (at its crest):

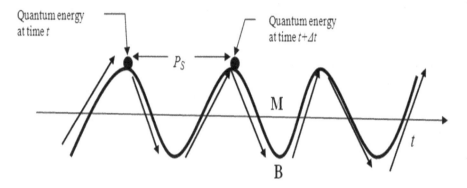

How is it so? At the time t, the quantum energy is at the top of the crest. Trigonometry says that the top of the crest is at a height of $A\cos(2\pi)=A$ the first time—the term A being the amplitude. Then the top of the crest occurs at every multiple $2\pi\theta$ of 2π such that $A\cos(2\pi\theta)=A$. In our situation here, $\theta = ft$. For instance if the frequency is $f=6$ crests per second, the the first crest occurs at $t=0$, then at $t=1/6$, then at $t=2/6$, etc. You can see that at each time t, the term $2\pi\theta = 2\pi ft$ is a multiple of 2π. The occurrences of the crests are cyclic.

Let Δt be the time span between two crests. Then based on the cyclic nature of waves explained in the previous paragraph, we have that

$$A \cos\big(2\pi f(t + \Delta t)\big) = A \cos(2\pi f t) \quad (1)$$

or the left-hand side may be changed to

$$A\cos(2\pi ft + 2\pi f\Delta t) = A\cos(2\pi ft) \quad (2)$$

There exists a trigonometric property that says that $A\cos(x+2\pi)$ = $A\cos(x)$. Don't you find that the equation (2) expresses the exact same property? *Yes.* In our situation, the variable $x = 2\pi ft$, and $2\pi = 2\pi f\Delta t$. This last equality implies that $\Delta t = 1/f$. Next, we know that $c = \lambda f$, where λ is the distance between two crests of the wave and c is the speed of light. The speed of light is implicated here because energy travels at it. Note that the quantum particle itself moves much slower than that! From $\Delta t = 1/f$, we obtain

$$c = \lambda f = \lambda(1/\Delta t) = \lambda/\Delta t \quad (3)$$

This last formula $c = \lambda/\Delta t$ is of the same form as $c = P_S/P_T$ (recall that P_S and P_T are the Planck distance and Planck time respectively). So then $P_S/P_T = \lambda/\Delta t$. Consequently, $P_S = \lambda$ (i.e., the Planck length is the same as the wavelength of the energy emitted by the quantum particle, and P_T is the reverse of the frequency). Actually, it's not quite that simple: we have to take into account the speed of the particle where the wave energy is emitted from. This factor is dealt with in the section "An Absolute Frame of Reference" of Chapter 5.

Appendix 3

(This appendix goes with Chapter 5: What Makes Motion Possible?).

A Formula for Discrete Energy

In this appendix we'll search for a quantum energy formula. The mathematical manipulations in this section and the next two sections will lead to a formula marked (6), this one

$$E = 8\pi^2 E_p f_s$$

which shows that energy is emitted in quantum discrete amounts (such as photons).

Intuitively, we know that frequency is an important property of waves. So we need to involve the frequency in the formula $E = 2mv^2$ found in the section "Discrete Energy and Frequences" of Chapter 5. A full frequency f occurs with a full swing from T_1 to B and back up to T_2:

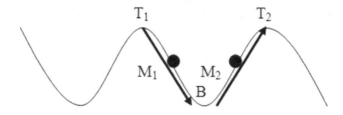

This happens in time duration $4t$. By definition a frequency is the reverse of that, so $f = 1/(4t)$ and $t = 1/(4f)$. Remembering that $d = A/2 = \frac{1}{2}at^2$, then $A = at^2 = (at)t = vt$. So then $v = A/t$ and the equation $E = 2mv^2$ becomes

$$E = 2mv^2 = 2m(A/t)^2$$

or

$$E = 2m(^A/_{(1/(4f))})^2$$

or

$$E = 2m(4Af)^2$$

or

$$E = 32m(Af)^2 \quad (1)$$

This is the energy for one wave. As can be seen, the higher the frequency, the higher the energy carried. If the wave has no mass, then it's pure energy. So in the equation above, the mass is imaginary because we are solely interested in an energy formula for energy, not matter.

This is where the equation $E = mc^2$ comes in handy. It not only says that matter can be transformed into energy, it also formulates equivalence between the two. So then a wave has an imaginary mass of $m = E_p/c^2$, where E_p is the energy of a *weightless* wave. Those waves are called *photons*. Let E_p be the lowest possible energy level for a photon. Therefore, it's a *constant*, and equation (1) becomes

$$E = 32m(Af)^2 = 32(E_p/c^2)(Af)^2 = 32E_pA^2(f/c)^2$$

Remembering that $f/c = 1/\lambda$ and the formula becomes

$$E = 32E_pA^2(f/c)^2 = 32E_pA^2(1/\lambda)^2 = 32E_p(A/\lambda)^2$$

In equation (1), we notice that the amplitude A and the frequency f have the same power in producing energy. But we are most interested in the frequency. Let $A/\lambda = f_s$ be the "standardized" frequency—this is the frequency taking into account the amplitude. The ratio A/λ is this one:

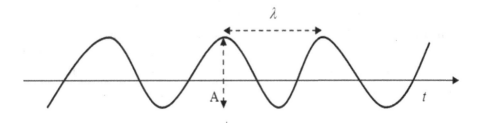

Why hide the amplitude by replacing it with this standardized frequency? Two reasons:

1. The unit of this ratio is *meters / (meters/wave)* = *wave*. So it's in units of number of waves. The higher that ratio A/λ is, the more waves there are that will be felt by nature as a higher frequency; the lower that ratio is, the lower the frequency.

2. Nature is more sensitive to frequency than amplitude. Think of a dim light. The dimmer the light, the less light there's, but the color doesn't change. A dim light changes the amplitude of the wave. So the higher the amplitude, the more photons are transmitted; the lower the amplitude, the less photons are transmitted. A prism splits the light into different colors with each color having its own frequency. The higher the frequency, the more energetic the photons transmitted. Note that the speed of the waves is always c regardless of amplitude or frequency.

The formula above becomes

$$E = 32E_p A^2 (f/c)^2 = 32E_p A^2 (1/\lambda)^2 = 32E_p f_s^2 \quad (2)$$

The standardized frequency is the one that ignores the amplitude and takes it as a contribution to the frequency. The reason that amplitude is irrelevant in this context is that nature uses solely the frequency to derive the distance traveled by a wave (see the section "An Absolute Frame of Reference" of Chapter 5. Also a light of higher amplitude gives a brighter light, hence more energy. But it is light of different frequencies that are of interest because they are the ones that interact with particles such as electrons. Indeed, an electron hit by a higher frequency light will react more than when hit by a lower frequency light. In fact, this phenomenon explains the colors of the rainbow with which we are all familiar.

Equation (2) says that energy of waves depends on the frequency squared, but this isn't quite true. Equation (6) in a section below will give the correct formula.

Toward Better Formulas

The formula (2) derived above although not the correct one is sufficiently accurate for us to see that the energy of a particle is indeed discrete. Nonetheless, let's find more accurate formulas. If you've learned trigonometry, you'll know that the distance d is given by the trigonometric equation

$$d(t) = A\cos(2\pi f t)$$

where A is the amplitude of the wave, f is its frequency, and t the time.

If you've learned some calculus, you'll know about a derivate of a function. The first derivate of the formula above gives the speed v upward and downward of the wave

$$\frac{d(t)}{dt} = -A2\pi f \sin(2\pi f t) = v$$

The second derivate gives the acceleration a

$$\frac{d^2(t)}{dt^2} = -A(2\pi f)^2 \cos(2\pi f t) = a$$

So then when the wave goes from top point T_1 to low point B on the graph, it accelerates from speed 0 to the maximum speed $A2\pi f$ at the middle path marked M of the line AB. At that point, the angle is $\pi/2$. So then in the cosines function, $2\pi f t = \pi/2$. Thus the time is translated from the frequency from $t = 1/(4f)$. So the equation of the section "Discrete Energy and Frequencies" of Chapter 5

$$E = F \cdot d = ma \cdot \left(\frac{1}{2}at^2\right) = \frac{1}{2}m(at)^2 = \frac{1}{2}mv^2$$

can be changed to

$$E = F \cdot d$$

or

$$E = ma \cdot d$$

or

$$E = m \left(\frac{d^2(t)}{dt^2} \right) \cdot d$$

or

$$E = m(-A(2\pi f)^2 \cos(2\pi ft)) \cdot (A \cos(2\pi ft))$$

or

$$E = -mA^2(2\pi f)^2 \cos^2(2\pi ft)$$

As the cosines can never be greater than 1 in the time range of 0 to t = $1/(4f)$, then the energy is no more than $E = mA^2(2\pi f)^2$; over an entire amplitude, it's twice that and then back on the top (crest) of the wave, another doubling of the energy. So the entire energy for one wavelength is no more than

$$E = 4mA^2(2\pi f)^2 \quad (3)$$

Note that the amplitude A is covered over the time $2t = 2/(4f) = 1/(2f)$. This gives an average vertical speed of

$$\text{distance/time} = A/2t = A/(1/(2f)) = 2Af = v$$

If we plug that in the formula (3), it gives

$$E = 4mA^2(2\pi f)^2$$

or

$$E = 4m\pi^2 A^2(2f)^2$$

or

$$E = 4m\pi^2(2Af)^2$$

or

$$E = 4m\pi^2 v^2$$

or

$$E = 2\pi^2(2mv^2)$$

This formula is of the same form as the one $E = 2mv^2$ found with the first approach in the section "Discrete Energy and Frequencies" of Chapter 5. Interesting!

Let's go back to the formula (3)

$$E = 4mA^2(2\pi f)^2$$

Remember that the imaginary mass is $m = E_p/c^2$, where E_p is the energy of a weightless wave of lowest energy. It's the photon of lowest energy. Then the formula above becomes

$$E = 4(E_p/c^2)A^2(2\pi f)^2$$

or

$$E = (4\pi)^2(E_p/c^2)A^2 f^2$$

or

$$E = (4\pi)^2 E_p A^2(f^2/c^2)$$

Remembering that $c = \lambda f$, this gives $c/f = \lambda$ or $f/c = 1/\lambda$, and the formula becomes

$$E = (4\pi)^2 E_p A^2(1/\lambda)^2$$

or

$$E = (4\pi)^2 E_p(A/\lambda)^2$$

So let $f_s = A/\lambda$ be the "standardized" frequency and the formula becomes

$$E = (4\pi)^2 E_p f_s^2 = 16\pi^2 E_p f_s^2 \quad (4)$$

So the overall energy of a wave is proportional to its frequency squared. Now this formula obtained with trigonometry and calculus looks awfully similar to this equation (2) of the previous section

$$E = 32E_p f_s^2$$

except that *16π² ≈ 32 x 4*.

Remember that E_p is a constant. Let's put the constants together as $g = 16\pi^2 E_p$, which I might call God's constant. So the formula becomes

$$E = g f_s^2$$

The Best Formula for Discrete Energy

The two formulas (2) and (4) of the previous sections only assume the vertical up and down wiggling motion that produce the wave, but don't really take into account the *shape* of the wave. This third approach will do that. A note: this section requires a good knowledge of calculus. If you're only interested in the final result, jump to equation (5) below.

The amount of energy contained in a wave is given by the integral function

$$\int_0^{A/2} 4F dy = Energy$$

where $F = ma$ is the force, dy the infinitesimal distance, and $y(t) = A\cos(2\pi ft)$. The factor 4 is because the integral covers only one-fourth of a full wave. Hence the first derivate of the formula gives the speed v upward and downward of the wave

$$y'(t) = -A2\pi f \sin(2\pi ft) = v$$

or another way to put it,

$$dy = -A2\pi f \sin(2\pi ft)dt$$

The second derivate gives the acceleration a

$$y''(t) = -A(2\pi f)^2 \cos(2\pi ft) = a$$

So the force is $F = ma = m(-A(2\pi f)^2 \cos(2\pi ft))$ and the integral function becomes (remember that at distance $A/2$, the time is $t=1/(4f)$). So then the integral is

$$\int_0^{1/(4f)} 4m(-A(2\pi f)^2 \cos(2\pi ft)) \cdot (-A2\pi f \sin(2\pi ft))dt$$
$$= Energy$$

or

$$\int_0^{1/(4f)} 4mA^2(2\pi f)^3 \cos(2\pi ft) \sin(2\pi ft)dt = Energy$$

Remembering the trigonometric identity $cos(x)sin(x) = \frac{1}{2}sin(2x)$, then the integral becomes

$$\int_0^{1/(4f)} 4mA^2(2\pi f)^3 \, \frac{1}{2}\sin\big(2(2\pi ft)\big)dt =$$
$$2mA^2(2\pi f)^3 \int_0^{1/(4f)} \sin(4\pi ft)dt$$

Let's focus on the integral portion solely for a moment and leave out the term $2mA^2(2\pi f)^3$

$$\int \sin(4\pi ft)dt = \big(\frac{1}{(4\pi f)}\big) \cdot (-\cos(4\pi ft))$$

With the boundaries 0 to $1/(4f)$, this integral evaluates to

$$\left(\frac{1}{(4\pi f)}\right) \cdot \left(-\cos\left(4\pi f \cdot \left(\frac{1}{(4f)}\right)\right) - -\cos(4\pi f \cdot 0)\right) =$$

or

$$\left(\frac{1}{(4\pi f)}\right) \cdot (-\cos(\pi) - -\cos(0)) =$$

or

$$\left(\frac{1}{(4\pi f)}\right) \cdot (-\cos(\pi) + \cos(0)) =$$

or

$$\left(\frac{1}{(4\pi f)}\right) \cdot (-(-1) + 1) =$$

or

$$\left(\frac{1}{(4\pi f)}\right) \cdot (1 + 1) =$$

or

$$\left(\frac{1}{(4\pi f)}\right) \cdot (2) = \frac{1}{(2\pi f)}$$

Now let's take back the term $2mA^2(2\pi f)^3$ that was left out a moment ago, and multiply it by $1/(2\pi f)$ giving

$$E = 2mA^2(2\pi f)^3 \cdot \frac{1}{(2\pi f)} = 2mA^2(2\pi f)^2$$

Replacing the imaginary mass $m = E_p/c^2$ by its energy, we get

$$E = 2\left(\frac{E_p}{c^2}\right)A^2(2\pi f)^2$$

or

$$E = 2E_p A^2 (2\pi(f/c))^2$$

or

$$E = 2E_p A^2 (2\pi(1/\lambda))^2$$

or

$$E = 2E_p (2\pi(A/\lambda))^2$$

We take again the standardized frequency $A/\lambda = f_s$ and this formula turns to

$$E = 2E_p(2\pi f_s)^2 = 8\pi^2 E_p f_s^2 \ (5)$$

So there you have it: three formulas (2), (4), and (5) that show the energy of a particle in motion with its frequency, with the latest equation (5) being the most accurate.

We're nearly done! All three formulas give the energy of *one wave*. But we're more interested in the energy of the wave per *frequency*, not just per wave. Consequently, we have to divide the formula by f_s

$$E = 8\pi^2 E_p (f_s^2 / f_s)$$

or

$$E = 8\pi^2 E_p f_s \ (6)$$

It turns out that the correct formula relating the frequency and energy of a wave was discovered by the physicist Planck in 1902, and is formulated as $E = h\nu$, where $h = 6{,}6 \times 10^{-34} \text{J} \cdot \text{s}$, and ν is the frequency. The energy unit is the joule. For instance with the Planck formula $E = h\nu$, the units are

$$E = \text{J} \cdot \text{s} \cdot \text{s}^{-1} = \text{Joule.}$$

This is correct because energy is measured in joule units. My frequency f_s has units $\text{m}/(\text{m} \cdot \text{wave}^{-1}) = \text{wave}$, and so my formula gives the units

$$E = \text{kg} \cdot (\text{m}^2/\text{s}^2) \cdot (\text{wave}) = \text{J} \cdot \text{wave.}$$

This "wave" unit isn't a real unit, but if the equation $E = h\nu$ made use of it, its units would become $\text{J} \cdot \text{s} \cdot (\text{wave} \cdot \text{s}^{-1}) = \text{Joule} \cdot \text{wave}$. This comes to the same units as my formula. Therefore, my formula is correct. Thank goodness—after all this work, it had better be right!

Now what's the real value of E_p? I looked up on the Internet for the lowest frequency of a wave to be known as about $v = 5$ x 10^{-3}/s. That gives a lowest energy wave of

$$(6.6 \text{ x } 10^{-34}\text{J.s}) \bullet (5 \text{ x } 10^{-3}/\text{s}) = 3.3 \text{ x } 10^{-36} \text{ J}$$

according to Planck's formula $E = hv$. That is also the lowest energy of a photon. So my lowest photon energy is $E_p = 3.3$ x 10^{-36} J•s. So then my constant in

$$E = 8\pi^2 E_p f_s$$

comes to

$$8\pi^2 E_p f_s \approx 78.9 \cdot (3.3 \times 10^{-36} J \cdot s) = 2.6 \times 10^{-34} J \cdot s$$

This is *very* close to the actual Planck's constant $h = 6.6$ x 10^{-34} J•s as discovered by Planck himself! Let $g = 8\pi^2 E_p$ be my constant. Then my energy formula is

$$E = g f_s$$

Planck's formula is expressed as $E = hv$. Keep my formula in mind (and Planck's formula too) because they'll be needed in the section "The Beginning of Nature" of Chapter 8.

For convenience's sake, I'll drop the s subscript:

$$E = gf \quad (7)$$

Appendix 4

(This appendix goes with Chapter 6: What is Gravitation?)

Formula for Centripetal Force

I show now how I derived the centripetal force formula in the section "Solar Systems in Equilibrium" in Chapter 6. I assumed a perfectly circular orbit of a celestial body:

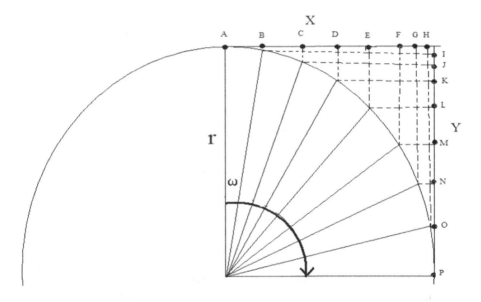

Suppose that the planet orbits in the direction of the arrow. To help illustrate the centripetal force at equal angular intervals ω, the positions of the planet are mapped onto the horizontal axis X, then the vertical axis Y, starting at point A then B, then C, all the way to point P. Any point on the X axis is given by the trigonometric formula

$$x = r \cos(\omega t)$$

and any point onto the Y axis is given by the trigonometric formula

$$y = r\sin(\omega t)$$

The speed on these axes is given by the derivates of these equations

$$v_x = dx/dt = -r\omega\sin(\omega t)$$

and

$$v_y = dy/dt = r\omega\cos(\omega t)$$

The acceleration onto the X axis and the Y axis is given by

$$a_x = \frac{dv_x}{dt} = -r\omega^2\cos(\omega t) \quad (1)$$

and

$$a_y = \frac{dv_y}{dt} = -r\omega^2\sin(\omega t) \quad (1)$$

So at any moment, the planet has the combined acceleration as depicted by the right-angled triangle

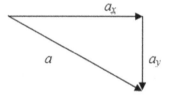

Thus, the combined acceleration is the square root of $a^2 = a_x^2 + a_y^2$
Using the equations (1) of these variables, this becomes

$$a^2 = (-r\omega^2\cos(\omega t))^2 + (-r\omega^2\sin(\omega t))^2$$

or

$$a^2 = r^2\omega^4\cos^2(\omega t) + r^2\omega^4\sin^2(\omega t)$$

or

$$a^2 = r^2\omega^4(\cos^2(\omega t) + \sin^2(\omega t))$$

Remembering that $cos^2(\omega t) + sin^2(\omega t) = 1$, the equation is reduced to

$$a^2 = r^2 \omega^4 \quad (2)$$

The angular variable $\omega = v/r$, where v is the speed of the planet. So then $\omega^4 = v^4/r^4$. Plugging that into the formula (2) gives

$$a^2 = r^2(v^4/r^4)$$

or

$$a^2 = v^4/r^2$$

Therefore the acceleration is $a = v^2/r$. As the force is given by the identity $F = ma$, where m is obviously the mass of the planet, it follows that the centripetal force is

$$F = \frac{mv^2}{r}$$

Appendix 5

(This appendix goes with Chapter 6: What is Gravitation?”).

From Quantum Physics to Newtonian Physics

In this appendix, I wish to show that my space-density gravitation model predicts the acceleration of celestial bodies around, say, the sun. An equation for that acceleration will be developed. (Note that you'll need an elementary knowledge of calculus.)

   Consider matter and light as their trajectories bend toward the sun. Imagine matter falling down and proceeding to cross the circle $C(r+dr)$ as shown below:

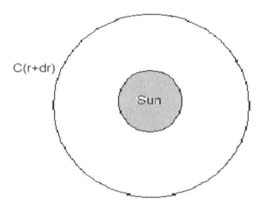

The parameter $r+dr$ designates the radius of the circle. Then matter and light get a distance of dr closer to the sun. So the matter and light are at the circle $C(r)$ as shown below:

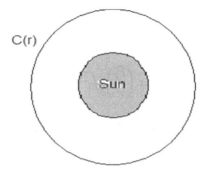

For now, let's ignore the light passing through and consider only matter. By the way, in this proof, energy density is to mean the KED of space, that is, the density that is the result from the effect of gravitation. Why do I use in this proof the KED rather than the PED of space? It's because I wish to prove that the PED will cause an acceleration that will, in turn, have an effect on the KED of space. This proof is about the consequence of my theory of PED of space.

It's fair to assume that the energy density on the edge of the circle $C(r+dr)$ is uniform, and that the energy density on the edge of the circle $C(r)$ is uniform as well. As matter moves using "floats." let's say that the circle $C(r+dr)$ crosses f "floats." My model says that the PED of space will cause matter to move toward the sun, and this is the gravitational force. Will acceleration result from that? According to Newton's Second Law of motion expressed by the formula $F = ma$, there certainly should be an acceleration. Let's investigate what that acceleration is.

. Onto the circle $C(r+dr)$, the value of the variable f is simply

$$f = \frac{2\pi(r + dr)}{P_S(r + dr)} \quad (1)$$

where $P_S(r+dr)$ is the Planck length between two "floats" at radius $r+dr$ from the sun. In the section "Discrete Energy and Frequencies" of Chapter 5, we determined that P_S is dependent on the speed of the particle expressed by the formula

$$P_S = P_s\sqrt{1 - (\tilde{V}/c)^2}$$

Another way to put it is

$$P_S = \left(\frac{P_s}{c}\right)\sqrt{c^2 - V^2}$$

Let's express V as $v(r+dr)$, which represents the speed of matter onto the edge $C(r+dr)$. So then we have

$$P_S(r + dr) = \left(\frac{P_s}{c}\right)\sqrt{c^2 - v(r + dr)^2}$$

With this expression, equation (1) becomes

$$f = \frac{2\pi(r + dr)c}{P_s\sqrt{c^2 - v(r + dr)^2}} \quad (2)$$

As all matter is subjected to by the same gravitational force, it's fair to assume that all the matter on the circle $C(r+dr)$ will reach the circle $C(r)$ at the same time.

Now let's assume that my model is unable to account for acceleration from gravitation, implying that we have to assume that all matter will reach the edge of the circle $C(r)$ at the *same speed* they had when they crossed the circle $C(r+dr)$. This implies that the number of "floats" did *not* increase while going from circle $C(r+dr)$ to $C(r)$ (the number of floats could increase only if the speed increased too, which we assume that it doesn't). So an equal number of f floats cross the circle $C(r)$ as onto the circle $C(r+dr)$. So we have a second equation for the variable f

$$f = \frac{2\pi r}{P_S(r)}$$

or put another way

$$f = \frac{2\pi r c}{P_s\sqrt{c^2 - v(r)^2}} \quad (3)$$

Note that the circumference of the circle $C(r)$ is obviously smaller than that of the circle $C(r+dr)$. Consequently the skips between the "floats" on the edge of circle $C(r)$ are closer together than those onto the circle $C(r+dr)$. As we discovered in Chapter 4, as a particle speeds up, its skips shorten. So we have that $P_S(r) < P_S(r+dr)$ or

$$P_s\sqrt{c^2 - v(r)^2} < P_s\sqrt{c^2 - v(r + dr)^2}$$

The only way that this can be is that $v(r) > v(r+dr)$. Consequently, we have to conclude that matter on the circle $C(r)$ gained speed (i.e., acceleration occurred). This proves that acceleration will result from my model of gravitation. The effect is like a body going down a hill. This is also what Einstein's model of space–time suggests, except that his model doesn't explain the motion in a purely physical way. Mine does!

Essentially, the matter that travels from circle $C(r+dr)$ to $C(r)$ has to squeeze in. This causes the matter to pick up speed. We can experience the same effect by making a hole in a box of juice and squeezing. The smaller the hole, the faster the juice comes out; the bigger the hole, the slower the juice comes out. Of course this is just an analogy: light and matter aren't liquids! Nonetheless, we'll see in Chapter 6 that there's some truth to this thermodynamic analogy.

Now that it has just been proven that acceleration occurs, let's attempt a derivation of a formula for that acceleration.

The equations (2) and (3) are equal

$$f = \frac{2\pi(r + dr)c}{P_s\sqrt{c^2 - v(r + dr)^2}} = \frac{2\pi rc}{P_s\sqrt{c^2 - v(r)^2}}$$

Some terms appear on both sides of the equality, so they cancel each other out, giving:

$$\frac{(r + dr)}{\sqrt{c^2 - v(r + dr)^2}} = \frac{r}{\sqrt{c^2 - v(r)^2}}$$

The square roots are an annoyance, so let's square everything

$$\frac{(r + dr)^2}{c^2 - v(r + dr)^2} = \frac{r^2}{c^2 - v(r)^2}$$

The denominators may be expressed as factors as follows

$$\frac{(r + dr)^2}{(c - v(r + dr)) \cdot (c + v(r + dr))} = \frac{r^2}{(c - v(r)) \cdot (c + v(r))}$$

Nothing can be faster than the speed of light, so $c + v(r+dr) = c$ and $c + v(r) = c$. The equality is then simplified to

$$\frac{(r + dr)^2}{(c - v(r + dr)) \cdot c} = \frac{r^2}{(c - v(r)) \cdot c}$$

Next the term c may be cancelled out

$$\frac{(r + dr)^2}{(c - v(r + dr))} = \frac{r^2}{(c - v(r))}$$

This is equivalent to

$$(r + dr)^2 (c - v(r)) = r^2 (c - v(r + dr))$$

or

$$(r^2 + 2rdr + dr^2)(c - v(r)) = r^2(c - v(r + dr))$$

The dr^2 may be neglected because it's extremely small compared to the other terms

$$(r^2 + 2rdr)(c - v(r)) = r^2(c - v(r + dr))$$

The term r may be factored out on the left side, giving us

$$r(r + 2dr)(c - v(r)) = r^2(c - v(r + dr))$$

Then dividing by r on both sides gives us

$$(r + 2dr)(c - v(r)) = r(c - v(r + dr))$$

A few manipulations later, we obtain

$$rc - rv(r) + 2cdr - 2v(r)dr = rc - rv(r + dr)$$

The term rc cancels out on both sides giving

$$-rv(r) + 2cdr - 2v(r)dr = -rv(r + dr)$$

Next let's place the term $rv(r)$ on the right side

$$2cdr - 2v(r)dr = -rv(r + dr) + rv(r)$$

On the left side, the term $2dr$ may be factored out, and on the right side the term r may be factored out as well, giving us

$$2dr(c - v(r)) = r(v(r) - v(r + dr))$$

The expression $v(r) - v(r+dr) = dv$, giving us

$$2dr(c - v(r)) = rdv$$

For all practical purposes, $c - v(r) = c$ because the speed of matter $v(r)$ in a gravitational field is much smaller than the speed of light. The equation is thus simplified to

$$2cdr = rdv$$

Putting the derivative dr on the right side and putting the term r on the left side gives us

$$\frac{2c}{r} = \frac{dv}{dr} \quad (4)$$

We're almost done. Note that *dv/dr* may be expressed as

$$\frac{dv}{dr} = \frac{dv/dt}{dr/dt}$$

The ratio *dv/dt* represents the acceleration, so let *dv/dt* = *a(t)*. The ratio *dr/dt* represents the speed of the matter subjected to gravitation, so let *dr/dt* = *v(t)*. With this, the equation (4) is expressed as

$$\frac{2c}{r} = \frac{dv}{dr} = \frac{dv/dt}{dr/dt} = \frac{a(t)}{v(t)}$$

or

$$\frac{2c}{r} = \frac{a(t)}{v(t)}$$

This finally gives the acceleration upon matter due to gravitation

$$\frac{2cv(t)}{r} = a(t) \quad (5)$$

You should be able to recognize that this expression is close to the correct one, $a = v^2/r$, for the centripetal force in a gravitational field. The reason that my equation doesn't match the correct one is in part due to the model used in this appendix that is only an approximation of reality. Indeed, my model assumes that objects fall straight down toward the sun, but this straight motion is most unlikely for two reasons:

1. There are 180 degrees of angle by which an object may approach the sun. A path straight down is at 90 degrees. For an object to fall straight down, it would have to approach the sun at *precisely* 90 degrees and stay at that angle. This is most unlikely.

2. Based on the argument (1), the object will approach the sun at an angle. Inevitably, the object will therefore approach the sun following an arc. At the quantum level, there are no arcs; there are only straight steps. A quantum particle (part of an object) would therefore follow the steps depicted below:

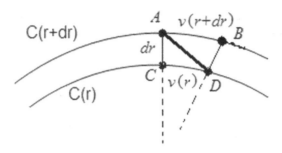

The dotted lines converge toward the center of the orbit, the sun. A particle in an *equilibrium* orbit around the sun would go from point *A* to point *B* onto the circle or orbit *C(r+dr)*. In the mathematical development so far, it was assumed that a particle falling toward the sun would go to point *C* from point *A*. From there, it would go to point *D* at speed *v(r)*. But in reality, the particle will go from point *A* directly to point *D*, following the diagonal *AD*, at which point, its speed will be *v(r)*.

I offer you a word of caution: the picture above seems to suggest that the *length* of the base, *CD*, of the triangle is *v(r)*. That's obviously not possible because a velocity isn't a length. The term *v(r)* is placed in the picture simply to indicate that the length of the base *CD* is associated with the speed *v(r)*.

As the diagonal *AD* is longer than the line *CD*, it follows that the actual real acceleration toward point *D* will be *smaller* than that one given by equation (5). Expressed mathematically, this implies that

$$a(t) = \frac{2cv(t)}{r} \cdot \frac{length\ CD}{length\ AD} < \frac{2cv(t)}{r} \quad (6)$$

We're searching for the correct expression for the acceleration $a(t)$ at point D. For that, we need to figure out the value of the ratio (*length CD*) / (*length AD*). At the moment, we only know that it's less than 1.

Note that the formula above is expressed in function of time while the picture associated with it is expressed in function of the radius. But we can't have both. As the acceleration is expressed in a function of time, let's consider the velocity in a function of time. So the picture above becomes:

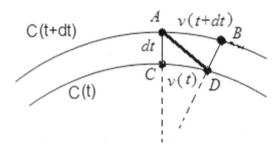

It's still the same picture, but everything is expressed as a function of time. Now you might notice that the right-angled triangle ACD looks very similar to this one from the section "A Better 'Tick' Formula" of Chapter 4:

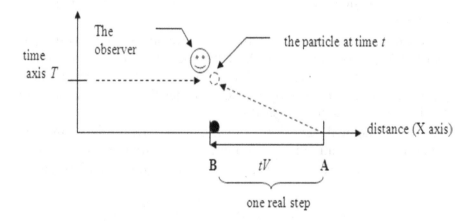

The two triangles are more than similar: they are identical! This triangle is the motion of a particle in space–time.

You might find this a contradiction that I use at this moment space–time motion in my development while stating in the section "Motion in Space–Time: The Plot Thickens!" of Chapter 5 that motion in space–time isn't possible. But no, there's no contradiction:

1. Einstein's space–time theory does describe well the motion of a body in a gravitational field. This appendix is indeed about determining motion based on my theory that gravitation has its roots at the quantum level.

2. As shown in the section "The Clock of the Universe" of Chapter 4, the distance traveled by a quantum particle is equivalent to the time span. This is also shown in section "Motion Finally Explained" of Chapter 7. Just like the distance by the hand on a clock is used to measure time, the same occurs at the quantum level. This is the reason that on the two pictures earlier of the orbiting particle, the distance dr on the right-hand triangle could be replaced by the time dt.

In the picture above, remember that the term c stands for the speed of light. The right-angled triangle has a hypotenuse of length ct, the base is of speed Vt, and the height of the triangle is the time on the time axis. In our case of gravitation, that height of time is dt as depicted in the picture. Based on this triangle, we conclude that the ratio in equation (6) is

$$\frac{length\ CD}{length\ AD} = \frac{Vt}{ct} = \frac{V}{c}$$

The term V is $v(t)$ in our case of gravitation. So the ratio is

$$\frac{length\ CD}{length\ AD} = \frac{v(t)}{c}$$

Placing this equation into equation (6) gives the correct expression of acceleration

$$a(t) = \frac{2cv(t)}{r} \cdot \frac{length\ CD}{length\ AD}$$

or

$$a(t) = \frac{2cv(t)}{r} \cdot \frac{v(t)}{c}$$

or

$$a(t) = \frac{2v(t)^2}{r} \quad (7)$$

We're not done yet with formula (7) because the actual acceleration is certainly less than that. The reason is that the length of the line CD is a Planck length that depends on the speed $v(t)$. The figure with the triangle ACD didn't take into account the acceleration. So the particle will reach point D with a faster speed than $v(t)$. As we learned in Chapter 4, the *faster* is the speed of a particle, the *shorter* is its Planck length. Therefore the line CD will be shorter than what the figure depicted. Given this realization, the correct picture is as follows:

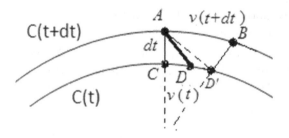

The line CD' is the old line CD from the previous picture. As a consequence of this new picture, we have

length of line CD < length of line CD'

So

$$\frac{length\ CD}{length\ AD} < \frac{v(t)}{c}$$

and

$$a(t) < \frac{2v(t)^2}{r}$$

The last step in this mathematical demonstration is to find the correct value of the ratio *(length CD)* / *(length AD)*. Let's see if the point *D* is exactly in the middle of the line *CD'*.

This is where it will come handy to remember that a quantum particle behaves like waves as we discovered in the section "An Absolute Frame of Reference" of Chapter 5. This behavior was depicted as follows in the section "Light with a Spin":

Another way to see the wavy motion is that the particle at point *A* will overspeed toward point *D*. The result will be that the centrifugal force will force the particle back onto the circle *C(t+dt)* at point *B* as shown here:

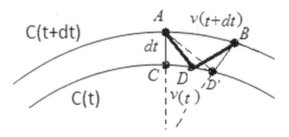

You might ask: why would the particle fall back at point *B* and not somewhere else on the orbit *C(t+dt)*? This is because the step *AB* is the shortest quantum step onto the orbit *C(t+dt)*. So the particle can't fall in between step *AB*, and it can't fall back farther than where point B is located because the centrifugal force is equal to the gravitational force that pushed the particle at point *D* in the first place. The effect is similar to a spring that's stretched at point *D*, then contracts back to the orbit *C(t+dt)* at point *B*. Essentially, the celestial body orbits the sun in a wavy motion such as was depicted in the section "Solar Systems in Equilibrium" of Chapter 6:

Of course the wavy motion here is greatly exaggerated, and this wavy motion takes place at the quantum level so it's not detectable at the macroscopic level. At the quantum level, the particle will move back and forth between the orbits $C(t+dt)$ and $C(t)$ as follows:

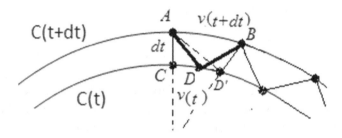

Because the crests of waves are evenly spaced, the point D is bound to occur at the middle of the line CD', and so *length CD* = ½ *length CD'*. It was mentioned earlier that

$$\frac{length\ CD}{length\ AD} < \frac{v(t)}{c}$$

Now we know that

$$\frac{length\ CD}{length\ AD} = \frac{1}{2}\frac{v(t)}{c}$$

So then from equation (6), we get the final expression for the acceleration

$$a(t) = \frac{2cv(t)}{r} \cdot \frac{length\ CD}{length\ AD}$$

or

$$a(t) = \frac{2cv(t)}{r} \cdot \frac{1}{2}\frac{v(t)}{c}$$

or

$$a(t) = \frac{v(t)^2}{r} \quad (8)$$

The correct centripetal acceleration in a gravitational field is indeed v^2/r. As I was able to derive that same centripetal acceleration, I feel very confident that my theory of gravitation definitely has a lot of truth to it. Also my theory appears to be equivalent to Einstein's space–time curvature theory as I made use of his space–time formulas that I derived in Chapter 4! Also my proof uses general relativity and quantum theory together to obtain a correct equation from Newtonian physics. Therefore, it appears that both theories can work together, with the reason being that gravitation takes place at the quantum level. This is made clear by the work in this appendix.

Appendix 6

(This appendix goes with Chapter 8: How Did the Universe Start?).

The Expansion of the Universe

Recall from the section "Fighting off the 'Void'" of Chapter 6 that my theory suggests that gravitation is a manifestation of nature's attempt to *neutralize* space density differentials. This effect takes place when an area of space becomes too thin of space density so that nature sends energy and matter there.

I theorize now then that the acceleration of the expansion started when the space density of the edge of the universe reached a critical minimal value. Indeed, as the universe expands, the surface of its edge increases, but the entire amount of energy available in the entire universe doesn't increase because the Big Bang ended a long time ago. My theory of expansion is itself based on my other theory of gravitation that suggests that gravitation is a manifestation of nature's attempt to *neutralize* space density differentials. Keeping that in mind, here is a picture of the expanding universe:

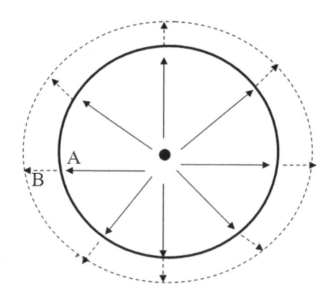

The inner sphere A is where the edge of the Universe is now, and the outer sphere B is where the edge will be dt time later. In the algebraic development that follows, calculus is used.

The area of the sphere A is $4\pi R^2$, and that of the sphere B is $4\pi(R+dr)^2$ where dr is the new radius of the universe at sphere B. So the speed of expansion at sphere B is

$$X = \frac{dr}{dt} \quad (1)$$

Note: usually the variable X is used to express distances. I use it in this appendix for the speed of expansion. Suppose that e is the amount of energy on the sphere A. Recall from Chapter 6 that I loosely defined space density as the amount of energy per unit surface (OK, it should be per unit *volume*, but I'll clarify that aspect later). So the space density on the sphere A is

$$\frac{e}{4\pi R^2} \quad (2)$$

This ratio turns out to be an important one and will be referred to often in this appendix. So let the term Σ express that ratio $e/4\pi R^2$.

Again recalling from Chapter 6 that nature will fight off space density that becomes too thin, there will be a time when the space density expressed by the equation (2) will reach an absolute minimum. At that moment the space density on sphere B will have to be *no less* than that of equation (2). Because the surface of sphere B is larger than that of sphere A, it's inevitable that nature will push some extra energy de onto the sphere B so as to keep the space density at that minimum. So let de be the increase of the amount of energy (including matter) on the sphere B so that sphere B has $e+de$ amount of energy on it. Thus, its space density is

$$\frac{e + de}{4\pi(R + r)^2} \quad (3)$$

The two equations (2) and (3) will have to be equal to maintain minimal energy density

$$\frac{e}{4\pi R^2} = \frac{e + de}{4\pi(R + r)^2}$$

This equation is easily reduced to

$$\frac{e}{R^2} = \frac{e + de}{(R + r)^2}$$

Squaring the denominator to the right gives us

$$\frac{e}{R^2} = \frac{e + de}{(R^2 + 2Rdr + dr^2)}$$

The term dr^2 is negligible, giving us

$$\frac{e}{R^2} = \frac{e + de}{(R^2 + 2Rdr + dr^2)} = \frac{e + de}{R(R + 2dr)}$$

Or reducing some more by eliminating the factor R

$$\frac{e}{R} = \frac{e + de}{(R + 2dr)}$$

With further manipulation, we obtain

$$de = e\left(\left(\frac{R + 2dr}{R}\right) - 1\right)$$

or

$$de = e\left(1 + \left(\frac{2dr}{R}\right) - 1\right)$$

or

$$de = e\left(\frac{2dr}{R}\right)$$

So the equation becomes

$$de = \frac{2e\,dr}{R}$$

Dividing both sides by dt gives us

$$\frac{de}{dt} = \frac{2e\left(\frac{dr}{dt}\right)}{R}$$

Remember the equation (1), $X = dr/dt$? It expresses the speed of the expansion. Using this, the formula above becomes

$$\frac{de}{dt} = \frac{2eX}{R} \quad (4)$$

Note that because the term Σ expresses that ratio $e/4\pi R^2$, this equation may be expressed as

$$\frac{de}{dt} = \Sigma 8\pi RX \quad (5)$$

Note that Σ is a constant. Let's see how the right-hand side ratio de/dt behaves. Will it increase or decrease as the expansion proceeds? If it increases, then the ratio increases, implying that energy reaches the edge of the universe at an accelerated rate. Let's now take the second differential of this latest equation (5). (Reminder: $d(ab)/dt = (da/dt)b + a(db/dt)$. Using this, then we have

$$\frac{d^2e}{dt^2} = \Sigma 8\pi \left(\left(\frac{dR}{dt}\right)X + R\left(\frac{dX}{dt}\right) \right)$$

or, as $dR/dt = X$,

$$\frac{d^2e}{dt^2} = \Sigma 8\pi \left(X^2 + R\left(\frac{dX}{dt}\right) \right) \quad (6)$$

The equations developed so far were focusing on the energy density, represented by the term Σ, on the *surface* area of the edge of the universe. Let's next concentrate instead on the ratio of the difference in energy over a *volume*

$$\frac{de}{dv} = \xi$$

This is the energy density in a volume. At the time that space density is at its lowest, this ratio de/dv will no longer go down, but rather will remain stable, so then ξ is a constant from that moment on. We may rewrite the ratio above as

$$de = \xi dv$$

The constant ξ may also be viewed as a conversion factor for when the space density is at its lowest. Dividing both sides by dt gives us

$$\frac{de}{dt} = \xi \frac{dv}{dt} \quad (7)$$

We've seen earlier the estimate $dv = 4\pi R^2 dr$, where dr is the change in radius. With it, the equation (7) becomes

$$\frac{de}{dt} = \xi 4\pi R^2 \frac{dr}{dt}$$

or

$$\frac{de}{dt} = \xi 4\pi R^2 X \quad (8)$$

because $dr/dt = X$. Note that this equation looks a bit similar to the equation (5)

$$\frac{de}{dt} = \Sigma 8\pi RX$$

where the term Σ expresses the ratio $e/4\pi R^2$.

Both equations (5) and (8) express the rate of change of energy over time. But that doesn't imply that the two equations are equivalent. The equation (5) expresses the rate of change of the energy over time on the *surface* of the edge of the universe, whereas the equation expresses the rate of change of the energy over time in a *volume*. So the differential of energy *de* isn't necessarily the same for surface and volume. To clarify this, let e_S be the energy on the surface of the edge, and let e_V be the energy over a volume at that edge. In addition, let de_S/dt be the rate of change of energy over a volume at that edge, and let de_V/dt be the rate of change of energy over a volume at that edge. So then the equations (5) and (8) are now expressed as

$$\frac{de_S}{dt} = \Sigma 8\pi RX \quad (9)$$

and

$$\frac{de_V}{dt} = \xi 4\pi R^2 X \quad (10)$$

Why do I feel the need to develop two differential energy equations: one over a surface, one over a volume at the edge? The answer is in what the constants Σ and ξ really mean physically. It's important to notice that the constant Σ reveals the energy density on the *edge* of the universe. As the universe expands, what do you think is first created at its edge? Understand that the edge can't be made of the three-dimensional space that we experience right here on earth. This is because before the three dimensions of space can be created, the *fabric* of space has to be created first. Only after the fabric of space is created can all three dimensions be created next.

So then the constant Σ gives the energy density of the *constitution* of space; essentially, it's the energy density of that *dark energy* as is the term used at times in this book. Once the fabric of space is created, then the three dimensions of space are created.

Then, once the three dimensions are created, energy can go into that brand new space, and that is what the constant ξ is all about. It represents the energy *volume* expansion at the edge of the universe. As it's brand new space near the edge of the universe, initially it's inevitable that it's the *minimal* amount of energy possible that moves into that new space. This is what the constant ξ is about: it gives the minimal energy density *in* space, whereas the constant Σ gives the minimal energy density *of* space.

Now let us take the second derivative of equation (10)

$$\frac{d^2 e_V}{dt^2} = \xi 4\pi \left(2R \left(\frac{dR}{dt} \right) X + R^2 \left(\frac{dX}{dt} \right) \right)$$

or since $dR/dt = X$,

$$\frac{d^2 e_V}{dt^2} = \xi 4\pi \left(2RX^2 + R^2 \left(\frac{dX}{dt} \right) \right)$$

Applying some more algebraic manipulations results in

$$\frac{d^2 e_V}{dt^2} = \xi 4\pi R^2 \left(2\frac{X^2}{R} + \left(\frac{dX}{dt} \right) \right)$$

Furthermore, we can factor out the term X

$$\frac{d^2 e_V}{dt^2} = \xi 4\pi R^2 X \left(2\frac{X}{R} + \frac{1}{X} \left(\frac{dX}{dt} \right) \right)$$

The formula $\xi 4\pi R^2 X$ happens to be de_V/dt as per the formula (10) above. Consequently, this differential equation may be expressed as

$$\frac{d^2 e_V}{dt^2} = \frac{de_V}{dt}\left(2\frac{X}{R} + \frac{1}{X}\left(\frac{dX}{dt}\right)\right) \quad (11)$$

This differential equation expresses the flow of energy over a volume near the edge of the universe. What about the flow of energy onto the surface of that edge? This has already been worked out and given by the equation (6), which may now be expressed as follows with the subscript S

$$\frac{d^2 e_S}{dt^2} = \Sigma 8\pi\left(X^2 + R\left(\frac{dX}{dt}\right)\right)$$

or when the term X is factored out

$$\frac{d^2 e_S}{dt^2} = \Sigma 8\pi R X\left(\frac{X}{R} + \frac{1}{X}\left(\frac{dX}{dt}\right)\right)$$

As $\Sigma 8\pi R X = de_S/dt$, the equation becomes

$$\frac{d^2 e_S}{dt^2} = \frac{de_S}{dt}\left(\frac{X}{R} + \frac{1}{X}\left(\frac{dX}{dt}\right)\right) \quad (12)$$

The differential equations (11) and (12) are almost identical! This observation will become handy later. The ratios d^2e/dt^2 and de/dt (regardless of the subscript V or S) express the energy flow at the edge of the universe (in the case of equation (11), over a volume; in the case of equation (12), over a surface). As all types of energy obey the Second Law of Thermodynamics, these two ratios reflect that law as well. Moreover, this equation includes the acceleration of the edge of the universe, dX/dt. These two equations suggest strongly that the acceleration of the universe obeys the Second Law of Thermodynamics. This will be proven later in this appendix.

Both equations (11) and (12) involve the term dX/dt, expressing the implication of an acceleration of the edge of the universe. But is it possible to deduce that the acceleration is positive or negative (deceleration)? As equations (11) and (12) are almost identical, we'll only use equation (12) to investigate.

It's obvious that the differential equation (12) can't be less than 0. If the differential equation is strictly positive, then de_S/dt increases at the edge of the universe, implying that the matter going outward accelerates too (note that this doesn't necessarily imply that the expansion of the universe accelerates). So then for d^2e/dt^2 to be positive, the expression in square brackets has to be positive

$$\frac{X}{R} + \frac{1}{X}\frac{dX}{dt} > 0$$

This would clearly be true as it's known by scientists that the universe expands so, $dX/dt \geq 0$. There's another way to see that the universe expands with acceleration. Suppose that the acceleration is 0. The acceleration of energy outward would then be

$$\frac{d^2e_S}{dt^2} = \frac{de_S}{dx}\frac{X}{R}$$

which is clearly greater than 0, so the energy will still accelerate outward even if the edge of the universe would *not* necessarily accelerate. But if the edge didn't accelerate, then energy at the edge would accumulate on the surface of the sphere, making the space density there increase above the minimal value of Σ. This violates the assumption made at the beginning of this section. Therefore, there has to exist some acceleration $dX/dt > 0$ at the edge of the universe.

But hold it! In the argument above, the assumption was made that the ratio $de_S/dt > 0$. Well, if de_S/dt is a nonzero constant, then d^2e_S/dt^2 has to be 0. But then the equation above becomes

$$0 = \frac{de_S}{dx}\frac{X}{R}$$

implying that $de_S/dt = 0$. But this contradicts the assumption that de_S/dt is a nonzero constant. So de_S/dt can't be a constant, and it can't be less than 0. If it were less than 0, then less energy would reach the edge of the universe as the radius increases. This would bring the energy density there below the minimal value Σ. Therefore, it has to be that the ratio $de_S/dt > 0$, and consequently the logical argument made earlier that $dX/dt > 0$ is correct.

Although we are able to deduce from differential equations (11) and (12) that $dX/dt > 0$, it seems impossible to derive an equation for dX/dt from them. Let's approach the problem by visualizing the energy flow at the edge of the universe. The differential equation (12) makes it clear that de_S/dt increases always. For this to be possible, new energy from behind the edge of the universe has to reach that edge. Here is the picture:

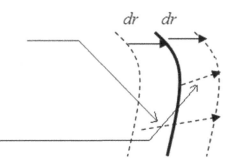

Energy from behind the edge catching up to the current energy traveling a distance outward 2dr.

Current energy from the edge moving outward dr

Both these dotted edges move outward in time dt. So the new energy that catches up with the edge does so twice as fast as the energy that's already on the edge. This will cause the new expanded edge to move faster, hence the acceleration. The average speed of expansion could be estimated as follows

$$\frac{e_S X + 2X de_S}{e_S + de_S} = \frac{X(e_S + 2de_S)}{(e_S + de_S)}$$

So the speed X increases slightly. Note that this formula is just an approximation of the speed, but a good enough one as the radius of the universe increases (as we'll see later).

The figure above suggests that the edge of the universe doesn't expand evenly, but rather more like a wave:

The crest of the wave would come from the energy from behind that caught up with the edge and hence is now moving faster. The section "Gravitons Revisited" of Chapter 8 shows that waves at the edge actually curl up inward causing photons to curl up as well, giving them a 2-spin rather than the normal 1-spin that light has. It's that change into 2-spin that makes the light at the edge turn into the dark energy making up the fabric of space.

Using the formula above, the change in speed X of the expansion is given by

$$\frac{X(e_S + 2de_S)}{(e_S + de_S)} - X = dX$$

or putting the left-hand side over the same denominator

$$\frac{X(e_S + 2de_S) - X(e_S + de_S)}{e_S + de_S} = dX$$

This reduces to

$$\frac{Xde_S}{e_S + de_S} = dX$$

Dividing by dt gives

$$\frac{X\left(\frac{de_S}{dt}\right)}{e_S + de_S} = \frac{dX}{dt}$$

or expressed another way

$$\frac{X}{e_S + de_S} \cdot \frac{de_S}{dt} = \frac{dX}{dt} \quad (13)$$

We now need an expression for $e_S + de_S$. Well, de_S is the energy addition on the *surface* of the edge of the universe that keeps the energy density constant there. This is precisely what the equation

$$\frac{e}{4\pi R^2} = \frac{e + de}{4\pi (R + r)^2}$$

used earlier expresses. It was found that this equation led to equation (4)

$$\frac{de_S}{dt} = \frac{2e_S X}{R}$$

or another way to put it,

$$de_S = \frac{2e_S X dt}{R}$$

Using this, then the expression $e_S + de_S$ is actually

$$e_S + de_S = e_S + \frac{2e_S X dt}{R}$$

or

$$e_S + de_S = e_S \left(1 + \frac{2X dt}{R}\right)$$

or

$$e_S + de_S = e_S \left(\frac{R + 2X dt}{R}\right)$$

Putting that into the equation (13) gives

$$\frac{XR}{e_S(R + 2X dt)} \cdot \frac{de_S}{dt} = \frac{dX}{dt}$$

As the value of R is so large, it's fair to assume that $R + 2X\, dt \approx R$. Note that this approximation will inevitably lead us to a somewhat incorrect equation for dX/dt, but a good approximation as the variable t becomes large. Using this approximation, the equation above is now reduced to

$$\frac{XR}{e_S R} \cdot \frac{de_S}{dt} \approx \frac{dX}{dt}$$

or

$$\frac{X}{e_S} \cdot \frac{de_S}{dt} \approx \frac{dX}{dt} \quad (14)$$

Plugging the equation (4)

$$\frac{de_S}{dt} = \frac{2e_S X}{R}$$

into equation (14), we obtain

$$\frac{X}{e_S} \cdot \frac{2e_S X}{R} = \frac{dX}{dt}$$

which is reduced to

$$\frac{2X^2}{R} \approx \frac{dX}{dt} \quad (15)$$

We finally have an equation for the acceleration of the universe! This equation proves that the expansion of the universe occurs with acceleration because clearly all terms are positive. This is the only good news, though. Sadly, when I worked out the solution to this differential equation, it gave a very poor estimate for dX/dt.

Let's think up a better solution. Why not use instead the energy differential equation (10)?

$$\frac{de_V}{dt} = \xi 4\pi R^2 X$$

by putting it into the formula (14)? Why would I wish to involve that energy equation? It's because it involves the constant ξ, which I have a hunch will turn out to reveal something interesting about space. Indeed it will as we'll see later. The trouble is that

$$\frac{de_V}{dt} \neq \frac{de_S}{dt}$$

So I may not put the term $\xi 4\pi R^2 X$ into the equation (14). Or perhaps, I may. Recall the differential equations (11) and (12) worked out earlier

$$\frac{d^2 e_V}{dt^2} = \frac{de_V}{dt}\left(2\frac{X}{R} + \frac{1}{X}\left(\frac{dX}{dt}\right)\right)$$

and

$$\frac{d^2 e_S}{dt^2} = \frac{de_S}{dt}\left(\frac{X}{R} + \frac{1}{X}\left(\frac{dX}{dt}\right)\right)$$

These two equations are so nearly identical that it suggests strongly that the ratios

$$\frac{de_V}{dt} \approx \frac{de_S}{dt}$$

Let's suppose that this is true. Recall that de_V/dt is expressed by the equation (9), and that de_S/dt is expressed by the equation (10). Let's see what we obtain by assuming that both expressions in equations (9) and (10) are approximately equal, that is

$$\xi 4\pi R^2 X \approx \Sigma 8\pi R X$$

or

$$\xi R \approx \Sigma 2$$

or

$$\frac{\xi}{\Sigma} \approx \frac{2}{R} \quad (16)$$

Later on in this appendix, we'll find estimate values for the two terms ξ and Σ, and discover that indeed this approximation is a good one! Therefore, the following is a good estimate

$$\frac{de_V}{dt} \approx \frac{de_S}{dt}$$

Using this approximation, we're now hopefully in a good position to find a better formula for dX/dt. From this approximation of the differentials, we may assume that equation (14) may be approximated by

$$\frac{X}{e_S} \cdot \frac{de_V}{dt} \approx \frac{dX}{dt}$$

Let's now put the expression of equation (10) into this formula

$$\frac{X}{e_S} \cdot \xi 4\pi R^2 X \approx \frac{dX}{dt}$$

or by rearranging the terms

$$\frac{\xi 4\pi R^2}{e_S} \cdot X^2 \approx \frac{dX}{dt}$$

Remember that $e_S/4\pi R^2 = \Sigma$, which is the space density on edge surface of the universe. Placing the term Σ into this approximation, we obtain

$$\frac{\xi}{\Sigma} X^2 \approx \frac{dX}{dt} \quad (17)$$

This equation proves that the expansion of the universe occurs with acceleration because clearly all terms are positive. So why did I choose to involve the ratio de_V/dt in the equation (14) rather than de_S/dt? Three reasons:

1. Both ratios are approximately equal.

2. We'll see later that this equation (17) gives much better estimates than equation (15) of the acceleration of the uiverse as the variable of time, t, becomes large.

3. Equation (17) involves two important constants ξ and Σ that are related in the sense that the energy density Σ determines how much energy may be sucked into the volume at the edge of the universe (i.e., the value of the constant ξ depends on the constant Σ). Equation (16) suggests a relationship based on the radius of the universe.

Although equation (17) is just an approximation, from now on, we'll assume that it's equality because it will be more convenient to work the equations that way. Also we'll see later that the approximations are pretty good ones.

Out of curiosity, if we use equation (17), note that the differential equation (11) becomes now

$$\frac{d^2 e_V}{dt^2} = X \frac{de_V}{dt} \left(\frac{2}{R} + \frac{\xi}{\Sigma} \right)$$

Given that $\xi/\Sigma \approx 2/R$ where R is the radius of the universe, then this differential equation becomes

$$\frac{d^2 e_V}{dt^2} = X \frac{de_V}{dt} \left(\frac{2}{R} + \frac{2}{R} \right) \approx 4 \frac{X}{R} \frac{de_V}{dt} \quad (18)$$

As for the differential equation (12), it becomes

$$\frac{d^2e_S}{dt^2} = \frac{de_S}{dt}\left(\frac{X}{R} + \frac{\xi}{\Sigma}X\right)$$

Given the ratio $\xi/\Sigma \approx 2/R$, the differential equation becomes

$$\frac{d^2e_S}{dt^2} = \frac{de_S}{dt}\left(\frac{X}{R} + \frac{2}{R}X\right) \approx 3\frac{X}{R}\frac{de_S}{dt} \quad (19)$$

We see that both second derivatives (18) and (19) have a nearly identical form. Therefore

$$\frac{d^2e_S}{dt^2} \approx \frac{d^2e_V}{dt^2}$$

which further supports our earlier assumption that

$$\frac{de_S}{dt} \approx \frac{de_V}{dt}$$

From these approximations, you might conclude that then both e_V and e_S have approximately the same energy density, that is, $\xi = \Sigma$. However, this isn't at all true for two reasons:

1. The differential equations above are over *time*, not over surface or over volume.

2. The units of ξ are J/m^3 (over a volume) whereas the units of Σ are J/m^2 (over a surface). As a surface is one dimension less than volume, we already can conclude that $\xi < \Sigma$. This is precisely what will be determined later in this appendix.

Gravitation Revisited

It's time now to solve the differential equation (17) of the previous section for the variable X, the speed of expansion. Let's take the integral of equation (17) then

$$\int \frac{\xi}{\Sigma} dt = \int \frac{dX}{X^2}$$

The integrals are easily solved as

$$\frac{\xi}{\Sigma} t = -1/X + C$$

when C is the constant of integration. This constant's value depends on some initial condition at the time that the acceleration dX/dt started at $t=0$. This condition is related to the radius of the universe and the speed of expansion at the time that the acceleration started. I have no way of knowing what that radius or the speed was, and it doesn't matter here because I'm interested in the acceleration of X, not X specifically. In other words, I'm interested in the derivative dX/dt of X. When the derivative is taken, the constant C disappears anyway. So then I may assume

$$\frac{\xi}{\Sigma} t = -1/X$$

from which the speed of expansion of the universe is simply

$$X = \frac{-\Sigma}{t\xi}$$

At first, this formula doesn't appear to make sense: the time t is in the denominator. So as time goes by, the speed X decreases, right? No! The negative sign makes the speed do the reverse: increase. The graph of this formula looks something like this:

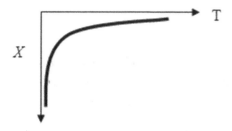

The negative sign suggests that the acceleration is in reverse—that is, it appears to be caused by some sort of antigravity! Interesting, because this is what physicists speculate.

Keep in mind throughout all mathematical work in this appendix that the variable t represents the time from the moment that the space density of the edge has reached and maintained its minimal value (i.e., when the term Σ has reaches its minimal value, as well as the term ξ). It's at that time, $t=0$, that the acceleration of the expansion of the universe started.

Now that we have a value for the speed of expansion, X, let's plug it in into the formula (17) of the previous section and see what expression for the acceleration we obtain

$$\frac{\xi}{\Sigma}\left(\frac{-\Sigma}{t\xi}\right)^2 = \frac{dX}{dt}$$

or

$$\frac{\Sigma}{t^2\xi} = \frac{dX}{dt} \quad (1)$$

This equation will be important later. Note that it was derived in the previous section from a few approximations, so it's only an *approximation* of the actual acceleration of the universe. This approximation is very bad for small values of the variable t, but gets better as the variable t increases in time. Nonetheless, I decided to keep this equation because it will lead me to values for the constants Σ and ξ that turn out to be very good estimates of energy densities related to space. Moreover, my equation will also lead me to Newton's famous gravitational equation

$$F = G\frac{m_1 m_2}{r^2}$$

I figured that if my theory of the acceleration of the expansion of the universe leads me to Newton's gravitational equation, there must be ring of truth to it.

Here's how my equation leads to Newton's gravitational equation. As $\Sigma = e_S/4\pi R^2$, the equation (1) may be expressed as

$$\frac{(e_S/4\pi R^2)}{t^2\xi} = \frac{dX}{dt}$$

or

$$\frac{e_S}{t^2\xi(4\pi R^2)} = \frac{dX}{dt}$$

Given the equivalence $e_S = mc^2$, this equation may be expressed with a mass that might be present at the edge of the universe

$$\frac{mc^2}{t^2\xi(4\pi R^2)} = \frac{dX}{dt}$$

or

$$\frac{m(c/t)^2}{\xi 4\pi R^2} = \frac{dX}{dt}$$

Newton's famous formula $F = ma$ will come in handy now. So as not to confuse the m in the above formula with this m in $F = ma$, let's use the capital M, so $F = Ma$. Let $a = dX/dt$, then the formula for the *force* applied at the edge of the universe may be expressed when the formula above for the acceleration is multiplied by M giving:

$$\frac{mM(c/t)^2}{\xi 4\pi R^2} = F$$

Arranged another way, this equation looks like this

$$\frac{(c/t)^2 mM}{(\xi 4\pi)R^2} = F \quad (2)$$

This is the force that matter is subjected to at the edge of the universe. Did you notice that this formula looks quite similar to Newton's formula for gravitation? That is,

$$G\frac{mM}{R^2} = F$$

Now, isn't this interesting! My own theory of gravitation outlined in Chapter 6 has led me to Newton's formula for gravitation! This is more than a coincidence. It looks like my theory might very well be an answer (an approximation, at least) regarding the mechanism that produces gravitation.

In my case, the gravitational constant G is

$$G = \frac{(c/t)^2}{(\xi 4\pi)} \quad (3)$$

A quick search on the Internet led me to a value of 6.673×10^{-11} m^3•kg^{-1}•s^{-2} for that constant G. Let's see if the units of my formula (3) check out. The ratio c/t has units m•s^{-2}. The term ξ has units J•m^{-3} or expressed another way,

kg• (m•s^{-1})^2•m^{-3} = kg• (m^2•s^{-2}) •m^{-3}
(Recall that the unit Joule = kg• (m•s^{-1})^2)

Putting those units into the formula (3) gives

$$\frac{(m \cdot s^{-2})^2}{kg \cdot (m \cdot s^{-1})^2 \cdot m^{-3}} = \frac{m^2 \cdot s^{-4}}{kg \cdot m^2 \cdot s^{-2} \cdot m^{-3}}$$

This may be reduced further to

$$\frac{s^{-2}}{kg \cdot m^{-3}}$$

or equivalently $m^3 \cdot kg^{-1} \cdot s^{-2}$. So the units do check out.

What's troubling is that my equation for G is *not* a constant—it's a function of time related to the expansion at the edge of the universe. Is Newton's gravitational constant G truly a constant? Is it the same value everywhere in the universe? Apparently not, according to this equation!

To be more precise, the "constant" G is proportional to the inverse of the time squared

$$G = \frac{c^2}{(\xi 4\pi)} \cdot \frac{1}{t^2} \quad (4)$$

This puzzling equation deserves more attention, and we'll come back to it in the next section.

In Newton's equation

$$G \frac{mM}{R^2} = F$$

the variable R is the distance separating the center of the two masses m and M:

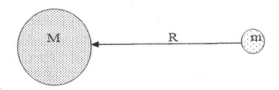

The force F makes the two masses attract each other. But the formula I discovered for the expansion of the universe

$$\frac{(c/t)^2 mM}{(\xi 4\pi)R^2} = F$$

is for this picture:

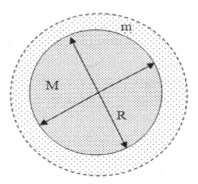

The mass M (minus the edge which has the mass m) would have to represent the mass of the entire universe with its center of gravity at the origin of the Big Bang. This picture is analogous to the following one where the force F makes the mass M push the mass m away.

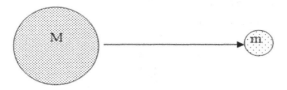

There's here a sense of an antigravity effect (as theorized by many physicists). So my equation does the opposite of Newton's equation. But is there truly an antigravity force? Yes—at the edge of the universe, there's energy only, and energy wants to spread outward creating an antigravity effect. I explained in detail my view on antigravitation in section "Space Density" of Chapter 6 and near the end of section "It All Started with a Bang!" of Chapter 8. So, the mass m is drawn toward the "emptiness" (no matter present) at the edge of the universe.

There's nothing mysterious about antigravitation. It may be viewed as gravitation in reverse. So, my formula (2) applies to bodies anywhere in the universe including at the edge where antigravity hides. It's only a matter of taking the picture of the edge of the universe inside out as follows:

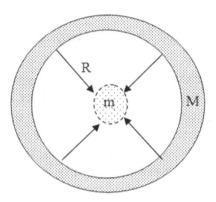

So the mass M can be seen as a ring attracted toward the mass m. The only difference is that in the first picture, the force is outward; in the second picture, it's inward. But in both cases, the same principle applies: nature sends energy where the space density is lower. So all the equations derived in the previous section apply. Viewed this way, it's gravitation that causes the acceleration of the universe, not antigravity. As an analogy, it's like looking at the negative of a photo. Nowadays photos are digital. But in the old days, a photo was created by taking its negative and reversing the colors. Nonetheless in both instances, it's the same picture in the photo.

In a way, it's all relative, as Einstein would like to say. Consider the earth and the moon:

and some asteroid attracted by the moon's gravity. As seen from earth, the asteroid moves away from earth, and this might be interpreted as some evidence of *antigravity pushing* the object away from earth. But for someone on the moon, it would be interpreted as *gravitation pulling* the object toward it.

The Second Law of Thermodynamics

There's more to be said about my theory of gravitation. As the universe expands, its energy tends to diffuse. This is due to the Second Law of Thermodynamics. If we can assume that the Big Bang is over, then no new energy comes out of the singularity point. Because of the First Law of Thermodynamics, no new energy may be created. Thus, the diffusion of energy implies that the overall density of energy of the universe decreases as it expands. Therefore

$$\frac{E}{V(t+dt)} < \frac{E}{V(t)}$$

where E is the entire energy of the universe. Also, due to the nature of diffusion, it's obvious that the density of energy at the edge of the expanding universe is even less than these two ratios. So then

$$\frac{de}{dv} < \frac{E}{V(t+dt)} < \frac{E}{V(t)} \quad (5)$$

where de/dv is the volume energy density at the edge of the universe. Now $V(t) = 4/3\,\pi R(t)^3$, and $V(t+dt) = 4/3\,\pi R(t+dt)^3$. Another way to express the volume formula is as

$$V(t+dt) = V + dv = {}^4\!/_3\,\pi(R+dr)^3$$

This formula is manipulated as follows

$$V + dv = {}^4\!/_3\,\pi(R^3 + 3R^2 dr + 3R dr^2 + dr^3)$$

or

$$V + dv = \frac{4}{3}\pi R^3 + \frac{4}{3}\pi(3R^2 dr + 3R dr^2 + dr^3)$$

or

$$V + dv = V + \frac{4}{3}\pi(3R^2 dr + 3R dr^2 + dr^3)$$

And that gives us an expression for *dv*

$$dv = \frac{4}{3}\pi(3R^2 dr + 3R dr^2 + dr^3)$$

The last two terms, *3Rdr²+dr³*, may be ignored as *dr* nears zero. So the formula becomes

$$dv = \frac{4}{3}\pi(3R^2 dr)$$

or simply

$$dv = 4\pi R^2 dr$$

Next we put that into the formula (5)

$$\frac{de}{4\pi R^2 dr} < \frac{E}{\frac{4}{3}\pi R^3(t)}$$

This reduces to

$$\frac{de}{dr} < \frac{3E}{R(t)}$$

or

$$de < \frac{3E dr}{R(t)}$$

Next dividing both sides by *dt* gives

$$\frac{de}{dt} < \frac{3Edr}{R(t)dt}$$

or

$$\frac{de}{dt} < \frac{3E\left(dr/dt\right)}{R(t)}$$

Remembering that $dr/dt = X$, we finish with

$$\frac{de}{dt} < \frac{3EX}{R(t)} \quad (6)$$

Let us bring back the equation (4) of the previous section

$$\frac{de}{dt} = \frac{2eX}{R} \quad (7)$$

Note that the inequality (6) is very similar to the equation (7). Moreover, notice what happens when the equation (7) is placed into the inequality (6). The inequality becomes

$$\frac{2eX}{R} < \frac{3EX}{R(t)}$$

The variable X and R cancel out, reducing the inequality to:

$$2e < 3E$$

This inequality is very evidently always true. So here are two observations:

1. The inequality (6) and the equation (7) are very similar.

2. Putting them together, we end up with a tautology.

This suggests that the equation (6) is a consequence of the inequality (7). I remind you that

1. The inequality (7) expresses a consequence of the Second Law of Thermodynamics (i.e., dissipation of energy).

2. Equation (6) expresses dissipation of the energy at the edge of the universe, starting at the moment that space density reaches a critical low. Recall that this idea was based on my own view of what causes gravitation.

As my view of gravitation led to essentially Newton's formula for gravitation, my view for the cause of gravitation is correct (i.e., it does describe gravitation). Furthermore the result shown in Appendix 5 proves me correct. Therefore, the two points above place together the Second Law of Thermodynamics and my own view of gravitation. Most inspiring! Consequently, I suggest to you that the effect of gravitation is a manifestation of the Second Law of thermodynamics. What a daring suggestion!

I believe there's a lot of truth to this because the Second Law says that energy tends to dissipate. It doesn't discriminate against the type of energy in question as *any* type of energy tends to dissipate. Because my view of gravitation says that an object is attracted to a body due to energy dynamics, this connection between gravitation and the Second Law should deserve some merit, as unbelievable as this may seem.

We have to realize that the Second Law of Thermodynamics is silent about the mechanism that causes dissipation of the energy. In fact, this mechanism depends on the type of energy. For instance, energy dissipation on a hot metal plate will occur differently than heat dissipation in the atmosphere. But all types of energy abide by that all-important Second Law of Thermodynamics.

In the section "A Formula for Discrete Energy" of Chapter 5, I found that waves emit energy in quantum units. In addition, energy is carried by waves. This inevitably includes gravitational energy. So

I propose the existence of gravitational waves. These waves would have to be of extremely low frequency because the gravitational force is so weak. Energy from those waves would be emitted by some massless particle. A quick search on the Internet led me to the Wikipedia Web site that says that physicists theorize the existence of gravitons. I discuss this topic in more detail in the section "Gravitons Revisited" in Chapter 8.

What is the Fate of the Universe?

During an Internet search, I found that the speed of galaxies extending away from each other is a mere 77 kms/sec per 3.26 million light-years. A simple calculation reveals an acceleration that averages 7.5×10^{-13} kms/sec^2 or 7.5×10^{-10} m/sec^2. This is some tiny acceleration as a snail accelerates much faster than that. There's another way to believe that the acceleration of the expansion of the universe is so tiny. Our own galaxy isn't immune to this acceleration, yet we humans don't feel any acceleration at all. Our galaxy appears to move along at a constant speed.

Now let's see what we can discover here. The equation (1) of the previous section

$$\frac{\Sigma}{t^2 \xi} = \frac{dX}{dt} \quad (1)$$

has for a graph the following:

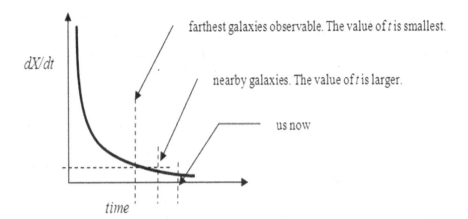

farthest galaxies observable. The value of t is smallest.

nearby galaxies. The value of t is larger.

us now

dX/dt

time

A physicist looking at my equation would surely mock it because it's nowhere near the correct equation for expansion. But I'm not deterred. Despite the inaccuracies, I decided to keep this equation because it leads me in the next section coming up to values for the constants Σ and ξ that turn out to be *very good* estimates of actual energy densities related to space. Moreover, my equation led me to Newton's famous gravitational equation

$$F = G\,\frac{m_1 m_2}{r^2}$$

I figured that if my theory of the acceleration of the expansion of the universe leads me to Newton's gravitational equation, there must be some truth behind my theory.

Note that because some of my mathematical derivations in this appendix are only approximations, inevitably the equation (1) is also an *approximation* of the actual acceleration of the universe over time. The next section coming up will use my formula to calculate an estimate to the acceleration for nearby galaxies (those with a large value of the the time t) and find that it's very accurate. As for galaxies farther away, I don't have any data to verify my equation against. Nonetheless observations obtained by astrophysicists tell that the acceleration *increases* the *farther* away the galaxies are. My equation does reflect that trend! The reason is that those galaxies are seen by us as they *were* billions of years ago. Thus, their acceleration detected here on earth right now is from very far into the past, and hence the variable t is small for those galaxies and larger for galaxies that are closer to us. My graph does reflect this inversely proportional relationship between acceleration and time.

This doesn't imply that my equation provides accurate values for the accelaration of those far away galaxies for which the variable t is small. It's abundantly obvious that the "spike" near $t=0$ isn't at all accurate because it is associated with accelations that would have generated speeds well beyond the speed of light—this is physically impossible. Clearly if there was a spike into the past, it was in reality not that pronounced when it happened. My equation obviously provides very poor estimates of the acceleration for small values of the variable t.

Note also that the estimation 7.5×10^{-10} m/sec^2 of the acceleration given by physicists is for the *observable* universe: its edge is obviously not observable. But the farthest galaxies observable are our best guess of what happens at the edge of the universe. The observable galaxies lie near the end of the curve of the graph where the curve has a nearly constant slope, reflecting the constant acceleration observed:

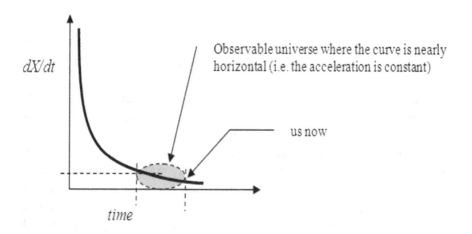

As the graph shows, the farthest galaxies are seen by us as having a high acceleration while nearby galaxies appear to have a lower acceleration. Note that near time $t=0$, the acceleration was extremely high. So the speed of matter going outward increased extremely rapidly many billions of years ago. In any case, the galaxies that happened to be near the edge of the universe that fateful day when the acceleration shot up (as shown in the graph) must have shattered under the strain of the acceleration. If there were any intelligent living beings then, their world must have collapsed as in the most horrific doomsday scenario.

The equation (1) expresses what happens at the edge of the universe. Can it give clues as to how the universe expanded after the Big Bang? The graph shows that the acceleration was extremely high at one point following $t=0$, then over time the acceleration slowed down considerably, but it will never reach zero. If right after $t=0$, matter near the edge of the universe was picking up speed very quickly, then quite possibly by now matter near the edge of the universe would be traveling faster than the speed of light! This is bizarre. All physicists agree that the speed of light is constant everywhere in the universe. Then either (a) this assumption is wrong; or (b) just before the time $t=0$ that this acceleration started, matter in the universe was expanding *very* slowly.

I think that the second option of a very slow expansion is correct and much more plausible. So has the universe always expanded with acceleration? I doubt it for two reasons: (a) the mathematical reason just given in the previous paragraph, and (b) the physical reason that now follows.

Let's assume that the entire energy that originated from the Big Bang all came out within a few million years. In other words, the Big Bang ended billions of years ago. (No physicists claim otherwise.) When the Big Bang ended, the radius of the Universe was obviously much smaller than it is now, yet it contained all of the energy the universe currently has. Consequently, the space density of the then-smaller universe was definitely much higher than what it is now and the edge of the universe had a very high space density as shown in the picture below:

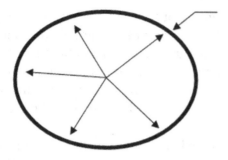

Space density of the edge is higher than the minimal possible

Because the space density of the edge was much higher than the minimal density, the edge didn't have to react the way it does now of quickly streaming more energy toward the edge. Consequently, all the equations that I worked out for the acceleration dX/dt hadn't taken effect yet. It was many billions years later before this acceleration happened. Because the entire energy of the universe was confined in a much smaller sphere than it has now, it must have taken many billions years for this acceleration to happen. I'm not suggesting that no energy was sent to the edge before this moment of acceleration as of course energy reached the edge all the time. It's just that the universe was in no hurry to send massive amount of energy at the edge.

So, the complete graph of the acceleration dX/dt versus time looks like this:

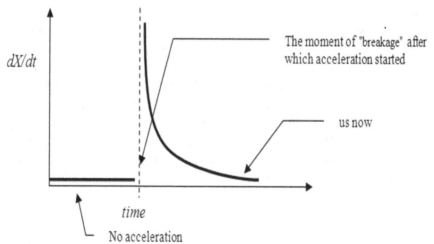

dX/dt

The moment of "breakage" after which acceleration started

us now

time

No acceleration

The vertical dotted line indicates the moment of "breakage" where mysteriously the acceleration started. At that point of "breakage," it probably was doomsday for much of the universe as I described earlier. Note that the graph above suggests that at the time of breakage, an asymptotic condition existed. This is a condition in which a function's value is infinite—in this case the acceleration. This condition is easily seen with the equation:

$$\frac{\Sigma}{t^2 \xi} = \frac{dX}{dt}$$

If $t=0$, the dX/dt is infinite. Understand that this asymptotic condition is physically impossible. The acceleration was very high near $t=0$, but clearly not infinite. So the graph depicted earlier should instead be depicted without a breakage as follows:

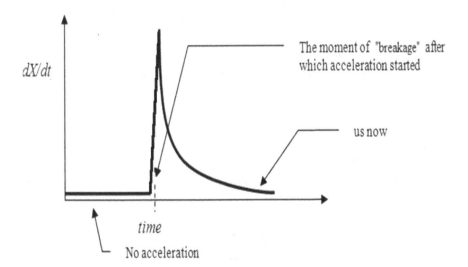

As I mentioned earlier, the equation for this graph is a very poor estimate of what happened near $t = 0$. Consequently this spike was in reality much less pronounced than shown here. Nevertheless, what caused the universe to suddenly expand with acceleration? What was the rush for? I can't help but notice that this graph looks very much like when a lightbulb bursts. At the moment of the burst, the luminosity of the bulb shoots up much higher than normal, and then quickly the light dies down.

The graph also looks like one in which a balloon is inflated at a constant speed, then at a critical point, it bursts, at which time the air escapes with a sudden sharp acceleration that slows down quickly. Should we conclude that perhaps just at the time the universe reached that "breakage," the space density of the edge of the universe had gone below the critical minimal space Σ density that I mentioned a couple of sections ago? Then did universe inevitably overreact by very quickly sending massive amounts of energy to the edge, thereby accounting for the spike in the acceleration dX/dt?

What could have caused the space density of the edge of the Universe to go below the critical minimal space density? Space density is measured by the amount of pure energy (that excludes matter) within a volume. As I mentioned before, the energy density must have been very high for a while after the Big Bang. I propose the following series of events that led to the spike of acceleration:

1. As the radius of the universe increased, that density decreased. At that time, the universe was made of subatomic particles. Then atoms started to form. This led to the creation of matter.

2. As matter is concentrated energy, that energy had to be "sucked" out of the surrounding environment. Consequently, the density of free-flowing energy *decreased* throughout the universe.

3. In turn, this led to less outward pressure to expand the universe. As a result, I suspect that the radius of the universe most likely *decreased* for a period of time in an attempt to keep the energy density from becoming too low.

4. As more and more matter was being created, this matter eventually was pushed outward so that the radius started to increase once again. This expansion increased the volume of the edge of the universe and thereby decreased the energy density on that edge.

5. This happened as matter was being created, which further decreased the free-flowing energy density.

6. These two factors (4 and 5) probably contributed in an exponential decrease of the energy density of the edge of the universe. Eventually the edge was too thin and so the universe reacted by sending a massive amount of energy toward it. This would explain the spike in the graph above.

Based on the previous paragraph, I propose that the acceleration of the expansion of the universe went something like this:

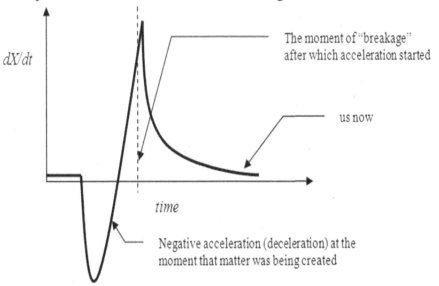

dX/dt

The moment of "breakage" after which acceleration started

us now

time

Negative acceleration (deceleration) at the moment that matter was being created

After the universe passed that "breakage" point, there was no return, and from then on, the universe has been essentially dying very slowly but all the while struggling to stay alive by accelerating energy toward the edge of the universe to maintain a minimal space density. Is this a sign that the universe is starting to run out of energy? I think so.

Now back to the equation (4) of the previous section

$$G = \frac{c^2}{(\xi 4\pi)} \cdot \frac{1}{t^2}$$

This yields the following graph for the gravitational "constant" G:

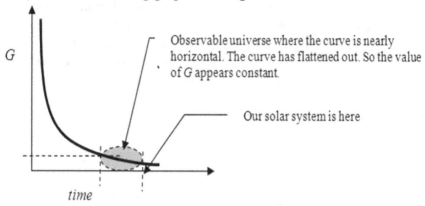

G

Observable universe where the curve is nearly horizontal. The curve has flattened out. So the value of G appears constant.

Our solar system is here

time

To suggest that the gravitational constant varies inversely with time seems ridiculous. Indeed, many scientists argue that if G varies, we should have plenty of evidence. For instance, the moon should move away rather rapidly from the earth, the planets and the sun would increase in size, etc. All these implications would be correct if the value of G did indeed vary rapidly. But does it? Obviously, it doesn't. Keep in mind that the graph above shows that the greatest variation of G would have occurred shortly after the acceleration of the universe started. According to the physicists, we're now some 7.5×10^9 years later. The changes in the value of G have now flattened out so much that G now appears constant.

OK—the purpose of this section is to investigate what might be the fate of the universe. I suggested a few paragraphs ago that the universe might be running out of energy. How bad is that?

According to the Web site www.space.com, the universe is 156 billion light-years wide and according to Wikipedia, the universe is 13.7 billion years old. There are 365 days in a year, 24 hours a day, and 3600 seconds in an hour. This calculates to 4.3×10^{17} seconds since the Big Bang. As for the size of the universe, the calculations come to 7.4×10^{26} meters radius. So the calculation of the *average* speed of expansion of the universe is only a matter of dividing

7.4×10^{26} meters by 4.3×10^{17} seconds, which gives 17.2×10^8 meters per second.

But this speed is faster than the speed of light! Could the speed of light been faster in the far past? Not so fast: there might be another explanation. The reason for this erroneous result might be that the age of the universe is calculated based on the *observable* universe. That number is obtained by multiplying how old we think the universe is by the speed of light. The reasoning there's quite straightforward: we can only see out to that distance from which light can have reached us since the universe began. So this might explain why I arrived at a speed of expansion in the range of the speed of light.

As the acceleration will never stop, the speed of expansion of the universe might eventually reach the speed of light. Oops—what will happen then? As the speed of light is the limit, the acceleration will stop dead all in a sudden! (I'm assuming here that the speed of light can't change and increase). What will be the consequences? One is that the edge will very quickly run out of energy, and this in turn will force the interior of the universe to send more energy toward the edge (this is based on my own theory of gravitation that was found earlier to be equivalent to the principle of dissipation of energy).

Essentially, the universe will eat itself to death. It's like if someone is asked to run nonstop for a whole week without eating or drinking. The person's body will have to find energy anywhere within the body to try to survive and the body will literally eat itself. I believe that this is precisely what would happen with the universe. Essentially the universe will fade away very much like clouds fade away if there's not enough humidity present in the atmosphere.

But as the universe will expand to no end, surely energy densities will reach numbers below the two constants $\Sigma = 8.1 \times 10^{16} \ J \cdot m^{-2}$ and $\xi = 2.2 \times 10^{-10} \ J \cdot m^{-3}$ (these constants are calculated in the next section). However, according to physicists, no matter what, these numbers can't be any lower. That means either (a) the radius of the universe will start to decrease despite my equation that states the contrary, or that (b) there's still energy coming out of the Big Bang to this day—and it will keep coming out forever!

I remind you that in the section "The Singularity Theorem" of Chapter 8, I stated that the singularity point was an *infinitely* dense black hole that contained the entire matter of the universe. If that is correct, then energy is still coming out of that black hole. I've expressed my doubts in that section about this possibility of an infinite amount of energy coming out of that singularity point.

If all energy created by the Big Bang has already been released, then instead of continually expanding, the edge of the universe will eventually stop expanding and will begin to shrink (decelerate) in order to maintain minimal energy densities. Then when the density of the edge is strong again, the expansion might resume again.

I theorize that the radius of the universe will shrink and expand like this forever. For the past sixty years, the popular theory regarding the end of the Uuiverse is that it will collapse into itself back into a singularity point due to gravitational pulls from galaxies. But physicists have never been able to prove this. Moreover, don't these physicists realize that it's *physically* impossible for the universe to shrink back into a singularity point? Indeed, to be reduced back to a singularity point again, the volume of the universe would have to become zero, meaning that the space dimensions would have to vanish! This is physically impossible.

Some physicists suggest that the whole universe will eventually be sucked into a single black hole. Perhaps—but keep in mind that a singularity point isn't the same as a black hole. The latter has a volume; the former doesn't. In a black hole, in principle the universe would still exist, only in a very small volume.

A singularity point means the absence of a physical universe. In Chapters 7 and 10, I showed that a system of thoughts created the physical universe. For the universe to shrink back into a singularity point would be the equivalent to undoing that system of thoughts. Needless to say, it's impossible to undo a thought, and some things in nature can't be undone (e.g., it's impossible to undo the birth of a baby). That same principle applies to the universe. I see nothing that can undo the Big Bang. Consequently, the universe will exist forever—that is, at any given moment in the future; the universe will still be there! The universe will *never* end!

You might recall that in Chapter 8, I claimed that the Universe isn't infinite and never will be. Am I not contradicting myself by suggesting that the universe will never end? No—in this case "never end" means that at any given x moment inthe future, the universe will still be there. As the universe had a beginning, at that moment x, the universe will have been in existence y number of seconds. Thus the number y will never be infinite. Infinity isn't a number; consequently the universe will never reach infinity even if it exists forever!

The universe will never come to an end! This seems so illogical because all things eventually come to an end, right? Plants, animals, insects, and humans eventually die off. Humankind will eventually disappear. The earth, our solar system, and our galaxy will eventually disappear. All galaxies eventually come to an end. The key word here is: *disappear*. If something has disappeared, it's obviously died. But the universe will never disappear even should it become a black hole because the universe consists of the space dimensions, the time dimension, and energy, and these will never disappear. It's simply impossible to undo God's initial Creation.

There's another argument to suggest that the universe will never end: time. As I argued in the section "What Created the Universe" in Chapter 7, God's thought of time triggered the creation of the universe. In the section "Our Virtual Universe" of Chapter 10, I showed that once a thought is generated, it never vanishes. Consequently, the thought of time will never disappear. Because that thought triggered the universe, its existence can never come to an end.

Near the end of section "The Expansion of the Universe and its Fate" of Chapter 8, more arguments related to the fate of the universe are proposed. You're invited to consult that section.

An Estimation of the Minimal Space Density

In this section, I'll continue to use the equations from previous sections of this appendix because they remain good approximations of the reality. The fate of the universe depends on the ability of space to retain minimal space energy at the edge of the universe. Throughout the mathematical development of the past few sections, the symbol Σ was used to symbolize the ratio $e_S/4\pi R^2$ (energy density) and the symbol ξ was used to express the ratio de_V/dv (energy density over a volume). A value for those two symbols may be calculated using the equations

$$G = \frac{(c/t)^2}{\xi 4\pi} \quad (1)$$

and

$$\frac{\Sigma}{t^2 \xi} = \frac{dX}{dt} \quad (2)$$

developed earlier. Let's find the value for the symbol Σ. From this second equation (2), we obtain

$$\frac{1}{t^2} = \frac{\xi}{\Sigma} \frac{dX}{dt}$$

Plugging that into the first equation (1) for G produces the equation

$$G = \frac{c^2 \left(dX/dt \right)}{\Sigma 4\pi} \quad (3)$$

Note that the value of G is proportional to the acceleration of the expansion of the niverse. As the acceleration is constant for the observable universe (as the graph earlier showed), then G is constant as far as what we can tell and feel. Keep this equation in mind as we'll come back to it shortly.

Given that physicists have estimated the acceleration dX/dt of the universe to be 7.5×10^{-10} m/s^2 and that G today is 6.673×10^{-11} m$^3 \cdot$kg$^{-1} \cdot$s^{-2}, we can plug those numbers into the equation above

$$G = \frac{(3 \times 10^8 \ m/s)^2 \cdot (7.5 \times 10^{-10} \ m/s^2)}{\Sigma 4 \pi}$$

or

$$G = \frac{(3 \times 10^8 \ m/s)^2 \cdot (6 \times 10^{-11} \ m/s^2)}{\Sigma}$$

or

$$= \frac{6.673 \times 10^{-11} \ m^3 \cdot kg^{-1} \cdot s^{-2}}{\Sigma}$$
$$(3 \times 10^8 \ m/s)^2 \cdot (6 \times 10^{-11} \ m/s^2)}{\Sigma}$$

So

$$\Sigma = \frac{(3 \times 10^8 \ m/s)^2 \cdot (6 \times 10^{-11} \ m/s^2)}{6.673 \times 10^{-11} \ m^3 \cdot kg^{-1} \cdot s^{-2}}$$

to obtain a value of 8.1×10^{16} $kg \cdot s^{-2}$ for Σ. Because one $kg \cdot m^2 \cdot s^{-2} =$ one *Joule* then the value of Σ comes to 8.1×10^{16} $J \cdot m^{-2}$, where J is one Joule, a measure of energy. We'll come back to this result in a few minutes.

In the first section of this appendix, I found that

$$\frac{\xi}{\Sigma} \approx \frac{2}{R} \quad (4)$$

So the ratio Σ/ξ is indeed not constant. Bad news! Does this mean we have to throw away the differential equation (17) obtained at the conclusion of the first section

$$\frac{\xi}{\Sigma} X^2 = \frac{dX}{dt}$$

from which the graph of the acceleration of the expansion of the universe was derived? Not so fast—my error of assuming the ratio Σ/ξ to be constant isn't quite all that bad for three reasons:

1. Notice that the ratio *2/R* changes *very slowly* as *R* becomes very large, and for two reasons:

   a. The rate of change of *2/R* is easily obtained by taking its derivative, $-2/R^2$. This formula flattens out very quickly after the universe expanded to just a few hundred kilometers. Keep in mind that all equations developed in this appendix are for after the expansion of the universe started to accelerate, which according to physicists was some 7.5 billion years ago. The radius *R* was most likely fairly large already by then. Thus, the ratio Σ/ξ most likely changed very little after that moment of acceleration up until today.

   b. The rate of change of the radius *R* is very small compared to the actual size of the Universe. So the ratio *2/R* indeed changes very slowly.

2. The equation is based on the confirmed assumption that the rate of dissipation of energy e_S is similar to the rate of dissipation of e_V. This was determined in the first section "The Expansion of the Universe."

Therefore, I'm confident that the results of this appendix remain fairly good approximations even if I assume the ratio Σ/ξ to be constant.

Next let's use the equation (4) and assume equality for simplicity sake

$$\frac{\xi}{\Sigma} = \frac{2}{R}$$

to solve for the value of the term ξ. Clearly from that formula

$$\xi = \frac{2\Sigma}{R}$$

The latest accepted value for R, the radius of the universe, is 78 billion light-years as reported at the Web site www.space.com. Given that a light-year is 9.5×10^{15} meters long, then the radius R is (78×10^9 light-years) x (9.5×10^{15} meters/light-year) $= 7.4 \times 10^{26}$ meters. So then

$$\xi = \frac{2\Sigma}{R} = \frac{2(8.1 \times 10^{16} \, J \cdot m^{-2})}{7.4 \times 10^{26} \, m} = 2.2 \times 10^{-10} \, J \cdot m^{-3}$$

This is an awfully low density of energy in space! Mathematician John Baez at his Web site math.ucr.edu/home/baez/vacuum.html used observations from a famous experiment, namely the Wilkinson Microwave Anisotropy Probe, to calculate a minimal energy density in space of 9×10^{-10} J•m^{-3} (material used with permission). My estimate with ξ is very close to that.

Note that when expressed in units of J•L^{-1}, constant $\xi = 2.2 \times 10^{-13}$ J•L^{-1}. Physicists call this number the *zero-point* energy of space. Note that the dark energy density, $\Sigma = 2.31 \times 10^{22}$ J•L^{-1}, is billions of times denser that the minimal energy in space. This suggests that the universe is made almost entirely of dark energy. But not quite: the term ξ specifies the *minimal* energy density in space. As we move away from the edge of the universe, this density inevitably increases. Physicists estimate that the universe is made of 73 percent dark energy. So 73 percent of the energy from the Big Bang is used to create the fabric of space.

Let's move to the other constant, $\Sigma = 8.1 \times 10^{16}$ J•m^{-2}. This is the density of energy at the edge of the universe, and it seems extremely high for a minimal space density of the fabric of space! When expressed in energy per liter, the number becomes a staggering 2.27×10^{22} J·L^{-1}. This conversion is based on the fact that 1 cm^3 = 1 ml, so (10 cm)^3 = 1 L, or (10^{-1} m)^3 = 1 L, or 10^{-3}m^3 = 1 liter. The square root of $\Sigma = 8.1 \times 10^{16}$ J•m^{-2} is 2.85×10^8 J•m^{-1}. That number cubed gives us 2.31×10^{25} J•m^{-3}. With the equivalence 10^{-3}m^3 = 1 liter, we obtain the space density of 2.31×10^{22} J•L^{-1}.

This number would be the energy density of *empty* space (i.e., before any energy is even moved in space). It's the energy necessary to the fabric of space itself. This is an extremely large space density considering that the energy density of the nucleus of helium-4 is 8.57×10^{30} J•L^{-1} (this is the *binding* energy number I obtained from Wikipedia). Something doesn't add up here ... or perhaps it does! Let's find out.

First, recall that this space density was derived from the equation (3)

$$G = \frac{c^2\left(dX/dt\right)}{\Sigma 4\pi}$$

All terms in that equation are known constants except for the acceleration of the universe dX/dt. But physicists claim to have calculated this acceleration accurately. So maybe the value of Σ is correct after all. We'll see in the next section that it is indeed so!

Second, recall that in the section "Can Anything Go Faster than Light" in Chapter 5, I speculated the existence of dark energy in space. That energy is inherent to space and is the mechanism that allows a quantum particle to go from one "float" to another. Essentially, dark energy allows motion to take place. Dark energy is the energy making up the fabric of space.

In the section "Motion in Space–Time: The Plot Thickens!" of the same chapter, I suggested that quantum particles skip in space by being carried by energy in space very much like when a person is carried above a crowd of people. The person temporarily is slowed down, and then reaccelerated, giving an impression of constant, smooth speed. Because of the principle of action–reaction, the force of reaction from space has to equal the force from the quantum particles.

Based on these thoughts, I have a hunch that the number $\Sigma = 2.31 \times 10^{22}$ J•L^{-1} representing the energy density of empty space means that empty space can support or carry any particle with an energy density of 2.31×10^{22} J•L^{-1} or less. What particles could that be? This was answered in the section "The Energy Dimension" of Chapter 8. It was determined there that the neutrino is the most likely candidate. Finally, I read on the Internet that physicists speculate that there's a connection between dark energy and neutrinos. My numbers bear that out!

Finally: what should we think of the gravitational "constant" G? We saw in the previous section that its value depends on the inverse squared of the time. How shocking: it's not a constant after all. But that equation

$$G = \frac{(c/t)^2}{\xi 4\pi}$$

doesn't at all make clear what physical phenomenon makes G vary. However, when the same equation is transformed into equation (3) that we saw earlier

$$G = \frac{c^2 \left(dX/dt \right)}{\Sigma 4\pi}$$

the value of G is expressed with the acceleration of the expansion of the universe, dX/dt. It suddenly makes more sense now. We have to remember that Newton's equation

$$F = G \frac{m_1 m_2}{r^2}$$

assumes two things:

1. That the two bodies are isolated (i.e., that there are no external forces upon them), and

2. That m_1 and m_2 are the masses at rest.

If the two bodies are being subjected to an acceleration coming from the expansion of the universe, these two assumptions no longer apply. Given this reasoning, it's therefore not surprising to find that the "constant" G depends on the acceleration of that expansion. In essence, this acceleration translates into an *additional force* applied on the two bodies.

Before finishing off the appendix, let's now plug the value for ξ found earlier into this formula

$$\frac{dX}{dt} = \frac{\Sigma}{t^2 \xi}$$

for the acceleration of the expansion of the universe to see if the equation does check out. After all, this equation was derived based on my theory of gravitation and my theory of the expansion of the universe. So it better work! We need an estimate for the value of the variable t, the time that universe started to expand with acceleration. The universe is estimated to be some 13.7×10^9 years old, but that isn't the number we seek. So I searched the Internet and found that physicists estimate that the universe started to accelerate some $t = 7.5 \times 10^9$ years $= 2.4 \times 10^{17}$ seconds ago. Let's plug the numbers into the formula

$$\frac{dX}{dt} = \frac{\Sigma}{t^2 \xi} = \frac{8.1 \times 10^{16} J \cdot m^{-2}}{(2.4 \times 10^{17} s)^2 \cdot (2.2 \times 10^{-10} J \cdot m^{-3})}$$
$$= 6.4 \times 10^{-9} m/s^2$$

Given that the estimated acceleration as observed by physicists is 7.5×10^{-10} m/s², the equation gives an acceleration that's not that far off the mark.

There are a few known facts that my equations have checked out:

1. Newton's gravitation formula could be derived from my theory.

2. The minimal space density $\xi = 2.2 \times 10^{-10}$ J•m⁻³ derived from my theory does correspond closely with the known estimate.

3. The minimal space density $\Sigma = 8.1 \times 10^{16}$ J•m⁻² derived from my theory does correspond closely to the energy density of neutrinos. See the section "The Energy Dimension" of Chapter 8.

4. My theory provides an estimate of the acceleration of the expansion of the universe that is close to the observed value for nearby galaxies.

I hope you'll agree that my modest theory of gravitation and of the acceleration of the universe should deserve some attention.

Index

602 A Personal Journey into the Quantum World